1001 Questions Answered About Hurricanes, Tornadoes and Other Natural Air Disasters

BARBARA TUFTY

Drawings by James MacDonald

Dover Publications, Inc., New York

Published in Canada by General Publishing Company, Ltd.,
30 Lesmill Road, Don Mills, Toronto, Ontario.
Published in the United Kingdom by Constable and Company,
Ltd.

This Dover edition, first published in 1987, is an unabridged and
enlarged republication of the work first published by Dodd, Mead &
Company, New York, 1970, under the title *1001 Questions Answered
About Storms and Other Natural Air Disasters*. A new Preface, a
Supplement to the Bibliography and a new Appendix comprising 27
updated answers have been specially written for this edition by the
author, who has also prepared a new hurricane map. The new material
has been indexed separately at the end of the book.

Manufactured in the United States of America
Dover Publications, Inc.
31 East 2nd Street
Mineola, N.Y. 11501

Library of Congress Cataloging-in-Publication Data

Tufty, Barbara.
    1001 questions answered about hurricanes, tornadoes and other
natural air disasters.

    Previous ed. published as: 1001 questions answered about storms and
other natural air disasters.
    Bibliography: p.
    Includes index.
    1. Storms—Miscellanea. I. Tufty, Barbara. 1001 questions about
storms and other natural air disasters. II. Title.
QC941.8.T84   1987   551.5′5   87-8935
ISBN 0-486-25455-0 (pbk.)

To

Harold

Christopher, Karen, Steven

Peejay

and

company

PREFACE TO THE DOVER EDITION

For this Dover edition I have provided new answers (updated through 1985) to 27 questions. A bullet (•) in the margin next to a question in the original text indicates that an updated version is to be found in the Appendix, which begins on page 351. A new Supplement to the Bibliography, listing six books and reports published since 1969, follows the original list, which begins on page 363. A Supplement to the Index, covering the material in the new Appendix, appears at the very end of the volume.

For help in preparing the supplementary material for the Dover edition, I wish to thank Henry Vigansky, Statistical Climatology Branch, NOAA (National Oceanic and Atmospheric Administration); James Campbell, Severe Weather Branch, NOAA; and Dr. Ken Bergman, National Weather Service, NOAA.

BARBARA TUFTY

ACKNOWLEDGMENTS

The slow growth and development of this book has come about from countless contributions of information, suggestions, and encouragement given by many people. To that vast array of individuals— scientists, researchers, authors, readers, librarians, secretaries, typists, clerks, friends, and moral supporters—I owe a great debt of appreciation. To list them all would fill another book, but I wish to express my thanks to several outstanding people, in particular to Dr. Helmut Landsberg, University of Maryland, who offered valuable suggestions for sections of the book in its early stages; and to Ann Ewing, Science Service staff writer, who gave helpful comments on the manuscript in final form.

My special thanks go to Herbert Lieb, Office of Public Information of Environmental Science Services Administration (ESSA), for his constant assistance to me and other authors in their never-ending search for information; and to Frank Forrester, meteorologist with the National Broadcasting Company, television, Washington, D.C. and information chief of the U.S. Geological Survey. Charles Thomas of ESSA's Weather Bureau Information Office and Sumner Barton formerly with the Weather Bureau have been most helpful in directing me to sources of weather information.

Among the many scientists and researchers who have contributed toward the accuracy and clarity of this book, in particular I wish to thank ESSA's Dr. Edwin Kessler, Neil B. Ward, and Dr. Gilbert Kinzer of the National Severe Storms Laboratory in Norman, Oklahoma; Dr. R. Cecil Gentry and Dr. Toby Carlson of the National

Hurricane Research Laboratory in Coral Gables, Florida; Robert E. Helbush and Samuel O. Grimm, Emergency Warnings Branch; Clarence Woolum, Eugene W. Hoover, Charles Roberts, Louis Harrison, J. Murray Mitchell, N. Arthur Pore, Maurice E. Pautz, Stuart G. Bigler, Alexander F. Sadowski, Maurice Arkin, Patrick E. Hughes, Donald C. House, Max M. Feinsilber, Harold Scott, and Bill West. Dr. Harold Glaser of the National Aeronautics and Space Administration was particularly helpful with the chapter on space storms.

Many people have contributed valuable information and services from other organizations, including The American Meteorological Society; American National Red Cross; Cold Regions Research and Engineering Laboratory; Crop-Hail Insurance Actuarial Association; National Academy of Sciences; National Center for Atmospheric Research; National Geographic Society; National Safety Council; National Science Foundation; Office of Disaster Relief of the Agency for International Development (AID), Office of Emergency Preparedness; Science Service; Smithsonian Institution; Travelers Research Center, Inc.; United Nations Educational, Scientific and Cultural Organization; U.S. Departments of Agriculture, Army, Navy, Air Force, Commerce, and Interior; and the World Meteorological Organization.

My continued thanks go to the many librarians who have tracked down publications from latest reports to out-of-print pamphlets—to Margit Friedrich and Mae Boyle of the Science Service Library; Marjorie A. Clark and Anne M. Tzarnas of the Weather Bureau Library; and William H. Heers and his staff at the Geological Survey Library.

I also wish to acknowledge Nadine Clement, Elizabeth Powell, Elizabeth Bakersmith, Princhessa Fields, Nancy Crist, and Helen Silver for their assistance in producing this book, and James MacDonald for his line drawings.

For any errors or discrepancies in this book, the author assumes full responsibility.

INTRODUCTION

Within the restless atmosphere that encircles the earth, catastrophic winds and storms are generated, many of which strike man and his communities with devastating force.

Sprawling hurricanes, vicious corkscrew tornadoes, torrential rains, crippling snows and ice storms, searing hot winds of the deserts, biting cold rivers of air from the polar regions—these ever-changing, ever-moving bodies of air are all natural phenomena that have swept the earth for eons, bringing destruction and anguish to mankind whenever they encounter his expanding societies. Yet these storms are an integral part of the life and ecology of our planet. By inducing exchanges of heat and cold on a vast scale from the equator to the poles, from upper to lower atmospheres, and by the continual releasing of energy, storms act as safety valves and essential balancers of the earth's climate. The atmosphere has persistently protected all living creatures of earth by forming a gaseous shield against destructive radiation and waves of solar and cosmic energy.

Man has long considered winds and weather as fickle, freakish, abnormal. Yet, like all natural phenomena, these events result from strict physical and chemical laws—created, sustained, and dispelled by an enormous profusion of complex but systematic factors. In constant response to effects of temperature, gravity, pressure, humidity, and topography, the atmosphere is cooled and heated, moistened and desiccated, churned and driven in ways that have often seemed mysterious and capricious. But behind the apparent vagaries and vicissitudes of the aerial elements, there is a grand global scheme at work, gradually becoming more understandable to those who probe its depths and seek to predict its ways.

As modern civilization becomes increasingly complex, it becomes increasingly sensitive and vulnerable to the various phenomena of weather. Events that are not meteorologically unusual may now have devastating impact upon expanding communities. A pre-spring warm air mass may trigger a deluge of melting snow water over a newly developing suburb; a burst of rain may bring houses, cars, lawns, and roads tumbling in a landslide; a lack of wind may stagnate death-dealing smog over a city; a pocket of clear air turbulence may jolt and drop a jet liner from the sky. Now that man is breaking out of the shell of earth's atmosphere into space, he is encountering new hostile forces in the solar system.

Meteorological research and application is a fast-moving field, in which new solutions to old problems and challenging controversies to established theories are constantly occurring. Changing theories and newly discovered facts often conflict with one another, and scientists may disagree with their fellow colleagues over their applications and interpretations.

By writing about these vibrant, complex, and technical matters in nontechnical language, the author has attempted to simplify many ideas for the layman and yet retain the necessary accuracy and detail. The author hopes this book will stimulate the curiosity of the reader and help him understand some of the causes, structures, locations, timing, and other patterns of winds and storms that ravage the earth.

CONTENTS

PHOTOGRAPHS

(Following page 112)

Hurricane waves in Florida, 1964

Satellite photograph of five hurricanes, 1967

Apollo 7 photograph of Hurricane Gladys, 1968

Hurricane-battered ships near Gulfport, Mississippi, 1969

Hurricane damage in Haiti, 1963

Tornado approaching the town of Vulcan in Alberta, Canada, 1927

Tornado funnel near Tracy, Minnesota, 1968

Aftermath of a tornado, St. Louis, Missouri, 1959

Aftermath of a devastating 1957 tornado

Spectacular flash lightning

Hailstones

Hail-shattered corn in Iowa

Hazardous fog

Ice storm in New England

Climate-changing industrial wastes

Iceberg in a North Atlantic shipping lane

Titanic at the start of her maiden voyage

Snow-blinded calf in a South Dakota blizzard, 1966

Ice rime in Wyoming

Aurora borealis

I. HURRICANES

Introduction. With powerful winds and torrential rains, more than 60 hurricanes rise from the tropical seas each year and spin in large curved paths across oceans and lands.

Towering ten thousands of feet high and covering several ten thousands of square miles, the hurricane rotates around its relatively calm central eye like a giant top, bringing death and destruction whenever it encounters segments of man's expanding civilization.

Hurricanes originate in warm sunny seas in two general areas north and south of the equator. The storms start innocently enough as mere disturbances of gently whirling winds and slightly lowering pressures. Given a spin from the rotating earth, young hurricanes travel leisurely at first, somewhat parallel to the equator. Then, fed by immense amounts of energy from warm moist air and nurtured by certain conditions of wind and pressure, the storms deepen in intensity and develop into full-fledged hurricanes that sweep toward the North and South Poles in great curves before losing their energy in the colder regions of the earth and dying out.

These storms have killed thousands of people, flattened buildings, destroyed towns, flooded vast regions of land, sunk armadas, and even changed the course of history. But they have also brought needed rain to parched regions and may prove to be an essential factor of the earth's meteorological system, transferring heat and energy between the equator and the cooler temperate regions toward the poles.

With instruments and equipment rapidly becoming more and more precise, scientists are exploring every dimension of these mighty storms, testing their vast reserves of energy and probing the complex mysteries of how they are created, develop, and die. As yet, man cannot control their powerful forces or change their direction—and he may not wish to, for more harm than good may come from tampering with such vast energies.

1. What is a hurricane? A hurricane is a large, rotating storm with strong winds blowing at speeds of 74 miles an hour or more around a relatively calm center called the eye. It blows counterclockwise in the Northern Hemisphere and clockwise in the Southern Hemisphere. The whole storm system may be some 5 to 6 miles high and 300 to 600 miles wide and moves forward, like an immense spinning top, at speeds about 12 miles an hour.

These storms, also called tropical cyclones and typhoons in certain regions of the world, can dominate the atmosphere and earth's surface over tens of thousands of square miles. Starting in certain seasons at low latitudes in tropical oceans near the equator, these hurricane systems usually move forward in a westerly direction parallel to the equator. They pick up speed as they develop and gradually swing toward the poles. Some hurricanes continue traveling toward the west; others recurve and move back toward the east before they die out. With the help of satellites, radar, airplanes, and other equipment, meteorologists are now keeping closer watch over these storms, each of which is given an individual name as it is carefully tracked.

Because of their considerable size and intensity, and because they last for several days or even weeks, hurricanes cause extensive damage with their heavy winds, rains, floods, high waves, and tides.

2. Where are hurricanes called cyclones? A hurricane is called a cyclone in the Bay of Bengal and the northern part of the Indian Ocean.

3. Where are hurricanes called typhoons? A hurricane is called a typhoon in the western area of the North Pacific Ocean, where they strike Japan and Korea with devastating regularity.

4. What causes a hurricane? Although scientists are coming closer to understanding the many factors that create hurricanes, they have yet to determine the exact processes of formation.

In essence, a hurricane must have the right ingredients of warmth and water vapor to supply its energy, as well as a certain amount of convection activity and vertical wind motion to bring in air from sea level and move it up through the storm system. Added to this, the wind system must be given the right amount of spin or twist, provided by the rotating earth. (For more detailed explanations, see Questions 78 through 91.)

5. What kinds of damage do hurricanes cause? Persistent winds and heavy gusts of hurricanes have caused enormous amounts of damage by blowing down buildings, toppling trees and telephone poles, hurling pieces of roofs, walls, branches, and other debris like ramrods against other structures, and crushing people beneath the weight of falling objects. At sea, thousands of boats and ships have been sunk or smashed to pieces on shores and many thousands of people drowned.

The major cause of destruction is the flooding that results from the large sea waves and tides driven ashore by hurricane winds, or from the torrential rains dropped from moisture-laden clouds.

● **6. What was the worst tragedy caused by a hurricane?** Probably the worst loss of life was caused by the tropical cyclone that swept up the north end of the Bay of Bengal on October 7, 1737, and over the mouth of the Hooghly River and the River Ganges delta—killing about 300,000 people. Strong gales and rough seas pushing up the Bay sank about 20,000 boats and seacraft of all descriptions and sizes. Along the coast, a wave some 40 feet high, driven by the wind and magnified by the configuration of the Bay, swept over the islands and lowlands, inundating some 6,000 acres.

A similar death toll of 300,000 was reported in 1881 when storm waves generated by a typhoon swept around the Haifong area in China. Facts and figures on this disaster, however, are somewhat vague.

7. What was the 1965 disaster in the Bay of Bengal? The year 1965 was a particularly terrifying one throughout the lowlands of India and East Pakistan, as cyclones and their enormous storm waves killed an estimated 60,000 people in three separate storms. Figures of the deaths are difficult to obtain and often vary, depending upon the information source. According to some reports, an estimated 17,000 people were killed by a cyclone in May; an estimated 30,000 were killed in June; and another 15,000 were killed in mid-December.

8. What other disasters have occurred in this area? In 1876, a cyclone-driven storm tide and waves killed an estimated 100,000 people in the Bay of Bengal area. A disease resulting from the inundation killed 100,000 more people shortly afterward.

Another catastrophe had taken place in 1864, when an estimated

50,000 people were killed by a cyclone and some 100,000 cattle and livestock were drowned.

More recent disasters in this area include the cyclones of 1942—on October 15 and 16 when 11,000 people died and on November 6 when 10,000 people were killed.

Another large storm on October 10, 1960, took the lives of 6,000 people, and on the 31st of the same month a storm killed 4,000 people.

9. Why do such terrible disasters occur in the Bay of Bengal? The area at the north section of the Bay of Bengal is particularly vulnerable to wind and wave disasters because the sea floor of the Bay slopes upward and thus huge masses of water are driven into shallow shores and over the low-lying land. The area here—part belonging to India, part to East Pakistan—is densely populated. Millions of fishermen, farmers, and their families live on the many islands at the mouths of the two great rivers, the Hooghly and the Ganges, and along the banks that border them. Calcutta, one of the most heavily populated cities in the world, is situated on the Hooghly River and is in the path of the inundations.

10. What famous typhoon damaged the U.S. Navy? One of the most destructive typhoons in the naval history of the United States rose out of the sea on December 17 and 18, 1944, during World War II and smashed into Task Force 38 of Admiral William Halsey's U.S. Third Fleet about 500 miles east of Luzon in the Philippines.

The storm caught the American fleet unprepared in the middle of refueling operations during naval encounters with Japan.

With winds blowing up to 150 miles an hour, the storm struck the fleet of some 90 ships spread over 3,000 square miles of the Pacific Ocean. Many of the ships had empty fuel tanks which caused them to be lightweight and ride high on the water, in no condition to fight the storm. Some of the captains ordered their empty oil tanks filled with seawater to act as ballast—a standing order for destroyers when encountering rough weather, but difficult to execute in a raging storm.

After the typhoon subsided, the damage was tallied: the fleet had lost 3 destroyers, 146 aircraft, and 790 men. Eighteen ships had suffered major damage.

11. Where did other death-dealing typhoons strike? In late September, 1959, typhoon Vera struck central Japan, causing a recorded 5,041 deaths and destroying some 40,000 homes. Damages were estimated to be about $1.2 billion.

Another Pacific typhoon, named Marie, struck Japan on September 26 and 27, 1954, and killed 1,700 people, sank 600 ships, and destroyed 200,000 buildings.

In September, 1953, a typhoon hit Vietnam and Japan, killing a recorded 1,300 people.

12. What was the worst hurricane disaster in the United States? On September 8, 1900, a violent hurricane crossed the coastline of Galveston Bay, pushing high hurricane tides 10 to 15 feet deep upon the city of Galveston, Texas. About 7:30 P.M. in only a few seconds, a sudden rise of water of about four feet surged through the city, and later a 20-foot wave added more water. During a 24-hour period, about 3,000 houses—half the homes of Galveston—were destroyed and 6,000 people drowned, thus making this the worst hurricane disaster in the United States in terms of people killed. The exact death toll may never be known, since whole homes were washed away and complete families died together. Property damage was estimated at $30 million.

● **13. What was another destructive hurricane?** Hurricane Flora was the name of a disastrous hurricane that rose out of the Caribbean Sea in 1963 and killed perhaps more than 6,000 people throughout the islands. (See Questions 35 through 40 for naming of hurricanes.)

First noticed on September 29, the gathering storm struck Haiti on October 3, killing an estimated 3,000 people as it ripped through villages and towns and inundated the island with floods. Flora then moved on to Cuba, where it was blocked from traveling farther westward by a strong high pressure area. For four days it moved back and forth over Cuba, killing uncounted thousands of people, devastating homes, and ruining the sugarcane and coffee crops. Then it drove on northward into the Atlantic Ocean, without touching the shores of the United States.

Winds of this powerful hurricane were recorded as averaging 160 miles per hour, with gusts up to 200 miles per hour.

14. What was the Lake Okeechobee hurricane? At 10:30 P.M. on September 16, 1928, hurricane winds and waves broke the eastern earthern dike built on the southern end of Lake Okeechobee, Florida, and inundated the flat farmland, drowning 1,836 people and injuring 1,849 in only a few hours. Most of these people were sharecroppers who had settled in the fertile area near the Everglades about 12 years earlier to raise eggplants, peppers, avocados, and other vegetables. Property damage in Florida was estimated to be $25 million.

The hurricane had blown from the Caribbean, where it had devastated the West Indies, killing about 2,000 people on the islands. Property loss in Puerto Rico alone was estimated at $50 million, with 300 people killed on that island.

15. What was the 1938 New England hurricane? The New England hurricane of 1938, lasting from September 10 until it blew itself out September 22, killed some 600 people, knocked down some 275 million trees, and created damages exceeding $250 million dollars. Most of the damage occurred in eastern Long Island, eastern Connecticut, Rhode Island, and southeastern Massachusetts, all of which were exposed to large surge waves.

This was the first hurricane to hit this region in 70 years and one of the very few on record to retain such strength as far north as 40 degrees latitude. It sometimes moved forward at unusually high speeds of 60 miles an hour.

16. What damage did hurricane Hazel do? No one believed that hurricane Hazel posed any threat to the Carolina coasts as it slowly lumbered around the West Indies for a week in early October, 1954. On the 12th of October, Hazel crossed southwestern Haiti, killing an estimated 400 to 1,000 people, including about 200 who were buried in landslides. It then swung out to sea, and people believed the United States was spared.

However, the hurricane veered west again and moved inland, sweeping over the beaches of South Carolina on October 15 with winds estimated as high as 150 miles an hour.

This was the most severe tropical storm in over 100 years in the Cape Fear area, and one of the most devastating tropical storms ever to reach northeastern United States and southeastern Canada. Total damage along the Carolina beaches was estimated at $61 million, as Hazel flattened business areas and homes, destroying every fishing pier from Myrtle Beach to Cedar Island, a distance of 170 miles. Yet because of radio and other warning systems only 19 people were killed in this area.

After the destruction along the coast, Hazel continued northward, creating an 800-mile-long path of destruction with nearly $89,000 in damages and 75 more fatalities. Some 14 hours later Hazel's center passed over Toronto, Canada, before breaking up and vanishing.

17. What devastation resulted from hurricane Camille? One of the most powerful hurricanes of the twentieth century swept across the Gulf of Mexico and slammed into the United States coastal areas August 17, 1969. With winds estimated at 190 miles an hour and gusts reaching 200, the hurricane Camille brought tides 20 feet above normal, dumped rainfalls of 6 to 8 inches, washed away towns, buildings, bridges, trees, and killed an estimated 500 people. Damages were estimated to run about $1 billion.

More tragedy occurred after the hurricane with weakened winds traveled inland into northern Mississippi, Tennessee, and Kentucky losing its hurricane status as it traveled north. Westerly winds suddenly pushed the waning storm center eastward, against the Appalachian mountains where it began unloading tons of water. More than 10 inches of rain fell over West Virginia and parts of Virginia, causing torrential floods to pour down valleys that swept away towns and caused loss of life and property. The wayward Camille crossed southern Virginia into the Atlantic Ocean where it picked up speed again, and finally dissipated August 22 over the ocean.

18. What is considered the most intense hurricane on record? Although the Labor Day hurricane of 1935 was small, it is considered the most powerful hurricane on record. It was only about 40 miles in diameter, yet had winds blowing as fast as 200 miles an hour and more. It had an extremely low pressure—at Matecumbe Key off Florida the barometer read 26.35, lowest on record. First sighted east and north of Turks Island in the Bahamas, the hurricane arrived at Florida Keys late September 2, with a storm wave estimated to be some 15 to 20 feet high. About 400 people lost their lives that night.

19. What were some other famous Atlantic hurricanes? One of the worst hurricanes in a long list of disasters was the storm that struck Santa Cruz del Sur, Cuba, in November, 1932, bringing a rise of the sea that drowned some 2,500 people, out of a population of about 4,000.

The Great Atlantic Hurricane of September 8 to 16, 1944, was aptly named, for it was one of the most violent hurricanes yet experienced. This storm traveled from northeast of Puerto Rico up the Atlantic coast into the vicinity of Chesapeake Bay, where it created more than $22 million in damage and killed 63 people. It continued toward New England destroying another $100 million worth of property and killing 390 more people before turning out to sea.

On June 27, 1957, hurricane Audrey swept up the Gulf of Mexico and smashed into Louisiana, killing 600 people and inundating villages and towns. The town of Creole was completely destroyed. Audrey traveled into Texas and continued northeastward through the Ohio Valley, into New York State, and then into Canada before losing its energy and disappearing.

20. What was the Great Hurricane of 1780? The year 1780 has long been known as the year of the Great Hurricane. At least three large Caribbean storms occurred about the same time and they have been confused together and sometimes considered as one. In analyzing eyewitness reports, historians claim three, perhaps five, storms that affected military forces and ships of England, France, Spain, and the Netherlands during the American Revolution. One storm destroyed the town of Savanna-la-Mar, Jamaica, in early October; another damaged the Spanish fleet assembled in the Gulf of Mexico to attack the British base at Pensacola; and a third storm damaged a British fleet and then a French fleet. Ordinarily, ships withdrew from the West Indies area in the hurricane season, but this year military forces lingered in the vicinity because of the Revolutionary War and were badly hit.

21. What was an account of the Great Hurricane? The following report gives an idea of the havoc raised by the 1780 hurricane.

The most terrible cyclone of modern times is probably that of the 10th of October, 1780, which has been specially named the "great hurricane." Starting from Barbados, where neither trees nor dwellings were left standing, it caused an English fleet anchored off St. Lucia to disappear, and completely ravaged this island, where 6,000 persons were crushed under the ruins. After this, the whirlwind, tending toward Martinique, enveloped a convoy of French transports, and sunk more than 40 ships carrying 4,000 soldiers; on land the towns of St. Pierre and other places were completely razed by the wind, and 9,000 persons perished there. More to the north, Dominique, St. Eustatius, St. Vincent and Porto Rico were likewise devastated, and most of the vessels which were on the path of the cyclone foundered with all their crews. Beyond Porto Rico the tempest bent to the northeast toward the Bermudas, and though its violence had gradually diminished, it sunk several English warships returning to Europe. . . .*

22. What war was averted because of a hurricane? A hurricane prevented war in March, 1889, when America, Germany, and England were on the brink of open hostilities. A German naval force had shelled a local village at Apia, Samoa, and destroyed some American property. Three American warships sailed into the harbor, where they encountered 3 German and 1 British warships. Before anyone could fire, a hurricane struck, sinking all 3 American and all 3 German ships, as well as 6 merchant ships. The British ship, *Calliope*, one of the few to be equipped with engines, barely managed to steam out of the bay and out of danger.

For lack of ships, the war was averted and a compulsory armistice was enforced.

23. When was the first hurricane of the New World recorded? Early reports have been lost, damaged, and misinterpreted, but historians believe that the first hurricane encountered by Europeans may have occurred in June, 1494, during the second voyage of Christopher Columbus. Journals record that a hurricane sank or blew ashore two ships left in the harbor at Isabella, Santo Domingo (now Dominican Republic). At the time of the tragedy, Columbus was sailing in another ship to the south of Cuba.

In October of the next year, 1495, Columbus himself experienced a hurricane in Isabella harbor. At this time 6 ships were reported sunk or destroyed, and the only ship to weather the storm was Columbus' *Nina*.

24. How costly are hurricanes? Estimates of damage in dollars vary, but approximate figures have been calculated on hurricane costs. They may cause millions and sometimes billions of dollars in damage

* Élisée Reclus, *The Ocean, Atmosphere, and Life* (New York: Harper & Brothers, 1873), p. 256.

each year. On the basis of damage inflicted by several large storms in the years 1964 and 1965, for instance, hurricanes were estimated to cost $6 billion for the two years. Between 1955 and 1964, U.S. hurricanes were estimated to have cost more than $10 billion.

In the Philippines, annual damages from typhoons may cost as much as $500 million.

25. Is property damage from hurricanes increasing? Property damage from hurricanes in the United States has been steadily rising in recent years, whereas deaths have been declining. Greater damage is inflicted as cities and industrial complexes continue to grow in areas subject to these storms. Yet the death rate is falling because of improved theories and weather instruments that permit better forecasting and warning systems.

26. What was one of the worst hurricane years? The year 1955 was a devastating year of hurricanes, with 11 storms destroying a total of $2 billion worth of property and killing some 1,500 people.

Hurricane Connie, third storm of the year, blew with 145-mile-an-hour winds across the North Carolina coast near Cape Lookout on August 12. It then moved across the Chesapeake Bay area and toward the north, over Lake Huron. The storm killed 25 people and caused over $86 million worth of damage.

Hurricane Janet, tenth storm of the year, was recorded with winds approaching 200 miles an hour. It hit the Windward Islands and the Grenadines, causing $3 million in property damages and killing 160 people.

27. What was the "billion-dollar hurricane"? Hurricane Diane of August, 1955, earned itself the name of "billion-dollar hurricane" as it moved through the eastern coastal areas. Until hurricane Betsy of 1965, Diane was considered to have caused the most property damage of any hurricane in the history of the United States.

Diane started to brew in the southeastern part of the north Atlantic Ocean on August 7. It was not considered a very strong hurricane, for the winds were not violent and did not whip up many sea waves. As it approached the coast of North Carolina, the winds subsided. Meteorologists tracking Diane classified it as just another "wind disturbance," and hurricane warnings were discontinued. Yet it was at this time that Diane caused some of the most damaging floods on record. An estimated 200 people were killed; mostly by drowning in flash floods.

Since that date, hurricanes have caused more economic damage than this, but Diane was the first to obtain this questionable renown and will always be remembered as the "billion-dollar hurricane."

28. Why was the dying Diane so destructive? Even though the winds of hurricane Diane diminished so much that the storm lost its official status as a hurricane, Diane kept moving, and began to pour record-breaking rains over the lands as it advanced over North Carolina, the middle Atlantic states, and parts of New England. Ordinarily, the flooding rains of the dying hurricane would have soaked harmlessly into the land or been sustained in lakes and rivers, but this hurricane followed the general track of hurricane Connie which had already saturated the region only a few days before. The added water load from Diane resulted in enormous flood damages and a death toll of some 200 people.

● **29. What U.S. hurricane caused the most property damage on record?** From August 27 to September 12, 1965, hurricane Betsy moved out of the western Atlantic and swept over southern Florida, across the Gulf of Mexico, and into Louisiana, then traveled northward, creating a record for property damage from a single hurricane—some $1.4 billion worth. Much of this damage was caused by flooding, particularly in Louisiana. With top wind speeds given as 136 miles per hour, Betsy caused 75 deaths in the United States.

● **30. What was another expensive hurricane?** On September 7, 1967, hurricane Beulah rose out of the sea to smash Texas and Mexico with damages costing more than a billion dollars. One of the strongest hurricanes in U.S. history, Beulah caused record floods over 40,000 square miles, giving Mexico the worst flood of the twentieth century. Riding with winds of 160 miles per hour, Beulah dropped 20 to 30 inches of rain, destroyed homes and businesses, and seriously damaged citrus and pepper crops. It killed 55 people, 12 of whom were in Texas.

● **31. What is it like to feel a hurricane approaching?** Reports that give an intense and immediate physical description of any natural disaster are difficult to find. Descriptions usually run to vague, emotional, and general wording such as "beyond belief," "terrible holocaust," "unimaginable terror" or to objective numerical reports made after the event has occurred, giving statistical death and destruction counts. The following excerpts give a more specific indication of the sounds, sights, and feeling experienced by a person as a hurricane draws near.

It is Wednesday morning; the sky is clear and the barometer is several hundredths of an inch higher than yesterday. The long, high swells are now definitely rolling in and, breaking, crash on the shore with a sound that can be heard nearly half a mile inland. . . . As the hours pass, the weather feels unusually oppressive and sultry, the winds are variable and fitful with occasional dead calms. . . . The sunset is red and orange, and very beautiful. . . .

It is after midnight when occasional gentle but rather gusty breezes begin to rustle through the tropical foliage, and the sound is somewhat different from usual. The moon still shines dimly through the high cloud. . . .

By 10:00 a.m. the wind is north, 45 to 50 mph, and increasing with every passing squall, and the barometer is down more than another tenth. The tree limbs are thrashing; the coconut palms bent from the constant northeasterly trades lean even more in the gale. One last look at the ocean before the door is closed and bolted—it is white with blowing spray and blends with the sky a short distance offshore. Within minutes the wind reaches hurricane force and continues to rise higher and higher, driving water in around the windows; water gurgles under the door, water everywhere.

During the next two hours the howling and screaming of the wind becomes indescribable, the house quivers and shakes and seems to want to rise off the foundation, torrents of rain gush all around. The wind is in excess of 100 mph and is now dealing destruction over the island. One hears splitting of limbs, the crashing of trees, the occasional thud of a coconut against the house, the noise of unseen objects flying through the air and hitting other objects. One is almost numbed by fear but work has to be done. . . . It seems impossible but the wind continues to increase, its pitch higher and higher. Verily it seems like a monstrous living thing trying to tear the house apart. . . .

32. Where did the word *hurricane* originate? The word *hurricane* comes from the Spanish word *huracan*, which may have originated

Gordon E. Dunn, Banner I. Miller, *Atlantic Hurricanes* (Baton Rouge: Louisiana State University Press, 1960), pp. 58–60.

from the vocabulary of Carib and other Indian tribes once inhabiting the Caribbean islands and Central and South America.

The Guatemalan Indians called the god of stormy weather *Hunrakan*. In the language of a former Haitian tribe, *huracan* meant evil spirit. The Galibi Indians of Dutch and French Guiana used the word *hyroacan* to mean devil, and the Quiche of southern Guatemala spoke of *Hurakan* as their god of thunder and lightning.

33. What other words are used for hurricanes? Other Carib words for hurricane include *aracan, urican,* and *huiranvucan,* which can be translated as big wind, powerful wind, and such terms.

There are many different words and spellings that have been used to connote high winds and storms—all sounding phonetically similar. Harrycain, jimmycane, furicano, and hurleblast are a few.

34. What are hurricanes called in other parts of the world? In Australia, hurricanes are called tropical cyclones or hurricanes in official reports on northwest Australia and the Timor Sea—and willy-willies in unofficial language.

A hurricane is called a baguio in the Philippines.

Tropical cyclones are sometimes called cordonazos in areas south and southwest of Mexico and Central America.

35. How have individual hurricanes been named in the past? For several hundred years, many devastating hurricanes in the West Indies were named after the particular saint's day on which they occurred. For instance, hurricanes were remembered as Hurricane Santa Ana, Hurricane San Felipe the First, and Hurricane San Felipe the Second.

Scientists then began identifying hurricanes and keeping track of them by using latitude and longitude and the dates on which they were detected. This system proved cumbersome and difficult because the storms traveled such complex and lengthy paths, over a period of many days and weeks.

● **36. How are hurricanes named today?** Meteorologists now use girl's names to identify hurricanes in the Atlantic Ocean and typhoons in the Pacific Ocean.

For hurricanes originating in the Atlantic Ocean, Caribbean Sea, and Gulf of Mexico, four sets of names are used in alphabetical order. The letters Q, U, X, Y, and Z are not included because there are not many names beginning with those letters. Every four years the cycle starts over again and names are reused.

If a major hurricane seriously affects the United States, the name assigned to it is retired for ten years and another name is substituted in the list.

Girls' names are also used to identify typhoons and hurricanes in the Pacific Ocean. Because the Pacific has a much larger number of tropical storms each year, the region has been broken down to include those occurring in the eastern North Pacific and those in the central and western North Pacific.

● **37. What are names of future Atlantic hurricanes?** Four quasi-permanent sets of hurricane names in alphabetical order have been prepared by the Weather Bureau to be used in future years in the Atlantic Ocean area.

For the years 1970, 1974, 1978—Alma, Becky, Celia, Dorothy, Ella, Frances, Greta, Hallie, Isabel, Judith, Kendra, Lois, Marsha, Noreen, Orpha, Patty, Rena, Sherry, Thora, Vicky, Wilna.

For the years 1971, 1975, 1979—Arlene, Beth, Chloe, Doria, Edith, Fern, Ginger, Heidi, Irene, Janice, Kristy, Laura, Margo, Nona, Orchid, Portia, Rachel, Sandra, Terese, Verna, Wallis.

For the years 1972, 1976, 1980—Abby, Brenda, Candy, Dolly, Evelyn, Felice, Gladys, Hannah, Ingrid, Janet, Katy, Lila, Molly, Nita, Odette, Paula, Roxie, Stella, Trudy, Vesta, Wesley.

For the years 1973, 1977, 1981—Anna, Blanche, C——*, Debbie, Eve, Francelia, Gerda, Holly, Inga, Jenny, Kara, Laurie, Martha, Netty, Orva, Peggy, Rhoda, Sadie, Tanya, Virgy, Wenda.

● **38. What are the names of future hurricanes (also called typhoons) in the eastern North Pacific Ocean?** In the eastern North Pacific Ocean, hurricanes are identified by a set of alphabetical listings of girls' names are four-year cycles, similar to the method used in the Atlantic Ocean.

---

* Camille was originally designated, but another name beginning with *C* will be assigned during these years, since the 1969 hurricane Camille was so devastating.

1970	*1971*	*1972*	*1973*
Adele	Agatha	Annette	Ava
Blanca	Bridget	Bonny	Bernice
Connie	Carlotta	Celeste	Claudia
Dolores	Denise	Diana	Doreen
Eileen	Eleanor	Estelle	Emily
Francesca	Francene	Fernanda	Florence
Gretchen	Georgette	Gwen	Glenda
Helga	Hilary	Hyacinth	Heather
Ione	Ilsa	Iva	Irah
Joyce	Jewel	Joanne	Jennifer
Kirsten	Katrina	Kathleen	Katherine
Lorraine	Lily	Liza	Lillian
Maggie	Monica	Madeline	Mona
Norma	Nanette	Naomi	Natalie
Orlene	Olivia	Orla	Odessa
Patricia	Priscilla	Pauline	Prudence
Rosalie	Ramona	Rebecca	Roslyn
Selma	Sharon	Simone	Sylvia
Toni	Terry	Tara	Tillie
Vivian	Veronica	Valerie	Victoria
Winona	Winifred	Willa	Wallie

39. What are the names of future hurricanes in the central and western North Pacific Ocean? In the central and western North Pacific Ocean, the entire list is used all the way through—the first name used each year is the name following the last one used in the preceding year. The entire list is estimated to last for a period of two to three years.

The list is repeated when the last name in Column 4 has been used.

Column 1	*Column 2*	*Column 3*	*Column 4*
Alice	Anita	Amy	Agnes
Betty	Billie	Babe	Bess
Cora	Clara	Carla	Carmen
Doris	Dot	Dinah	Della
Elsie	Ellen	Emma	Elaine
Flossie	Fran	Freda	Faye

Grace	Georgia	Gilda	Gloria
Helen	Hope	Harriet	Hester
Ida	Iris	Ivy	Irma
June	Joan	Jean	Judy
Kathy	Kate	Kim	Kit
Lorna	Louise	Lucy	Lola
Marie	Marge	Mary	Mamie
Nancy	Nora	Nadine	Nina
Olga	Opal	Olive	Ora
Pamela	Patsy	Polly	Phyllis
Ruby	Ruth	Rose	Rita
Sally	Sarah	Shirley	Susan
Therese	Thelma	Trix	Tess
Violet	Vera	Virginia	Viola
Wilda	Wanda	Wendy	Winnie

● **40. Why are hurricanes named for girls?** Experience shows that
the use of girls' names in written and spoken communications is
shorter, quicker, and causes fewer mistakes than any other hurricane
identification yet used. The names are clearly pronounced, quickly
recognized, and easily remembered.

These assets are especially important when detailed storm information is sent over long distances to many weather stations.

The idea of using girls' names may possibly have originated in the
novel *Storm*, by George R. Stewart, published in 1941, in which a
meteorologist referred to active storms as girls.

**41. What other methods have been suggested for naming hurri-
canes?** Suggestions have been made to designate hurricanes by numbers (1–2–3), by English letters (A–B–C), by Greek letters (alpha–
beta–gamma), by boys' names (Arthur–Ben–Christopher), by the
phonetic alphabet used by U.S. military services during World War
II (able–baker–charlie), by the International Civil Aviation Organization's phonetic alphabet (alfa–bravo–coca), by the names of
animals (antelope–bear–coyote), or by descriptive adjectives (annoying–blustery–churning). Many more suggestions include little-known
mythological characters and historical personalities as well as famous
personalities, places, and things.

42. What is the structure of a hurricane? The seemingly chaotic and violent hurricanes actually have a definite structure, with a well-defined pattern of winds. Scientists are finding more about this as they probe closer into these mammoth storms.

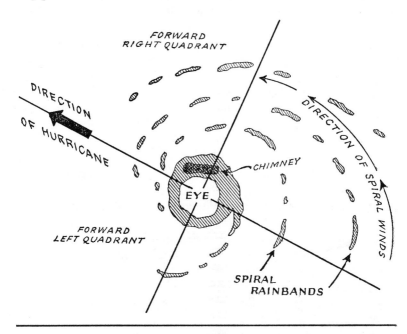

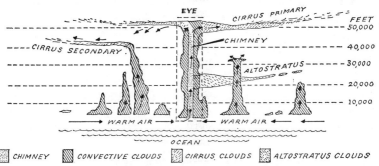

Structure of a hurricane

In essence, a hurricane has an area of relative calm in the center called the eye, toward which winds and rain clouds spiral in enormous bands. (See Questions 43 through 47.) Around this eye blows a bank of clouds—the region of strongest winds. Within the wall of clouds, the chimney or hot tower of the hurricane is located. This is a primary energy cell of a hurricane, through which moist heated air moves upward from the ocean surface. (See Question 49.)

As the winds spiral into the center of the storm, they bring in moist air in bands of precipitation called rainbands. Towering sometimes as high as 50,000 feet around the center of the storm, these rainbands may extend outward from the storm's center for several hundred miles. (See Question 50.)

Around the edge of the whole rotating storm, some 200 to 300 miles from the hurricane center, moderate winds blow in short flurried gusts.

43. What is the eye of a hurricane? The eye of a hurricane is the innermost portion of the storm, a zone of surprisingly light breezes or even of almost windless calm. Within this eye, skies are often clear, and sunlight or starlight may stream all the way to the earth's surface, while winds and clouds continue raging around the edge of the eye.

On the average, eyes are about 14 to 20 miles across. In the Pacific Ocean, where typhoons are often larger than other such storms, the eyes are usually larger, sometimes as large as 50 miles in diameter. The cross section of the eye may vary in shape, ranging from circular to elliptical.

Within this eye are found the lowest pressures, the highest temperatures, and the lowest relative humidities of the storm.

The eye is not always in the center of the storm. Sometimes it turns or moves in various directions within the storm itself, which continues to move forward on its own course.

44. How does the eye appear on radar? When viewed on radar, the eye is usually readily identified. In mature hurricanes it appears clearly as an echo-free area.

45. Why have eyes been called treacherous? Many people, not understanding the structure of a hurricane, have been deceived by the sudden appearance of the serene eye as the storm passes overhead. Some venture out into the sunny calm, believing the storm to be over—only to be killed or injured when the other half of the hurricane arrives, a few minutes to half an hour later, with powerful winds and heavy rains blowing in the opposite direction.

46. How have eyes helped storm-tossed victims? To some people and creatures, the hurricane eye has been an unexpected blessing. Sea captains often welcome the calm as an opportunity to secure ship and repair damages before the other half of the storm arrives.

Birds, butterflies, and insects caught in the eye of a storm have managed to fly within it in front of the second half of the storm, dropping to rest on a ship or land when it appears. Birds in particular have been known to travel great distances in the eye of a hurricane. Tropical birds have been found in New England after a hurricane from the Caribbean Sea passed through—more than 2,000 miles from their home.

47. What does it feel like to ride through the hurricane's eye? On August 15, 1951, a U.S. Weather Bureau reconnaissance plane approached the eye of typhoon Marge raging in the Pacific Ocean. The following account describes the experience.

Soon the edge of the rainless eye became visible on the [radar] screen. The plane flew through bursts of torrential rain and several turbulent bumps. Then, suddenly, we were in dazzling sunlight and bright blue sky.

Around us was an awesome display. Marge's eye was a clear space 40 miles in diameter surrounded by a coliseum of clouds whose walls on one side rose vertically and on the other were banked, liked galleries in a great opera house. The upper rim, about 35,000 feet high, was rounded off smoothly against a background of blue sky. Below us was a floor of low clouds rising to a dome 8,000 feet above sea level in the center. There were breaks in it which gave us glimpses of the surface of the ocean. In the vortex around the eye the sea was a scene of unimaginably violent, churning water.*

48. How do wind forces move within a hurricane? A hurricane has sometimes been likened to a simple heat engine, with warm moist air supplying the driving force.

As the sun beats down over tropical seas, the moist air becomes

* From "Hurricanes" by R. H. Simpson. Copyright © 1954 by Scientific American, Inc. All rights reserved.

heated and starts to rise. More moisture-laden air is sucked into this warm core of the storm center, and the whole system is given a spiral twist by the rotation of the earth.

As the air ascends within the column, it expands in the reduced pressures of the upper atmosphere, and cools by expansion. By this cooling, moisture is condensed, which in turn releases heat to the surrounding atmosphere of the storm core, thus regenerating the heat cycle. This acts to intensify the storm.

The rising air flows out from the system at various heights and at the top, mixing with the cooler surrounding atmosphere where it starts to sink. This condition results in lower pressures in the center of the column and higher pressures outside, creating the kinetic mechanical energy—wind.

49. What are the hot towers of a hurricane? Hot towers are the primary energy cells of a hurricane, through which rise warm moist air currents, generating enormous amounts of energy that drive the hurricane. Located on or within the cloud wall that surrounds the storm's eye, these hot towers, also called chimneys, reach as high as 50,000 feet from the surface of the ocean, perhaps higher. These hot towers are prime targets for researchers trying to modify the storm's energy by dropping chemicals into the storm from airplanes. (See Questions 149 through 152.)

50. What are hurricane rainbands? Most of the heavy rain from a hurricane falls from the series of rainbands, or bands of precipitation, that spiral in toward the center of the storm. Long and narrow, they vary in width from 3 miles to as much as 23 miles, while the length of these bands may extend more than 300 miles. Some of them are 45,000 feet high, some are only 20,000 feet, and others are even less high at greater distances from the storm center. The outermost fringe of a rainband sometimes may extend several hundred miles in advance of the eye itself.

Essentially, these rainbands indicate the paths of winds bringing in warm moist air to feed the storm. The appearance of the spiral band structure changes gradually as the storm progresses and as individual showers move within the band, more or less along the spirals. Meteorologists can track the movements of a storm from the rainband pattern.

Rainbands preceding a hurricane are a potential flood threat, with heaviest rainfall usually occurring in the right semicircle of the approaching storm.

51. How many rainbands does a hurricane have? The number of rainbands around an eye varies from one to several. As many as 10 bands may exist in a large storm. Between them, rainfall is relatively light.

52. When were rainbands detected? Until only a few years ago, after the development of radar during World War II, men were unaware that such rainbands existed around the hurricane eye. With the raindrops reflecting the echo back to a satellite or land station, radar signals have clearly shown the rainband pattern.

53. How does a hurricane obtain its energy? Great amounts of energy are created when warm water is evaporated from tropical seas. In a hurricane this energy is stored as latent heat in the water vapor which, upon condensing, forms the towering clouds circling the hurricane eye. About 90 percent of this heat energy is released when the vapor expands, cools, and condenses into rain.

54. How much energy does a hurricane contain? As water vapor of a hurricane condenses into rain and about 90 percent of the total energy is released as heat, only as little as 3 percent may be converted into mechanical energy, or energy of the circulating winds. This relatively small amount of mechanical energy amounts to an equivalent of 360 billion kilowatt-hours per day. The total electrical energy generated in the United States each day is only about 2 billion kilowatt-hours. Thus, the mechanical energy of a hurricane in one day is equal to about a six-month supply of electrical energy for the entire United States.

55. How has the idea of this energy been translated into layman terms? Various calculations have been made to translate the idea of a hurricane's energy so the layman can understand:

One second of hurricane energy is equivalent to ten Hiroshima atom bombs.

One hour of hurricane energy is equal to all the electric power generated in the United States during an entire year (this figure is based on hurricane energy also released as heat—an estimated 16 trillion kilowatt-hours for one day).

The hurricane that hit Galveston, Texas, in September, 1900, had sufficient energy to drive all the power stations in the world for four years.

56. What is the direction of wind flow within a hurricane? In the Northern Hemisphere, the winds spiral inward in a counterclockwise direction. Within the core of the hurricane, they continue in the same direction, but change to clockwise circulation after flowing away from the center at high altitudes.

57. What are the general regions of wind circulation in a hurricane? Wind circulation within a hurricane can be divided into three general regions: the outer portion where winds are blowing at gale forces of about 40 miles an hour, extending from a diameter of 350 to 400 miles to within 20 to 60 miles of the center; the second region is of winds blowing at hurricane forces of 74 or more miles per hour, with a diameter of 100 miles extending to within 15 miles of the eye; and the third region is that of maximum winds averaging about 120 to 150 miles per hour. Some of these have never been exactly clocked because instruments are often broken in hurricanes. These winds blow in a circular band 5 to 30 miles wide immediately surrounding the eye of the storm. From this point inward, the winds fall off abruptly and the calm eye is formed.

58. What are some of the high speeds of rotating hurricane winds? A windstorm is not classified as a hurricane until its rotating winds reach speeds of 74 miles per hour or more. As the winds spiral inward, the closer they get toward the eye the harder they blow. Winds of 90 to 100 miles per hour are fairly common in mature hurricanes. Often they can reach 150 miles per hour and more. Brief gusts attain even greater speeds of 200 miles per hour.

59. How large are hurricanes? Small hurricanes exist with diameters of only 25 miles, while some of the greater hurricanes may extend 400 to 500 miles. The largest hurricanes, which build up across the great expanses of the Pacific Ocean, have diameters of more than 1,000 miles.

Some storms rise into the atmosphere 25,000 feet or more. Large storms may reach as high as 50,000 feet above the surface of the ocean.

60. How fast does a hurricane travel? Like a top spinning swiftly around its axis and "walking" slowly across the floor, the forward movement of the whole mass of a hurricane is relatively slow compared to the speeds of its rotating winds.

In its early stages of formation, while it may be only a small tropical storm, a hurricane moves forward very slowly—about 10 to 15 miles an hour. Sometimes these storms can remain stationary over one spot, while their winds continue to rotate at faster speeds as they gain energy from the hot moist air over tropical waters.

As the hurricane matures and moves poleward out of tropical waters, it picks up speed and may travel about 20 to 30 miles per hour. At latitudes of about 25 degrees north, the point of recurvature, hurricanes sometimes slow down again and even remain stationary for a while. Then they may pick up speed and move forward as fast as 60 miles per hour in the higher latitudes, before they begin to decrease in intensity, and die out.

61. What is the hurricane's point of recurvature? As a hurricane develops, matures, and eventually dies, it travels a curved path generally moving toward the west, then northward. In the North Atlantic Ocean, hurricanes often travel northward, then curve back toward the east, as they move from one prevailing wind stream to another. The area where the hurricane tends to turn is often called the point of recurvature. This point is usually near latitude 25 degrees, the boundary between the tropical easterlies and the middle-latitude westerlies.

62. How far do hurricanes travel? An average hurricane may travel about 300 to 400 miles a day—or about 3,000 miles before it dies out.

Some tropical cyclones have been tracked from east of Australia around the northern and western part of the continent to New Zealand—a distance of about 9,000 miles.

Rarely, a long-lived hurricane travels as far as 10,000 miles across the Atlantic and back again.

63. What specific hurricanes traveled long distances? The hurricane of August and September of 1900 was tracked from its beginnings in the Atlantic Ocean to the Texas coast, where it caused much devastation to Galveston. (See Question 12.) It then moved northeastward toward the Great Lakes region, then eastward across the Atlantic Ocean, across Europe, and into Siberia.

Hurricane Faith of August and September, 1966, journeyed a possible 10,000 miles as it crossed the Atlantic Ocean twice before dying out near the North Pole.

64. How long does it take a hurricane to pass over a point on earth? The time it takes a hurricane to pass an observer at one particular place on land or sea depends on how large the hurricane is and how fast it is moving.

Some storms, about average in size, have taken about 10 hours to pass a location. The first part of the storm may last some 4 hours. Then comes the quiet time of the eye—about 20 to 30 minutes. It takes another 4 or 5 hours for the other half of the storm to pass a spot, for the winds to die down, and for the rains to stop.

65. What are the signs of an approaching hurricane? As one stands near the ocean shore, one of the first definite signals of an approaching hurricane is a change in the sea. Along the coast where the storm is approaching, the level of the sea begins to rise slowly—several hours before the storm arrives. Great long waves called sea swells begin to pound with a long slow rhythm that is quite different from that of ordinary waves. As the storm approaches, the waves become heavier and rougher, and the sea rises higher in hurricane tides.

66. How does the sky look as a hurricane approaches? Signs of an oncoming hurricane can also be seen in the sky. Upon the horizon and spreading gradually over the sky appear cirrus, or high feathery, clouds, which often seem to converge at one point of the horizon— possibly indicating the location of the storm center. At sunset or sunrise the clouds on the outer border of the storm become highly colored, with deep tones of brilliant red and orange.

67. What other signs indicate a hurricane? The falling pressure in a barometer is a sure sign of a hurricane's approach, as the low pressure of the storm moves closer. Barometer pressures begin to drop about 12 hours before the arrival of the storm center. At the outer limits of the storm, rain begins to fall in squally showers, increasing in frequency and intensity. As the center of the storm approaches, the rains are heavy and steady; the winds continue rising and the low moaning of the wind can be heard through tree branches, telephone wires, and around corners, becoming higher in pitch and intensity as the storm draws near.

68. What happens to the barometric pressure during hurricanes? In normal weather, in tropical and subtropical regions, the weather barometer reads about 30 inches at sea level. This means that a normal pressure of about 15 pounds per square inch is being exerted at the earth's surface or about 2,160 pounds per square foot.

As the hurricane approaches, the inwardly spiraling and rising winds of the center cause a reduction of air pressure. The barometer falls, slowly at first, then more rapidly as the eye draws near. The rate of fall depends upon the depth of the barometric depression in the storm and the speed at which it approaches.

For each inch drop indicated by the barometer, about 70 pounds has been lifted off every square foot.

In fully developed hurricanes, the barometer nearly always falls below 29 inches at sea level. The lowest sea-level barometer readings for the entire world have all been recorded during hurricanes.

69. What were some of the lowest pressure readings? The lowest sea-level barometric pressure reading for a hurricane in the United States was 26.35 inches, recorded at lower Matecumbe Key, Florida, as the Labor Day hurricane passed over on September 2, 1935. The second lowest hurricane pressure was recorded with Camille in August, 1969—26.61 inches.

Another reliable record was taken aboard the S.S. *Sapoeroea* caught in a Pacific Ocean typhoon some 460 miles east of Luzon on August 18, 1927. This was the record low of 26.18 inches.

70. How many hurricanes form each year? In general, about 60 full-fledged hurricanes and typhoons rise out of the tropical seas each year to spin in wide curved paths toward the poles.

The exact number of such hurricanes cannot be stated with precision, mainly because such information is lacking in little-populated areas and also because the numbers vary considerably from year to year.

71. How many hurricanes form in the Northern Hemisphere? Some experts have calculated that in an active year for hurricanes, perhaps 50 such storms form in the Northern Hemisphere alone. This number seems relatively small when compared to about 20 cyclones formed each day outside the tropics in the winter season.

72. How many hurricanes form in the Pacific Ocean each year? Over a period of years, records of hurricanes average out with the following numbers:

About 21 hurricanes a year are formed in the southwestern part of the North Pacific Ocean. Many tropical cyclones form in the eastern North Pacific area.

In the South Pacific Ocean, more than 20 severe storms form, many of which reach hurricane status. About 6 hurricanes a year sweep up the Bay of Bengal, and about 5 form each year in the southern Indian Ocean and head toward Madagascar.

73. How many hurricanes form in the Atlantic each year? In the North Atlantic Ocean, about 10 hurricanes have been forming, on an average, per year for the past 20 years. The largest number of hurricanes and tropical cyclones recorded in the Atlantic in any 1 year was 21 in 1933. The smallest number was only 1 recorded in 1890 and in 1914; and 2 in 1925 and 1930.

74. Do many hurricanes form on the west coast of America? Fewer than 3 hurricanes a year develop in the Pacific Ocean off the west coast of Central America and Mexico each year.

75. How many hurricanes form in other parts of the world? One or 2 hurricanes form per year in the Arabian Sea region. About 1 hurricane a year forms northwest of Australia.

76. Can there be more than one hurricane at a time in an area? Usually there is only 1 hurricane at a time brewing in any specific area. However, 2 or 3 may be at different stages of development at the same time. In 1893, 1950, and 1967, as many as 3 hurricanes were reported in raging conditions at the same time in the regions of the Gulf of Mexico, the Caribbean Sea, or the Atlantic Ocean.

77. What are the general trends in the numbers of hurricanes? Over the past 70 years, an average of 8 hurricanes per year was recorded in the North Atlantic Ocean. This average increased to 9 per year over the past 40 years, and to 10 per year over the past 20 years.

This increase in hurricane numbers may be due to better detection methods, but meteorologists have noticed a significant increase of hurricanes beginning around 1930 when the general atmosphere seemed to be warming up. Some scientists believe there is a general decrease of hurricanes beginning around 1965 when the general atmosphere seemed to be gradually cooling.

78. What forces help form a hurricane? Several complex forces help generate a hurricane. These exist only in a few places in the world and at certain seasons of the year.

In warm moist tropical seas to the north and south of the equator, small atmospheric disturbances and depressions (see Questions 88 and 89) move westward in the easterly wind belts somewhat parallel to the equator. Here the disturbances begin to feel the effect of the Coriolis, or deflection, force of the rotating earth. (See Questions 79 through 82.) Three important environmental factors are also necessary for hurricane formation: the intertropical convergence zone or region where trade winds meet and converge, the easterly pressure waves, and the intrusions of cool air from the polar vortex. (See Question 83 through 91.)

79. What is the Coriolis force? The Coriolis force is a deflecting force set up by the rotation of the earth.

As the earth rotates on its axis from west to east, all trees, mountains, and everything else closely attached to it by gravity moves with it. But the free-moving air is not attached so closely by gravity, and may continue to blow in a certain direction while the earth "slips" beneath it.

For instance, suppose a mass of cold air starts to move southward from the North Pole toward a point on a meridian. In the time necessary for it to reach that point, the point will have moved to the east

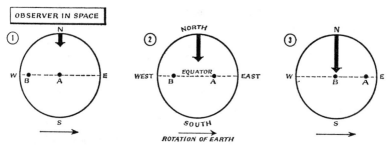

1) Body of air starts from north toward point A. 2) Air continues moving. Points A and B move east with rotating earth. 3) By the time air mass reaches latitude where point A originally was, and where point B is, point A is far to east. Hence, body of air appears to have curved to right of its direction of motion, in relation to point A.

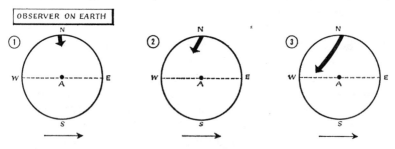

Drawing rotates with observer on earth at point A. 1) Air starts toward point A. 2) Air continues moving. From point A's view, air appears to have veered westward (see 2) above). 3) By time air reaches latitude of A, air seems to have blown from north toward west in curved path.

with the rotating earth, and the body of air will reach another point that has moved in from the west. Hence, to an observer on earth, the air seemed to have blown from the north toward the west in a curved path. To an observer in space, the body of air actually followed a straight line; it was only the earth that slipped beneath it to give an appearance of a path curved to the right of its direction of motion.

80. How does the Coriolis force affect cold air masses in the Southern Hemisphere? Masses of air from the South Pole are affected in much the same way as those from the North Pole. With the earth slipping beneath, the body of cold air moving from the

South Pole toward the equator seems to undergo deflection to the left of its direction of motion.

81. Where does the Coriolis force become most effective? The earth's spin has only a small effect upon moving air at the equator; however, the deflecting force becomes most effective from about 6 to 15 degrees north and south of the equator.

82. Where does the name Coriolis force come from? The name of this deflecting force comes from a French mathematician, Gaspard Gustove de Coriolis, who lived in the nineteenth century. Coriolis noticed that an object moving across a turning surface such as a turntable veered toward the right or left of its path of motion depending on the direction of rotation. From these and other relatively simple observations he evolved some basic principles of force for the rotating earth.

83. What is the intertropical convergence zone? The intertropical convergence zone is the area where the prevailing trade wind systems of the Northern and Southern Hemispheres meet and converge— near, but usually not on, the equator. This zone does not stay in one location throughout the year, but tends to follow the sun, with a lag of 2 to 3 months. In late summer, when the sun is still north of the equator, the zone can be found at its northernmost location of about 12 degrees north latitude. In winter it tends to move south—although in the Atlantic Ocean it does not move so far south as in the Indian Ocean and the western part of the Pacific.

84. How wide is the intertropical convergence zone? The dimensions of the zone of converging winds is highly irregular. Sometimes it is strong and is concentrated in a band 50 to 100 miles wide. At other times it is so weak it is hard to locate. It does not entirely circle the earth, and often exists only in segments.

85. By what other names has this converging wind zone been called? The intertropical convergence zone, called ITC or ITCZ for short by meteorologists, is also called the equatorial front, although it is not a true front since there are no great differences in tempera-

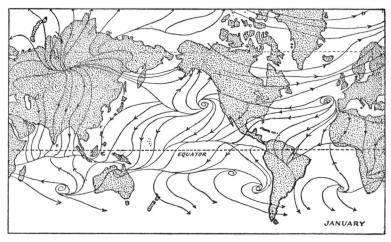

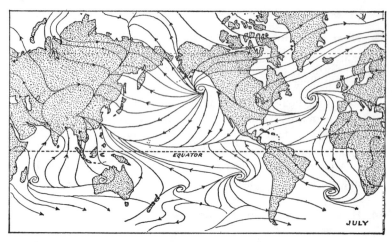

Winds of Northern and Southern Hemispheres meet at intertropical convergence zone

ture. The region is also erroneously called the doldrums, an area of little or no winds. (See Question 568.) Scientists are finding that the ITC zone moves in and through the doldrum area on a separate path.

86. What are easterly pressure waves? Easterly pressure waves are a series of rising and falling pressure centers or waves embedded in easterly winds traveling westward near the equator. These centers of pressure changes, known as isallobaric centers, are factors necessary for formation of hurricanes.

87. How do hurricanes originate in these pressure waves? The easterly pressure waves over tropical oceans are often the spawning regions of hurricanes.

Hurricanes seem to develop from wave troughs of low pressure, in areas where the pressure is falling. These low-pressure troughs may move one after another in series from east to west. Within them winds may begin to circulate, developing into what meteorologists call tropical disturbances, depressions, and storms. (See Questions 87 through 90.) A low pressure area may travel more than 2,000 miles before any indication of a storm is observed. It may travel another 2,000 miles before a wind circulation develops into hurricane intensity.

88. What are tropical disturbances? Tropical disturbances are low pressure areas, where rotating winds are blowing at gentle to moderate speeds of 8 to 18 miles per hour at ocean level, or even faster at higher altitudes. This type of disturbance occurs frequently throughout the tropics.

89. What are tropical depressions? Tropical depressions have more definite circulating winds and show a closed, circular system of lines of equal pressure. The winds are considered strong, blowing as fast as 38 miles an hour.

90. What are tropical storms? When circulating winds are blowing at speeds between 39 and 73 miles per hour, meteorologists call the system a tropical storm.

91. How do trails of the polar vortex help create hurricanes? Trails of troughs of the circumpolar vortex, which is the flowing mass of cold air around the poles, sometimes become mingled with the prevailing westerly winds of the earth's middle latitudes at high levels, and occasionally penetrate far into the tropics, into the convergence zone, often producing very bad weather.

Scientists do not yet fully understand the function this intruding cold trough of polar vortex plays, but they believe its presence is a contributing factor in hurricane development.

92. What are some phases of a hurricane's life? Opinions differ when it comes to defining the various phases of a hurricane's life—its birth, development, maturity, and death. Most authorities seem to agree, however, on four basic stages:

1) The formative state, which begins in a widespread condition of instability when an organized wind circulation develops in an easterly pressure wave or in the vicinity of the intertropical convergence zone.

2) The immature stage, when the young vortex deepens and reaches its maximum intensity. During this time, the storm is most symmetrical and covers a relatively small area.

3) Maturity, when the storm area is spreading out without further lowering of the central pressure. Actually its intensity is decreasing.

4) The decay and dying stage, when winds are slower than 74 miles per hour, the eye is filling with clouds or cold polar air, and the whole storm is collapsing and spreading out. Hurricanes often die as they move poleward over land and are cut off from their sources of moisture and heat.

93. How long does a hurricane take to generate? The formation of a hurricane may take as long as 9 or 10 days—or it may happen swiftly in 6 to 12 hours.

94. How long do hurricanes last? Some hurricanes have lasted only a few hours, while others have traveled around for several weeks before they weakened and died.

The average life of an Atlantic hurricane is about 9 days. Storms brewed in the month of August normally live the longest, with an average life of 12 days. This is because they develop in the eastern part of the North Atlantic and travel the long water route for several days before recurving northward.

95. Why do hurricanes formed in the Caribbean Sea or Gulf of Mexico endure only a short while? Hurricanes formed in the Caribbean Sea and Gulf of Mexico in June or in November generally have shorter lives since they strike land soon after formation and lose their energy source of moisture. They also slow down because of greater friction over land.

96. What causes a hurricane to dissipate and die? Hurricanes begin to weaken and die when either source of their enormous energy

—warmth or moisture—is diminished. This happens when the storm moves over land or into higher latitudes over colder water. Over land, the storm cannot obtain enough moisture; while in latitudes higher than 40 or 45 degrees, it cannot obtain enough warmth. Once the amount of moist warm air is reduced, the supply of available latent heat of condensation is also reduced. With the large source of energy cut off, friction causes the winds to decrease in speed and gradually the eye fills with clouds. Thus the storm's force is dissipated.

97. Where do Atlantic hurricanes usually die? Many Atlantic hurricanes die out in the central or northern part of the ocean, dwindling into sea storms in lonely areas not frequented by ships or airplanes.

98. How far can a hurricane travel over land? Over land, a hurricane that has traveled 200 or 300 miles may still have enough force to blow down power lines and drop heavy rains. Yet the farther inland the storm moves over dry land, the weaker it becomes. By the time it has traveled 600 to 800 miles inland it has usually lost its punch and become an ordinary storm.

99. What are some phenomena that accompany hurricanes? As a hurricane travels across the earth's surface, it is accompanied by devastating phenomena such as hurricane waves, swells, tides, and surges of sea water that are driven onto shore by the powerful winds (see Questions 100 through 109); torrents of rain dropped from clouds (see Questions 112 through 114); tornadoes (see Question 115); and lightning (see Questions 116 and 117).

100. What are hurricane waves? Winds of a hurricane act upon water in such a way that large waves are generated. These waves, called hurricane wind waves, while in the influence of the storm travel outward in all directions from the storm center. The largest waves are generated on the right side of the storm, where they have the increased wind power of the storm's forward motion.

101. How large are hurricane waves? Winds of a hurricane can generate some of the highest ocean wind waves known to man. Hurricanes have lashed waves to heights of 40 to 50 feet. Some of the greatest heights of hurricane waves on record were reported by ships of the main squadron of the Imperial Japanese Fleet suddenly caught in a typhoon on September 26, 1935. Waves were reported as high as 45, 54, 75, and 90 feet.

102. What determines the size of hurricane waves? The size or height of hurricane waves depends on the speed of the wind that caused them, the length of time the wind blows on them, and the fetch, which is the distance over which the wind blows in a relatively straight line.

103. What are hurricane swells? As hurricane waves move out of the influence of the storm's generating winds into calmer, less windy regions, they lose height and become relatively low, smooth, undulating movements known as swells.

Swells may persist for a long period of time, in directions different from that of the hurricane. Some swells have been found thousands of miles from their point of origin.

104. How fast and far do waves travel ahead of a hurricane? It is generally agreed that sea waves move with a speed that is somewhat slower than that of the wind that produces them.

In a hurricane, the forward sea waves produced by high-speed rotating winds travel slower than these winds, yet travel faster than the whole storm system. For instance, the wave-producing, rotating winds of a hurricane may average about 100 miles an hour and generate waves that travel about 30 to 50 miles an hour. These waves travel faster than the whole hurricane system, which moves forward on an average of about 12 miles an hour.

These waves, or swells, can travel 1,000 or 2,000 miles ahead of the approaching hurricane.

105. Why does the slowing down of pounding waves on shore indicate a hurricane? Ordinary wind-driven waves may have wave periods of about 6 to 10 seconds. (A wave period is the length of time for one entire wave to pass a certain spot.) These waves fall on shore at a rate of about 6 to 10 waves per minute.

Since the speed of a traveling wave depends on its length, the longer waves outrun the shorter waves which are left behind.

A hurricane generates swells that are much longer than ordinary, with a period of about 12 to 15 seconds, breaking on the shore at the rate of about 4 or 5 waves per minute.

In general, the longer the distance the waves travel from the storm, the greater will be the period of waves observed on shore. Also, the more intense the storm at sea, the longer will be the period of swells breaking.

The slow, heavy breaking of these advance swells on shore when the wind is still light and gentle is one ominous sign of an approaching hurricane.

106. What are hurricane tides?
A hurricane tide is the abnormal rise of sea water along a coast toward which a hurricane is moving. It is a combination of the regular true tide and additional water brought ashore by the hurricane.

It is not a true tide, which is the periodic rise and fall of the sea level caused by the gravitational pull of the sun and moon.

The rising of a hurricane tide can result from the gradual pile-up of water, pushed for long periods of time over long distances by the approaching storm. The tide may begin to rise many hours before the storm arrives while the storm center is several hundred miles away.

Hurricane tides may rise 6 or 7 feet above normal tide—sometimes as much as 20 feet—and last for several hours. These tides can be greater at times of lunar high tide. A hurricane tide may raise the sea-water level along several hundred miles of coastline.

Hurricane tides are also known as storm tides or meteorological tides.

107. What is a storm surge?
A storm surge is part of a hurricane tide. It is an increase of water piled up by forces of an oncoming hurricane—principally the effect of winds pushing on the water, the decrease of atmospheric pressure as the storm center approaches, and the water transported by the action of oncoming waves.

Meteorologists define a storm surge as the difference between the actual storm tide and the normal tide which would have occurred had there been no storm.

108. Why are storm waves and tides so destructive?
Flooding that results from hurricane surges and tides is a major cause of death, injury, and destruction.

Hurricane-induced waves carry enormous force and can destroy many objects, especially when debris such as tree trunks or heavy beams are picked up by the waves and used as battering rams. They can knock down houses and even thick concrete installations in a short length of time. Moving at speeds up to 50 or 60 miles per hour, they can break down piers, bridges, and walls and hurl boats and other movable objects several miles inland.

The erosive powers of hurricane-driven waves are great. They can scour out 30 to 50 feet of beach within a few hours and undermine highways, buildings, and sand dunes with their forceful pounding.

109. What makes some waves particularly destructive? Waves are particularly destructive when there is deep water a short distance offshore, so they build up to powerful heights before spilling over the land. Reefs and shallow water tend to break up this wave action before the waves hit the shore.

110. What is the tragedy of the U.S. East Coast beaches? Beaches all along the U.S. Eastern Coast are being gradually washed away by constantly driving winds and pounding waves—particularly those accompanying hurricanes.

Each year, some 20 million cubic yards of sand disappear permanently from Florida's beaches—much because of hurricane winds and waves.

Beaches at Miami have almost completely disappeared except at low tide. Normally these beaches would have been replenished by sediment and sand washed down from the coasts of Georgia and the Carolinas, but conservation projects in those states have been trying to hold the coastal soil and sand in place.

111. Why are salt water floods so destructive? Much damage can be created by the inundation of fields, plains, and shoreline properties with floods of salty water for a long period of time. Salt water kills vegetation and can leave salt and other chemicals in the soil that are harmful to new crops. Salt water can damage fresh water wells and underground water supplies near the shore by making them brackish and unfit to drink until they are flushed out by recurring rains or by man-engineered washing.

112. How much rain falls from hurricanes? Some of the world's heaviest rainfalls are connected with hurricanes, for these storms pick up enormous quantities of water vapor—about 2 billion tons per day —from the oceans and release it as rain, often over land.

The amount of rainfall on a particular area depends on a number of factors, including winds, forward speed of the storm, and topography of the land. The exact amount may not be known, since after winds reach 50 miles per hour, sometimes only 50 percent of the rain is caught in rain gauges.

In general the average amount of rain a hurricane drops varies from 3 to 6 inches in any one location, but rainfalls in excess of 20 inches are not rare.

113. What were some of the heaviest rainfalls from hurricanes? The greatest amount of rainfall recorded over a 24-hour period was 73.5 inches on March 15 through 16, 1952, at Cilaos, Reunion Island, in the Indian Ocean. Typhoon Gloria dropped 49.13 inches on September 10 through 11, 1963, at Paishili, Taiwan, and a storm in Baguio in the Philippines dropped 42 inches in 1911.

For comparison, the average annual rainfall of Paris is about 22 inches, of London 24 inches, of Honolulu 22 inches, and of San Francisco 19 inches.

114. What were some record hurricane rainfalls in the United States? The record rainfall for a U. S. hurricane was 23.11 inches at Taylor, Texas, in 1921.

The Cedar Key hurricane of September, 1950, poured nearly 39 inches of rain in one day and night on Yankeetown, Florida, off the Gulf Coast. This 9-day hurricane traced an unusual double loop in the Cedar Keys area, and the coast from Sarasota northward suffered extensive wind and flood damage. The coastal area inland from Yankeetown to Tampa was flooded for several weeks.

Record rainfalls occurred in the northeast region of the United States when the dying hurricane Diane dropped more than 12 inches of rain in a 24-hour period on already saturated soil. (See Questions 27 and 28.)

Hurricane Beulah of 1967 dropped 20 to 30 inches of rain over 40,000 square miles of Texas and Mexico, giving Mexico the worst flood of the twentieth century.

115. Do tornadoes form along with hurricanes? Several tornadoes have been observed forming, traveling, and collapsing along the edge of a hurricane. They have been observed most often in the front part of the hurricane, at almost any time of day. Meteorologists report that these tornadoes are usually less severe than the single tornadoes of the Midwest and have paths that are relatively shorter and narrower.

For a long time people did not notice the occurrence of hurricane-generated tornadoes—possibly because they were so preoccupied with the hurricane itself they did not observe other accompanying phenomena. Also, after a hurricane has passed, tornado destruction can be difficult to distinguish from any other kind.

Tornadoes have been observed most often in Florida, Cuba, the Bahamas, and along the coasts of the Gulf of Mexico and the South Atlantic Ocean. Two were noticed over Florida, accompanying hurricane Betsy, on September 9, 1965. Hurricane Carla in 1961 released 26 tornadoes. One hundred and fifteen were counted spinning off hurricane Beulah in September, 1967. More than a hundred were counted with hurricane Camille in 1969.

116. Does lightning appear with hurricanes? In the midst of the winds, rains, and swirling clouds of a hurricane, meteorologists have observed frequent and continual lightning. Electrical activity is intense during the formative stages of a hurricane. This activity grows less during the mature phases, and then increases again as the hurricane decays. By means of radio direction-finding equipment, researchers have found electrical discharges they call sferics, which is a contraction of the word *atmospherics*, the natural electrical phenomenon detected by radio methods.

Researchers have thought that possibly these sferics could be used to track hurricanes—but so far, they have been only marginally successful.

117. What is the firefly effect? Sometimes people along the coast have noticed what seem like swarms of millions of tiny fireflies blinking and flying in hurricane winds. Actually these lights are tiny electrostatic discharges generated by friction as sand particles are picked up and driven by the winds.

118. Why is most damage caused in the hurricane's forward right quadrant? Meteorologists have generally divided a hurricane into four quarters of cross section, with the main axis parallel to the direction in which the storm is moving.

Since the force of the forward movement of the whole hurricane is added to the force of those winds blowing around the eye, the heaviest and strongest winds will be found in the right front sector. Here also are found the heaviest rainfalls, as well as the highest and strongest storm waves.

As a hurricane moves westward or northwestward and crosses the coastline, areas to the north of the storm center will be harder hit by wind, rain, and sea waves than those to the south. The weakest winds are usually found in the back left quadrant of the storm, since the battering ram effect of the wind and water is lessened because the main bulk of the storm is moving away from it.

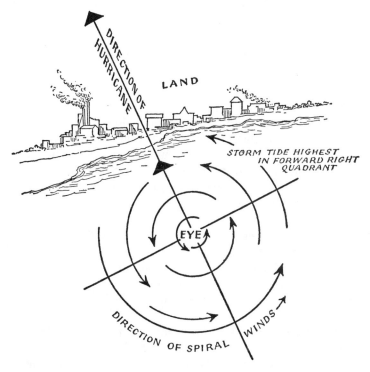

Storm winds and waves highest in forward right quadrant

119. Where do hurricanes originate? There are about eight regional areas in the world where hurricanes tend to develop—all within two general bands on either side of the equator. Each band is about one thousand miles wide.

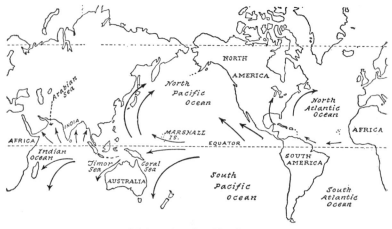

Origin and paths of hurricanes

120. Where are most hurricanes formed? More hurricanes originate in the southwestern part of the North Pacific Ocean than any other place on earth. These storms are usually generated between the Marshall Islands and the Philippines and move in a clockwise curve, first westward toward the China coast, then northward and northeast over the Philippines, Korea, and Japan.

121. Why do most hurricanes form in the North Pacific Ocean? Most hurricanes generate in the western part of the North Pacific Ocean for several reasons. For one thing, the intertropical convergence zone is often very strong in this area and far north of its average position. Also, sea surface temperatures are high.

122. Where do hurricanes form in the Atlantic Ocean? Many hurricanes generate in the southern portions of the North Atlantic Ocean around the Cape Verde Islands, or in the western Caribbean Sea and the Gulf of Mexico. These storms move westward toward the West Indies and the Southern and Eastern Coasts of the United States.

123. Where else in the Northern Hemisphere do hurricanes form? Forming in the northern Indian Ocean, some hurricanes push into the Bay of Bengal, where the configuration of the Bay and land contributes toward making these storms among the worst natural disasters in the world. (See Questions 6 through 9.)

Other hurricanes occur in the Arabian Sea.

124. Where in the Southern Hemisphere do hurricanes originate?
Another area spawning a large number of hurricanes is the southern Indian Ocean, south of the equator, from which intense hurricanes sweep toward Madagascar and southeast Africa.

The Timor Sea, north of Australia, generates hurricanes that curve southwestward over the Australian coast.

Many hurricanes are formed in the western part of the South Pacific Ocean.

125. Why don't hurricanes form in the South Atlantic Ocean?
As yet, no hurricane has been recorded as originating in the South Atlantic Ocean, or in the eastern part of the South Pacific Ocean. This dearth of storms occurs mainly, meteorologists believe, because the intertropical convergence zone tends not to move across the equator in a southerly direction over these areas.

126. What prevents hurricanes from forming on the equator?
Hurricanes tend not to form on the equator, and only rarely a few degrees north and south of it, because that area is not sufficiently affected by the Coriolis force of the rotating earth to give the generating winds the spin necessary to form hurricanes.

Hurricanes tend not to form farther north and south of the 25 degree latitudes because the climate is too cold to provide the heat energy they need.

127. Can hurricanes form over land? For centuries people believed that hurricanes could be formed only over water. Scientists now know, however, that the forerunning disturbances from which hurricanes develop sometimes occur over land.

128. Which hurricanes were land-generated? Wind disturbances of 2 1966 hurricanes formed over land: the early arrival, hurricane Alma, began as a moderate swirling disturbance in the area of Panama in June; and the northbound traveler, Faith, generated as a moderate disturbance over west Africa in the middle of August.

129. What is the season for hurricanes? Hurricanes are formed in the Northern Hemisphere most often during the late summer and early autumn, at a time when the prevailing trade winds of the 2 hemispheres converge north of the equator. At this time the sun is traveling southward toward the equator, and the surface of the sea has reached its maximum temperature—about 80 to 86 degrees Fahrenheit.

130. What is the hurricane season for North Atlantic hurricanes? The official hurricane season for the U.S. Weather Bureau extends from June 1 through November 30.

August, September, and October are the months when most Atlantic Ocean hurricanes form, with September the month when the majority occur. Some storms have been generated in June or July, and some even as early as May. Most early storms originate in the western Caribbean Sea, moving into the Gulf of Mexico and crossing the coastline into Mexico or the Gulf states.

During August and September, and less frequently in July and October, most hurricanes develop over the eastern section of the North Atlantic Ocean near the Cape Verde Islands and move westward toward the coastal areas of the United States.

In late September, October, and November, hurricanes again tend to originate in the western Caribbean Sea. Most of these late ones turn northward and northeastward in the lower latitudes, crossing Florida or the Greater Antilles.

131. When do hurricanes occur in the Southern Hemisphere? In the Southern Hemisphere, the hurricane season occurs from November through April—the equivalent of summer and autumn for that hemisphere.

Typhoons of the South Pacific Ocean and Indian Ocean, south of the equator, generally form in the autumn of the year—from January through April.

132. What were some Atlantic hurricanes that appeared early in the year? One of the earliest hurricanes on record was hurricane Alice, which appeared in the vicinity of the Lesser Antilles during January, 1955. Another early hurricane developed in March, 1908, in that same location.

One of the earliest hurricanes on record to strike the United States was hurricane Alma, which moved across the Gulf of Mexico and crossed the top part of Florida in the first week of June, 1966. A few days later it dissipated off the East Coast.

Severe storms are often formed over tropical waters during the winter months, but rarely do they reach hurricane intensity.

133. What are the general paths of hurricanes? Since tropical storms originate in regions dominated by trade winds blowing in a westerly direction, their initial movement is westward, and slightly toward the poles. After a while, many drift far enough north or south to pass out of the area dominated by easterly winds and move into regions dominated by west winds—the prevailing westerlies of the middle latitudes. These winds tend to reverse the direction of hurricanes so that they may move in an eastward direction. The entire path of a hurricane is roughly parabolic.

Many variations occur in these general paths, changing from one month to another, and depending upon the pattern of areas of high and low pressures.

134. What is the general direction of North Atlantic hurricanes? Most Atlantic hurricanes are formed about 900 miles southeast of the Bahama Islands and begin moving westward and slightly north, skirting the edge of the permanent high-pressure area known as the Bermuda-Azores high. These hurricanes begin to curve toward the northeast at about 25 or 30 degrees north latitude.

135. What makes some hurricane paths so erratic? The paths of hurricanes are subject to pressure systems of the surrounding atmosphere, as well as the influence of prevailing winds and of the spinning earth.

While some hurricanes travel in a general curved or parabolic path, others change courses quite abruptly. They have been known to reverse direction, zig-zag, turn back toward the equator, make loops, stall, return to the same area, and move in every direction of the compass.

These changes usually occur as the storm passes through areas of light and variable winds between the prevailing easterlies and westerlies. They also are influenced by the presence of highs and lows.

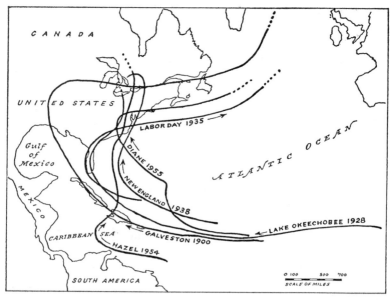

General paths of some devastating Atlantic hurricanes*

High-pressure areas act as barriers, and if a high is well developed, its outward-spiraling flow will guide the hurricane around its edges. Low pressures, on the other hand, can tend to draw the hurricane system toward their slowly inward rotating winds. With all these pushes and pulls, the hurricane can follow a seemingly erratic route.

136. What happens to hurricanes as the climate becomes warmer?
During a gradual warming cycle over the Northern Hemisphere, when the glaciers and the permanent snow line show signs of retreating, several changes take place to force hurricanes that normally curve over the open Atlantic Ocean inland into the coastal areas of the United States.

For one thing, the Bermuda-Azores high-pressure area moves farther north during a warm cycle and becomes abnormally strong. This high tends to block the hurricanes, forcing them farther to the west and north. With warmer climates, more hurricanes are generated, and these are able to retain their force for longer periods of time and push farther north.

*A similar map, showing paths of major Atlantic hurricanes from 1969 through 1985, is given in the Appendix, p. 361.

137. What are some benefits of hurricanes? Hurricanes are not all bad, in spite of their destructive power. In fact, they may even be essential in many ways to maintain certain environmental factors necessary for life on earth. Hurricanes help maintain the heat balance throughout the world, act as safety valves to release excess energy, and bring torrents of fresh water to replenish crops and ground water.

138. How do hurricanes help keep the heat balance? Hurricanes help keep the balance of heat and cold by transferring heat accumulated in the tropics and subtropics from the sun.

In these tropical areas more heat is being received than is being reradiated, while in the North and South Pole regions, more heat is being radiated out into space than is being received and absorbed. Circulating air currents, aided by the spinning of the earth and giant storms, help carry heat from the equatorial regions toward the polar regions, thus distributing the sun's radiant energy.

139. What are some historical events in hurricane research? In the early days of U.S. history, only meager information on hurricanes was obtained when ships traveling to and from the New and Old Worlds reported storms they had encountered.

In the early 1800's, accounts of storms and hurricanes were made in the United States, the Caribbean islands, and India. Gradually the idea evolved that storms were whirlwinds moving forward, and that hurricanes traveled from tropical waters in a curved track.

In 1857 members of the religious Jesuit Order of the Catholic church established the Observatory of the College of Belen in Havana, Cuba, under the direction of the Reverend Benito Vines who set up the first system of hurricane warnings.

In 1870 a Federal Weather Service was established under the U.S. Army Signal Corps.

At the outbreak of the Spanish-American War, in 1898, Weather Bureau officials pointed out to President William McKinley that throughout history more ships had been sunk by weather than by war. After McKinley's statement: "I am more afraid of West Indian hurricanes than I am of the entire Spanish Navy," the Hurricane Warning Service was set up under the U.S. Weather Bureau, which had been established in 1880.

In 1905 the first wireless radio report was sent from a ship at sea.

A 24-hour hurricane watch was instituted in 1935 and teletype communications were set up between Jacksonville, Florida, and Brownsville, Texas.

In 1943 the first reconnaissance plane flew into the eye of a hurricane, and the next year aircraft reconnaissance was undertaken by the U.S. Air Force and the Navy.

140. How are hurricanes spotted and tracked today? For centuries, hurricanes developed and remained undetected until they slammed onto land, taking inhabitants by surprise and causing much damage.

Today an elaborate network of instruments, men, and equipment at the National Hurricane Center in Miami, Florida, part of the Environmental Science Services Administration, search out potential hurricanes in their early stages of birth, and track them through their life cycle until they decay and die.

Hurricanes are now observed from the air by satellites and research and commercial planes, at sea by buoys and ships, and on land by radar and other equipment on land stations. (See Questions 141 through 148.)

141. What satellites spot hurricanes? A new era opened in April, 1960, when TIROS I (Television Infrared Observation Satellite), the experimental television weather satellite, detected for the first time and photographed a fully developed typhoon about 800 miles off Brisbane, Australia. This satellite flew in orbit 450 miles from the earth and circled the earth once every hour and 39 minutes.

Since then successive TIROS satellites have been sent up to send pictures and data earthward in unceasing streams, from which hurricanes can be located as they first start to form.

In August, 1964, a $100 million satellite, Nimbus, was sent up on a near-polar orbit that was designed to monitor the weather over every portion of the globe each 24 hours. This satellite circles the earth once every 98.3 minutes, with one of its camera systems transmitting continuous pictures of cloud formations at the rate of more than 1,000 pictures a day. Infrared photography enables the satellite to take continuous pictures even in darkness.

Latest eye on the weather is the ATS satellite—the Applications Technology Satellite—which flies 22,300 miles above earth, transmitting pictures of cloud covers and disturbances and yielding new insights into global weather.

142. What is the TOS system? The TIROS Operational Satellite system or TOS as it is called, financed by Environmental Science Services Administration and operated by the National Environmental Satellite Center, provides both worldwide and local cloud cover pictures daily to aid in weather forecasting and warning.

Seven such satellites had been launched by late 1968, equipped with instruments such as arrays of atmospheric radiation sensors and automatic picture transmitters (APT).

143. What is Project Stormfury? Project Stormfury is a joint effort of the Environmental Science Services Administration and the Navy to explore the structure and dynamics of hurricanes in order to understand them better, to improve prediction of where they may travel, and to test methods of possibly modifying some aspects of these destructive storms.

Originating in 1961, Project Stormfury consists of experimentally seeding hurricanes to determine the aftereffects.

To achieve their aims, more than 200 scientists, technicians, and flight crews fly into and around a hurricane at altitudes of 1,000 feet to more than 60,000 feet. They use as many as 17 aircraft—including DC-6s, C-54s, WB-57s, and other aircraft supplied with equipment for seeding, monitoring, and photographing hurricanes.

144. What are Hurricane Hunters? When a storm is observed by satellite, ship, or plane, in the tropical oceans of the Caribbean and Atlantic, a reconnaissance plane is sent by the Air Force, Navy, or the National Hurricane Center of the Environmental Science Services Administration to make a thorough investigation. These planes and their crews, officially called Airborne Early Warning Squadron FOUR, are nick-named Hurricane Hunters. They are equipped as flying meteorological laboratories. Once the men locate the storm, they begin taking continual readings on its temperature, pressure, cloud structure, speed of winds, extent of storm area, and other pertinent data.

With bases at Naval Air Station, Jacksonville, Florida, and at Ramey Air Force Base in Puerto Rico, these flights are made at frequent intervals to keep constant check on the storm's intensity, speed, and direction. Reports are radioed back to the National Hurricane Center in Miami, Florida. Here the reports are plotted, analyzed, and used for forecasting.

145. What instruments are used by the hurricane planes? As Hurricane Hunters scan the skies for storms and monitor them, many instruments are used for precise records. With the advent of powerful, long-range airborne radar, areas of some 200,000 square miles can be observed with one sweep. In one weather flight, information on some 1,500,000 square miles can be gathered.

Another valuable instrument is the dropsonde, a radio transmitter attached to a small parachute that is dropped in the storm center. As it falls, it transmits such information as temperature, relative humidity, and pressure back to the plane.

146. How effective is radar in the search for and study of hurricanes? Radar emerged from World War II as a powerful new tool for detecting and tracking hurricanes. With this equipment, which reflects electromagnetic waves from water droplets and ice particles in thunderstorms, tornadoes, or squall lines within the hurricanes, scientists have learned much about hurricanes, and are able to determine the direction and rate at which a storm is moving.

Current radar equipment at ground stations has a range of 150 to 250 miles. This is the maximum range because any point beyond that lies beyond the horizon as a result of the earth's curvature.

A network of weather radars extends along the Atlantic and Gulf Coasts from Brownsville, Texas, to Nantucket, Massachusetts, and forms a protective screen through which no hurricane can slip unobserved. Where one radar station leaves off, another takes its place as the storm moves up the coast.

147. How are earthquake instruments used to detect hurricanes? One method of tracking a hurricane is with seismic equipment that records tremors passing through the earth. As air pressures become lower in the storm center, water in the center of the hurricane rises to form a sort of hill or hump, and pressures of the water upon the sea floor beneath it are increased. As the storm and hill of water move forward, the shift of water pressure can be detected by delicate in-

RAIN SHIELD

SPIRAL BAND

EYE

Tracking a hurricane with radar

struments called microseismographs. Two or more such seismographs in different coastal towns working together can be used to plot the path of the oncoming hurricane with a certain amount of accuracy.

148. What land equipment is used to track and study hurricanes?
In general, during the hurricane season, reports are constantly received from land stations in the eastern and southern United States, Mexico, Central America, islands in the Antilles, and Bermuda, and, more irregularly, from the Azores and the Cape Verde Islands.

At some of these stations, information from the upper air is supplied by balloons tracked from ground stations which determine the wind directions and velocities up to 30,000 feet. Radiosonde, small radio transmitters, are sent aloft with helium balloons, to relay back data on the pressure of the atmosphere, temperature, and humidity. Some of the stations also scan the skies by radar.

149. Can hurricanes be controlled or modified? Scientists have considered many methods of trying to reduce the destructive force of the hurricane.

Some of the ideas proposed include seeding the turbulent storms with silver iodide crystals in an effort to rapidly release the latent heat energy. Another idea is to spread a chemical film one molecule thick over the sea surface where the storm is beginning to generate. Evaporation would be somewhat slowed down and some scientists believe this diminished amount of water vapor would remove the storm's vital source of energy.

150. What is cloud seeding? Cloud seeding is the dropping of artificial nuclei into a cloud of supercooled water droplets to start formation and precipitation of ice crystals or rain. Supercooled water droplets are those that are at temperatures below the freezing point but that have not turned to ice. In order for these drops to condense into a raindrop, condensation nuclei or freezing nuclei must be present to form the raindrop core. (See Question 734.) By adding thousands of these condensation nuclei to a cloud, researchers hope to trigger the formation of ice or raindrops that would precipitate to earth, hence modifying the cloud.

Pioneer work on the artificial seeding of clouds began in the 1890's when scientists of the United States and Germany worked on devices for projecting dry ice into clouds. Current theories and experiments were developed from laboratory experiments in 1946 when Vincent Schaefer, Irving Langmuir, and Bernard Vonnegut dropped dry ice into a box of supercooled water droplets.

Since then experiments have been made on dropping or shooting nuclei into clouds in efforts to decrease the energy of destructive storms, such as hurricanes or hail-bearing thunderclouds, or to bring rain to parched areas.

151. What chemicals are used as seeding nuclei? Two seeding agents are commonly used in experiments—silver iodide, and solid carbon dioxide or dry ice.

Silver iodide, whose crystals form extremely effective freezing nuclei, is usually vaporized in ground generators and carried into the clouds with rising currents of hot air.

The chemicals can also be shot into cloud formations by cannons or by rockets or dropped in smoking canisters from airplanes, dispersing nuclei as they fall through the cloud.

Lead iodide has also been used as a seeding agent in some experiments with detonating nose cones. The lead iodide is a relatively easy material to pack into the cones, and it disperses easily within the cloud.

152. How might seeding help alter a hurricane? In the laboratory, scientists have found that by dropping or seeding crystals of dry ice or silver iodide into supercooled water vapor in cloud chambers, water drops freeze into ice and snow crystals and precipitate, releasing latent heat. Experiments have been conducted with many cloud formations also.

Some meteorologists believe that by dropping silver iodide particles into the band of clouds next to the eye of a hurricane, the supercooled water droplets should be transformed into ice crystals, hence releasing the heat energy into the storm system near the warm core.

The effect of this additional heat should reduce the atmospheric pressure adjacent to the low-pressure center of the storm and reduce the difference in pressure across the cloud wall—the steep pressure gradient which produces winds of hurricane force. Scientists believe the smoothing of the pressure gradient should cause the hurricane winds near the center to diminish and cause the spiral bands of the storm to expand. This should decrease the hurricane's strength.

153. In what areas can hurricanes be seeded safely? Because of the unknown effects that may result from man's tampering with a hurricane, great care is taken to conduct experiments in hurricane seeding far enough away from inhabited land so no damage can be done.

For several years, an isolated area, somewhat in the shape of a football, was designated as the seeding laboratory. This area was located astride the Tropic of Cancer, south of Bermuda and north of the West Indies. Here any hurricane passing through cannot reach a populated area within 36 hours. It seemed an ideal spot for scientific experiments, and scientists were ready each hurricane season to test their theories. Unfortunately, hurricanes kept skirting this area for several years, and no hurricane was seeded there.

Precautions were subsequently modified in 1967 and 1968 to permit seeding of any suitable storm in the southwest North Atlantic, including the Gulf of Mexico and the Caribbean Sea, if there is a 10 percent or less probability that the storm center will come within 50 miles of a populated area within 24 hours after seeding.

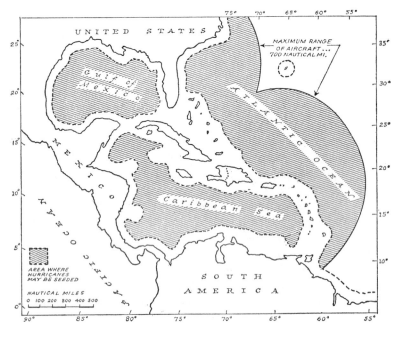

Hurricane seeding areas

154. What hurricanes were seeded? The seeding of hurricanes has been undertaken with extreme caution as scientists attempt to tamper with the immense energies of the storms.

In 1961, research planes seeded hurricane Esther, and in 1963 hurricane Beulah. Both experiments resulted in minor changes in the cloud system that offered no conclusive information but were enough to warrant continuing the experimentation.

In 1962 and 1968 the hurricane season was mild, and in 1965, 1966, and 1967 all hurricanes skirted the area designated for the tests.

In 1969 hurricane Debbie was seeded on two days—August 18 and 20, when it was several hundred miles north-northwest of San Juan.

No dramatic changes in winds or the cloud structure were observed, and the hurricane continued to move in its track past Bermuda and out to sea. Weather scientists considered the seeding an operational success and accumulated large amounts of data for further study.

155. What other seeding experiments are under investigation? In the belief that hurricane rainbands may constitute an important link in the chain between relatively simple cumulus convective activity and a mature hurricane, scientists are investigating the possibility of seeding hurricane rainbands and tropical cumulus cloud lines over the ocean. The lines of maritime cumulus clouds are similar in many ways to the rainbands. Scientists believe that by seeding these cloud lines, information will be obtained to assist in designing new hurricane experiments.

156. Could a nuclear bomb affect a hurricane? Because of the enormous amount of energy contained in a full-blown hurricane, a nuclear bomb might have little or no effect on diverting it, one way or another. Most scientists strongly consider such a control method a useless notion since it not only violates the atmospheric test ban, but it could, by radioactive fall-out, do more damage than good.

157. How are warnings of a hurricane transmitted? When there are definite indications that a storm may be growing to hurricane size, the Weather Bureau puts into operation a vast Hurricane Warning System.

Established in 1938 by joint action of the Weather Bureau, the U.S. Air Force, and the U.S. Navy, this system has three principal functions: to collect necessary observational data, to prepare forecasts and warnings, and to distribute warning information as rapidly and efficiently as possible to the public in the potential path of the hurricane.

To aid in the distribution of hurricane forecasts and warnings during the hurricane season, a special hurricane teletype circuit is installed connecting all Weather Bureau city and airport offices on the Gulf and South Atlantic coasts from Brownsville, Texas, to Charleston, South Carolina. A second teletype circuit serves all other offices along the Atlantic seacoast from Miami to Portland, Maine.

All stations on these circuits immediately receive observations and forecasts placed on the line by another station. Atlantic Hurricane Centers are Boston, New Orleans, Washington, Miami, and San Juan. Warnings and advisories are prepared and distributed by these centers as soon as a tropical storm or hurricane is discovered and for as long as it remains a hurricane or a threat to life or property in an area south of latitude 50 degrees north and west of longitude 35 degrees west.

The San Francisco and Honolulu Weather Bureau offices and the Guam Typhoon Center are the Pacific Hurricane Warning centers.

158. What warnings are given as winds increase in speed? Small craft warnings are given if a hurricane moves within a few hundred miles of the coast. Small craft usually are warned to take precautions and not to venture into the open ocean.

Gale warnings are issued for coastal areas whenever winds of more than 39 to 54 miles per hour are expected to occur.

Storm warnings are issued when winds blow at speeds of 55 miles per hour and above. Weathermen are using the term "storm" to replace that of "whole gale" in indicating winds that are blowing at speeds of 55 to 73 miles per hour.

When a gale warning or storm warning is issued as a hurricane approaches the shore, the warning may be changed to a hurricane warning if the hurricane continues moving toward the coast. Gale or storm warnings are then extended to other areas on either side of the expected hurricane path—areas close enough to experience such winds, but not close enough to feel the full effects of a hurricane.

159. What are advisories? Advisories are formal messages issued each six hours concerning tropical storms and hurricanes. They give warning information on where a storm is located, how intense it is, where it is moving, and what precautions should be taken.

When a hurricane is near the U.S. mainland, more frequent messages, called bulletins, are issued to supplement the advisories.

160. What is a hurricane watch? As the hurricane continues to approach the mainland and is considered a threat to coastal and inland regions, meteorologists issue a hurricane watch for the regions in the calculated path.

This watch does not mean that the hurricane is definitely going to strike. It means that everyone in the area covered by the alert should watch more carefully for the hurricane and be prepared to act quickly if definite warnings are issued that the hurricane will strike.

161. What is a hurricane warning? A hurricane warning is issued to coastal areas where winds of 74 miles per hour or higher are definitely expected to occur. A warning also may include coastal areas where dangerously high water or exceptionally high waves are predicted, even though the winds expected may be of less than hurricane force.

When the warning is issued, all precautions should be taken immediately.

The warnings are seldom issued more than 24 hours in advance. If the hurricane path is unusual or erratic, the warnings may be issued only a few hours before the onset of hurricane conditions.

162. What flags are used to warn of hurricanes? Many coastal areas use a system of flying flags by day and hanging lanterns by night to give warning of rising winds to small ships at sea without radio. A northeast storm warning, for instance, is indicated by a red pennant above a square red flag with a black center by day—or two red lanterns, one above the other, at night.

A southeast storm warning is indicated by a red penant below a square red flag with black center by day, or one red lantern at night.

A southwest storm warning is a white penant below a square red flag with black center by day, or a white lantern below a red lantern by night.

A northwest storm warning is indicated by a white penant above a square red flag with black center or white lantern above a red lantern at night.

Hurricane warnings are shown by two square red flags with black centers, one above the other by day—or two red lanterns with a white one in between at night.

163. What are some precautionary actions to be taken before a hurricane? The key to hurricane protection is preparation. By taking sensible measures before, during, and after a hurricane, many lives can be saved and property damage averted.

First advice is to keep well informed by listening to the latest warnings and advisories on radio or television.

Remove children, and other young or helpless people and livestock, to safe ground. Secure all boats and items left loose on piers or boat houses. Leave low-lying beaches or other areas that may be swept by high tides or storm waves. Remember, roads to safer areas may become flooded before the main portion of the storm arrives.

Plan to stay home during the hurricane if the house is out of danger from high tides and is well built. Board up windows or protect them with storm shutters or tape. Secure anything that might be blown away or torn loose. Garbage cans, garden tools, toys, porch furniture, and other objects become weapons of destruction when they are picked up by the winds and hurled against objects.

Other advice includes such things as: store drinking water in tubs and jugs; buy food that needs little or no refrigeration or cooking; check flashlights and battery-powered equipment; and make certain there is enough gasoline in the car.

164. What should be done during a hurricane? It is important to remain indoors during a hurricane. It is extremely dangerous to travel or move about when the winds and tides are whipping your area.

Keep track of the storm's progress through the Weather Bureau's advisories on radio and television.

Avoid the eye of the hurricane. If the calm storm center passes directly over your region, there will be a lull in the wind and sudden calm that may last for a few minutes to half an hour or more. Stay in a safe place during this time, and remember the second half of the circular storm will sweep over the region. On the other side of the calm eye, the wind will rise very rapidly to hurricane force, coming from the opposite direction.

165. What should be done after the hurricane has passed? After the storm has passed, one should seek necessary medical care at Red Cross disaster stations or hospitals.

It is most important to stay out of disaster areas, since the presence of many nonqualified people often hampers the first-aid and rescue work.

Drive carefully along debris-filled streets. The roads may be undermined by hurricane waves or floods and collapse under the weight of a car.

Avoid loose or dangling wires and report them to the power company or nearest law enforcement officer. Report broken sewer or water mains.

Check refrigerated foods for spoilage, for often the power has been cut off during the storm.

Stay away from river banks and streams, as they are unsafe after a deluge.

166. What organizations are involved in hurricane research and warning? The U.S. Weather Bureau of the Environmental Science Services Administration and the Department of Defense are very active in hurricane research and warning systems. The Administration maintains several divisions for hurricane research: the Institutes for Atmospheric Sciences, Experimental Meteorologic Branch; the National Hurricane Research Laboratory, including the National Hurricane Center in Miami, Florida; and the ESSA Weather Bureau Hurricane Warning Offices at San Juan, New Orleans, Washington, and Boston. Weather Bureau offices in San Francisco and Honolulu keep watch for Pacific tropical disturbances that could develop into hurricanes and typhoons in the Pacific.

The National Meteorological Center in Maryland keeps track of hemispheric activity.

The United Nations Economic Commission for Asia and the Far East and the World Meterological Organization are active in international hurricane research.

II. TORNADOES

Introduction. With winds rotating so fast they cannot usually be measured, tornadoes pack more concentrated violence into a small area than any other windstorm on earth. The twisting funnel of a tornado dips down from the base of a storm cloud toward the ground for only a short time, yet wherever it touches earth it can cause terrible destruction, especially in populous areas.

Whirling downward in corkscrew fashion from a turbulent overcast sky, the funnel-shaped cloud of a tornado is frequently accompanied by lightning and a frightening roar. As the funnel moves over the ground, it literally explodes houses, splintering and scattering them into broken bits and jumbled debris. During their brief existence, tornado funnels damage everything they encounter. They have been known to pick up freight cars and drop them to earth again hundreds of feet away, twist and uproot houses and trees, mangle automobiles and airplanes. They often kill any human being or animal unfortunate enough to be caught in their paths. These powerful winds can also play freakish tricks such as sucking all water from a well, plucking feathers off a live chicken, or carrying a cabinet of dishes for hundreds of yards and setting it down again without breaking a dish.

These storms are difficult to predict and as yet cannot be controlled. With the development of modern instruments—radar, lasers, specially equipped aircraft, satellites, computers, and other electronic devices—scientists are improving techniques for gathering data and issuing warnings far enough in advance for people to seek adequate protection.

Considerable amounts of effort and research are under way to develop techniques for forecasting tornadoes with greater accuracy and for sending faster warning communications. Research divisions are also engaged in weather-modification studies trying to find techniques for "dampening down" some of the furious energy potential in tornado-producing clouds.

167. What is a tornado? A tornado is a relatively small but extremely powerful vortex of winds in the shape of a funnel dipping toward earth from a storm cloud and rotating at tremendous speeds

Possible size (diameter)	Possible speed of forward motion	Possible speed of rotating winds	Possible time, duration of storm

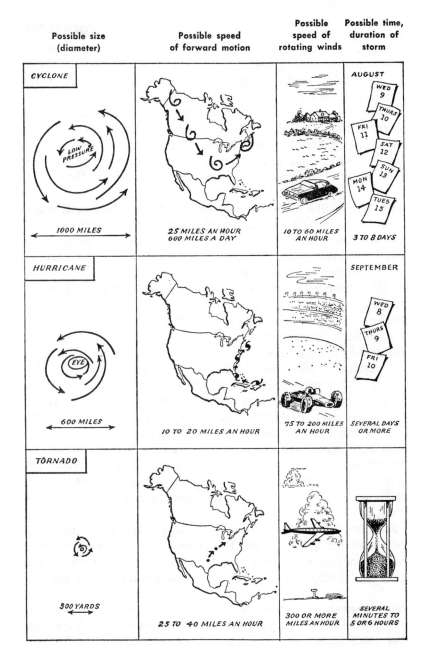

CYCLONE			AUGUST
LOW PRESSURE			WED 9 / THURS 10 / FRI 11 / SAT 12 / SUN 13 / MON 14 / TUES 15
1000 MILES	25 MILES AN HOUR 600 MILES A DAY	10 TO 60 MILES AN HOUR	3 TO 8 DAYS
HURRICANE			SEPTEMBER
EYE			WED 8 / THURS 9 / FRI 10
600 MILES	10 TO 20 MILES AN HOUR	75 TO 200 MILES AN HOUR	SEVERAL DAYS OR MORE
TORNADO			
300 YARDS	25 TO 40 MILES AN HOUR	300 OR MORE MILES AN HOUR	SEVERAL MINUTES TO 5 OR 6 HOURS

Difference between a cyclone, hurricane, tornado

around a center of low atmospheric pressure. The dynamic force of wind currents spiraling inward and upward creates a partial vacuum at the "eye" or center of the whirl which exerts an explosive effect as it passes over buildings and other objects.

Tornadoes are some of the smallest and yet most violent wind storms known to man. They affect relatively small areas—on an average, paths are less than a mile wide and several miles long—but they strike faster and with more concentrated violence than any other storm, shattering and uprooting nearly everything in their paths.

168. What is the difference between a cyclone, a hurricane, and a tornado? People constantly are confusing the three kinds of storms—cyclones, hurricanes, and tornadoes. These three wind systems are all whirls of winds within which air spirals in toward a low-pressure center and then moves upward.

They differ from one another in size, velocity of rotating winds, speed at which they travel, and length of time they exist. In general, the faster the winds spin, the smaller the storm and the shorter the time it lasts. For purposes of clarity, the essential differences are summed up in a general way as follows:

A cyclone is any atmospheric system of rotating winds around a center of low barometric pressure. It is known as the low of the weather maps. A cyclone has winds up to 60 miles an hour, can spread out more than 1,000 miles in diameter. It travels about 600 miles a day, and may last for a week, sometimes several weeks.

A hurricane or typhoon has winds exceeding 74 miles an hour up to 200 miles an hour, perhaps more. This wind storm can spread out with a diameter of 600 miles, moves at an average speed of 10 to 20 miles per hour, and lasts for several days, sometimes more than a week.

Scientists believe that tornado winds may reach speeds of nearly 500 miles per hour or more. This storm is usually less than a mile wide. It advances at average speeds of 25 to 40 miles per hour and may last for only a few minutes. Some have lasted for a few hours.

169. What causes a tornado? Tornadoes are formed under specific unstable atmospheric conditions involving layers of air with contrasting temperatures, moisture contents, densities, and wind flows. Complicated energy exchanges within these unstable air systems produce the vortex, which is the whirling funnel of rotating winds.

Many theories have been suggested to explain exactly how a tornado is generated. Researchers seem to agree that tornadoes are caused by the combined effects of two kinds of forces: thermal, or those involving heat transfers; and mechanical, or those involving rotating winds and forces other than heat. Both forces working together can produce tornadoes, but generally one or the other is the stronger agent. (See Questions 170 and 171.)

170. How do thermal forces help produce a tornado? Complicated heat transfers occur whenever a cold layer of air encounters a warm layer of air. In most cases the cold air mass, being heavier than the warm, moves along close to the ground in a wedge-shaped formation that helps push the warmer air upward.

At certain times of the year, however, and in certain locations, a set of conditions exists wherein a mass of cold air moves over a warm air mass, producing unstable and turbulent atmospheric conditions that can generate tornadoes.

171. How do mechanical forces help produce a tornado? Mechanical forces acting with energies other than thermal can help produce a tornado in areas where the air is already unstable.

These forces include air flowing in from the sides, rotation of the earth, and the sweep of high-powered jet wind streams aloft. These factors slowly constrain the rotating air currents already set in motion by differences of temperature, density, and other things.

As the radius of circulating winds is decreased, the speed of rotation increases, just as an ice skater spins faster by drawing in his arms, thus contracting his body radius as much as possible.

These converging, rotating, accelerating winds all help set the tornado vortex in motion.

172. What are general conditions for tornado formation in the United States? The exact process of tornado formation is still in doubt, but in general this is how it works in the central area of the United States:

A layer of cool, dry air moving high in the sky from the west or northwest rides over warm, moist surface air coming from the south. The warm air, sometimes a few thousand feet deep, is thus constrained from its natural tendency to rise by the lid of cold air pushing down on it. This sets up a condition of imbalance that is adjusted by a breakthrough, occurring most likely where the potential energy is greatest. At this breakthrough, the warm lower air forces its way upward and the heavier cool air drops down. When this movement is accompanied by strong winds blowing in from the sides, a rotary flow of air is created that may form into a tornado vortex.

173. What part does the jet stream play in tornado formation?
Meteorologists have known that the long swift river of air called the jet stream (see Question 522) has helped breed tornadoes in conjunction with masses of warm air and of cold air. They have long suspected the arctic jet stream that sweeps down from Canada as a major factor in producing U.S. tornadoes, but photographs from a space satellite in 1968 provided conclusive evidence that the subtropical jet streams sweeping out of the Pacific Ocean through Mexico play a major role in the creation of tornadoes.

174. What did satellite photographs show about tornadoes?
Alerted by the Weather Bureau that tornadoes were likely to be formed in the Midwest on April 23, 1968, specialists programmed the camera on the Applied Technology Satellite 3 (ATS–3) to take a series of pictures at 15-minute intervals over North and Central America. These photographs, 50 in all, were made into a film and surprised meteorologists by showing a 300-mile-wide river of air—clearly marked by clouds—rushing at 100 miles per hour out of the Pacific Ocean from a point 1,000 miles southwest of Baja, California. Moving at altitudes of some 40,000 feet, the jet stream crossed central Mexico, central Texas, Oklahoma, and Missouri. Near the Mississippi River, the stream snaked its way through an eastward-moving cold front. When it swept across the forward edge of the front, the jet stream caused thunderstorms to swell up to enormous size. More than a dozen tornadoes were sent spinning across Michigan, Ohio, Indiana, Kentucky, Tennessee, and Mississippi.

Scientists have called this photographic series perhaps the most useful meteorological discovery of 1968.

175. Might electricity help cause a tornado? Lightning and other electrical processes have long been suggested as possible sources of the energy required to initiate and maintain tornadoes.

As early as 58 B.C. the Roman philosopher and poet Lucretius suggested such an idea. In 1456 the Italian statesman Niccolò Machiavelli noticed and described the brilliant illuminations that accompanied tornadoes. Scientists in the mid-1800's believed that tornadoes could be electrical winds. Today scientists continue to study the unusual electrical effects associated with tornadoes. The suggestion that electricity might be a cause of tornadoes has not had much scientific support to date, but continues to arouse interest. Data on the relationship between electrical activity and tornadoes are continuing to be collected by scientists and eyewitnesses.

176. What was the worst single tornado disaster? The Tri-State tornado of March 18, 1925, was the most destructive single tornado known in history. This tornado formed in southeast Missouri, cut a swath across Illinois, and ended 219 miles away in Indiana, 3½ hours later. It killed 689 people, injured 1,980, and caused about $17 million damage.

The rampaging Tri-State twister holds several records to date. It caused the highest death toll, had one of the longest continuous tracks, lasted the longest time, and moved forward at an average speed of 62 miles per hour, surpassed only by 2 other tornadoes on record. It reached a peak of 73 miles per hour during the last 36 miles of its journey. Another feat of this tornado was the fact that it moved in an exceptionally straight line over much of its journey—for the first 3 hours as it traveled on a east-northeastern path for a distance of 183 miles, it did not vary more than one compass degree off course.

177. Why were so many people killed by the Tri-State tornado? A significant aspect related to the high death toll was the fact that the funnel of this Tri-State tornado was difficult to see, at times almost impossible. People in its path did not realize the tornado was coming until just a few seconds before it hit. The base of the storm cloud was so close to the ground, the sky so dark, and the amount of dust and debris so massive and confusing that the tornado funnel was completely obscured until there was little or no time to escape.

178. When did a tornado almost wipe out an entire town? On Good Friday evening, between 5 and 6 P.M., March 26, 1948, a tornado struck suddenly in Indiana, nearly wiping out the town of

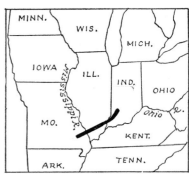

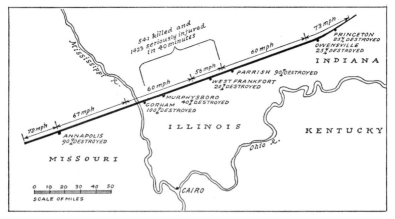

The Tri-State Tornado, March 18, 1925

Coatesville with a population of 500. It destroyed 80 percent of the town, killed 16 people, and injured 150 others.

The tornado formed at Terre Haute and moved across the state, with an 880-yard-wide path, finally ending at Red Key. Total damage of this tornado was more than $3 million.

179. What destructive series of tornadoes occurred in 1965? A cluster of 37 tornadoes struck the Midwest on Palm Sunday, April 11, 1965, killing 271 people, injuring 5,000, and causing $300 million damage to property. For 9 hours, this series of tornadoes churned throughout a 6-state area, devastating portions of Iowa, Illinois, Wisconsin, Michigan, Indiana, and Ohio.

180. Why did the Palm Sunday tornado inflict so many deaths?
Tornado warnings had been issued by the Weather Bureau throughout
the area in which the April 11, 1965, tornadoes struck, but people
either ignored the warnings or did not hear them. The Midwestern
communities were wholly unprepared for tornado emergencies. To
prevent another tragedy such as the Palm Sunday catastrophe, the
Weather Bureau has been making concerted efforts to inform people
in tornado areas as to what these storms are, when to expect them,
and what to do when they strike.

**181. What has been the highest death toll from one series of
tornadoes?** A series of 60 tornadoes, dropping down from one parent
cloud system, struck the southern and southeastern United States on
February 19, 1884, and took the highest death toll on record. Reports
listed 800 persons killed in one day as the tornadoes raged through
Mississippi, Alabama, North and South Carolina, Tennessee, Kentucky, and Indiana.

● **182. What were some other outstanding tornado disasters?** On
May 7, 1840, at Natchez, Mississippi, 317 people were killed, houses
were burst open, and an undetermined amount of damage was caused
by one single tornado.

A tornado that traveled only a short and narrow path and did not
blow with especially violent winds happened to cross the heavily
populated city of St. Louis, Missouri, on May 27, 1896. It killed 306
people and caused almost $13 million in damage.

Another outstanding disaster, aftermath of a series of 31 tornadoes,
occurred on March 21 and 22, 1952, when 208 persons were killed
in the 6 states of Arkansas, Tennessee, Missouri, Mississippi, Alabama, and Kentucky. About 1,154 people were injured, and over
$14 million in damage was caused.

Eighteen tornadoes spun a 175-mile path of destruction across
northern Illinois and 7 other states including Iowa, Wisconsin, Indiana, and Michigan on April 21, 1967. During the storms, 60 or more
people were killed, and some 9,000 more were injured or otherwise
affected. The series of tornadoes caused many millions of dollars of
property damage, including more than $20 million in Illinois and
$3 million in the area of Grand Rapids, Michigan.

183. What was an account of a tornado's aftermath? On Friday, May 8, 1840, the day after the Natchez, Mississippi, tornado, an extra edition of the *Weekly Courier and Journal* ran the following article:

HORRIBLE STORM!!
NATCHEZ IN RUINS!!!

Our devoted city is in ruins, and we have not a heart of stone to detail while the dead remain unburied and the wounded groan for help. Yesterday, at 1 o'clock, while all was peace, and most of our population were at the dining table, a storm burst upon our city and raged for half an hour with most destructive and dreadful power. We look around and see Natchez, yesterday lovely and cheerful Natchez, in ruins and hundreds of our citizens without a shelter or a pillow. Genius cannot imagine, poetry cannot fill up a picture that would match the ruins and distress that every where meets the eye. 'Twas the voice of the Almighty that spoke, and prudence should dictate reverence rather than execration. All have suffered, and all should display the feeling of humanity and the benevolence of religion!

"Under the Hill" presents a scene of desolation and ruin which sickens the heart and beggars description—all, all, is swept away, and beneath the ruins still lay crushed the bodies of many strangers. It would fill volumes to depict the many escapes and heart-rending scenes. . . .*

184. What other disastrous tornadoes have occurred? On Sunday night, April 5, 1936, a tornado spun through the residential section of the city of Tupelo, Mississippi, toward the northeast and disintegrated about 20 miles away. Even though the storm was only 400 yards wide, it killed 216 people, injured 700 others, and caused $3 million in destruction to property.

Causing a death toll of 153 people and property damage of more than $5 million, a family of 4 tornadoes moved through the mountains of West Virginia and Maryland and into Pensylvania on June 23, 1944.

The tornado that struck Louisville, Kentucky, on March 27, 1890, was reported to be one of the most damaging in the country up to

---

* "The Press Reports the Natchez Tornado—7 May, 1840," reprinted in *Weatherwise*, The American Meteorological Society, Boston, April, 1966, Vol. 19, No. 2, p. 69.

that time. In its travel through Louisville into Carroll County, Kentucky, the storm killed 106 people and caused more than $3 million in damage.

One of the most economically destructive tornadoes struck Will Rogers and Tinker Air Fields near Oklahoma City on March 20, 1948, causing a $10 million loss to buildings and aircraft. A second tornado struck Tinker Field 5 days later, and added another $6 million to the damages. No one was killed, and only 9 people were injured.

185. What does it feel like to be directly under a tornado? Only a few people have had the frightening privilege of looking up into the interior of a tornado cloud and living to tell the tale. One vivid description of such an experience came from farmer Will Keller who lived near Greensburg, Kansas:

On the afternoon of June 22, 1928, between three and four o'clock, I noticed an umbrella-shaped cloud in the west and southwest and from its appearance suspected there was a tornado in it. The air had that peculiar oppressiveness which nearly always precedes the coming of a tornado.

I saw at once my suspicions were correct. Hanging from the greenish black base of the cloud were three tornadoes. One was perilously near and apparently headed directly for my place. . . .

Two of the tornadoes were some distance away and looked like great ropes dangling from the parent cloud, but the one nearest was shaped more like a funnel, with ragged clouds surrounding it. It appeared larger than the others and occupied the central position, with great cumulus clouds over it.

Steadily the cloud came on, the end gradually rising above the ground. I probably stood there only a few seconds, but was so impressed with the sight it seemed like a long time. At last the great shaggy end of the funnel hung directly overhead. Everything was still as death. There was a strong, gassy odor, and it seemed as though I could not breath. There was a screaming, hissing sound coming directly from the end of the funnel. I looked up, and to my astonishment I saw right into the heart of the tornado. There was a circular opening in the center of the funnel, about fifty to one hundred feet in diameter and extending straight upward for a distance of at least half a mile, as best I could judge under the circumstances. The walls of this opening were rotating clouds and the whole was brilliantly lighted with constant flashes of lightning which zig-zagged from side to side. . . .*

● **186. On the average, how many people have been killed each year by tornadoes?** In the years from 1953 through 1968, the death toll from tornadoes has averaged 123 people per year in the United States.

In the period 1916 to 1953, an average of 230 people were killed each year.

187. What year had the highest death toll? In 1925, 842 people were killed by tornadoes. This is the highest annual death toll to date.

Other years with high tornado death figures were 1936, when 552 people were killed; 1927, when 540 were killed; 1953, when 516 died; and 1917, when 509 people were killed.

● **188. When was the death rate the lowest?** The death rate from tornadoes was lowest in 1962 and 1937, when only 28 and 29 people respectively were killed.

189. Why are tornadoes so destructive? Destruction from these violent storms is caused by the combined actions of the extraordinary high-speed winds blowing around the vortex and the partial vacuum and swift updraft in the center of the funnel. These factors create sudden and great atmospheric pressure differences within small areas.

With some of the highest-speed winds in the world within a circular storm, tornadoes strike with enormous force against any object in their path.

Pressures exerted by winds increase approximately as the square of the velocity. In other words, a 500-mile-per-hour wind exerts a pressure 25 times that of a 100-mile-per-hour wind. These high winds can turn normally harmless inert objects into dangerous missiles. Sand and gravel are driven like shotgun pellets, and straws of wheat can turn into penetrating arrows, lodging as much as an inch deep within tree trunks.

* A. A. Justice, "Seeing the Inside of a Tornado," *Monthly Weather Review*, May, 1930, p. 205.

190. How do tornadoes explode houses? In a tornado system, upper air is removed outward in a horizontal direction through the middle and upper part of the funnel faster than the air flows in from below. This depletion results in a partial vacuum within the center of the funnel at its base near the ground.

The low atmospheric pressure in this center may cause a sudden drop of air pressure by 8 percent or more in a matter of seconds as it passes over objects. The air pressure within the house thus becomes suddenly greater than that outside, and the walls and roof literally are pushed outward in an explosive manner.

For instance, if a house is at sea level pressure, a normal pressure of about 15 pounds per square inch exists on all sides of the house, inside and out. If a tornado moves over the house, the pressure on the house beneath the funnel may suddenly drop to 14 or 13 pounds per square inch. Since the pressure inside the house drops relatively slowly, especially if doors and windows are closed, the internal force exerted outward suddenly is enough to push out the walls, just as if there had been an explosion from inside. Meteorologists compute that if a tornado passes over a house with a ceiling space of 20 by 40 feet in area, the internal force suddenly exerted outward would be about 68 tons.

191. What other kinds of damages are caused by the suction power of a tornado? The abrupt pressure drop and the accompanying violent updraft winds are powerful enough to lift heavy objects such as trains and houses and carry them for considerable distances. The vacuum effect of a tornado has ripped concrete surfaces off highways, pulled corks out of bottles, exploded chests, peeled bark off trees.

192. What is the lowest recorded air pressure of a tornado? The lowest officially recorded air pressure in a tornado occurred in May, 1896, when a tornado funnel passed over Lafayette Park in St. Louis, Missouri. The pressure recorded was 26.94 inches, lowest tornado record to date. Hurricane pressures are lower. (See Question 69.)

Another report of even lower pressures was made by two men on August 20, 1904, as a tornado struck Minneapolis. They stated the barometer needle dropped to 23 inches as the storm passed over, then returned immediately to its former reading. The barometer had been checked for accuracy by the Weather Bureau just before the storm.

193. What are other names for tornadoes? Tornadoes have long been called by their popular name, twisters.

In many Midwest states they are erroneously called cyclones. Even though technically tornadoes are cyclonic storms in that the winds spiral inward toward a center of low pressure, they are not cyclones. (See Question 576.)

The word tornado comes from the Latin word *tornare*, which means to turn or to twist. The Italian word *tornare* means to turn, as does the Spanish word *tornear*. The French word for tornado is *tornade*.

194. What does the funnel cloud of a tornado look like? The most distinctive characteristic of a tornado, making it different from all other storms occurring over land, is the pendant cloud dangling from the base of the overhead cloud. This cloud often has a funnel shape— wide at the top and tapering to a small diameter toward earth.

The funnel often twists and stretches into other strange forms. People have described it as a trunk of a huge elephant swaying back and forth over the land, sucking up objects as an elephant sucks up water. Other people have reported that it looks like a long trailing rope, or a snake dangling and writhing from the sky. Sometimes as it hangs straight down, the diameter of the base expands and grows as large as the top, making the whole tornado look like a huge thick pillar.

Because swift upper winds blow the higher clouds, the top part of the funnel cloud often moves faster than the lower, while the bottom end trails behind. Friction at the spot where the funnel touches earth also makes it drag. Sometimes the top and bottom are stretched so far apart the cloud looks like a long horizontal string or rope.

195. How big are these funnels? The length of the funnel cloud dangling from the sky may be anywhere from about 800 to 2,000 feet. The diameter of its destructive tip at the earth's surface varies quite a bit—ranging from about 9 feet to slightly over a mile. The average diameter is about 200 to 300 yards.

196. Why does a funnel appear different shades of gray? A tornado funnel is a cloud loaded with condensed water vapor that appears different shades of gray because of the diffraction of light on the water droplets. Sometimes when the moisture content is fairly low, the cloud appears light colored, at times almost invisible.

Whether a funnel appears light or dark depends mostly, however, upon where the source of light is shining from and upon the background of the tornado. If an observer is standing between the funnel and the sun, he sees the funnel as white. If the sun or source of light is behind the funnel, the funnel appears dark.

As a tornado moves over land like a giant vacuum cleaner, it shatters and sucks up a great variety of material such as dust, soil, and debris. This material makes the funnel darker, sometimes almost black. On the other hand, one tornado funnel turned almost white from snow that was swirled into it as it passed over a snow field on the east slope of the Wasatch Mountains in Utah.

197. How does a tornado funnel behave? Formed several thousand feet above the earth's surface, usually in conjunction with a severe thunderstorm, tornadoes seem to drop down literally from the skies. Some funnels never reach the ground. Others may touch the ground in one spot, then rise into the air again and dissipate. Sometimes a twister may burrow downward, plow a path several yards wide, lift up in the air to leave an undamaged area for several miles, and then lower again and continue its destructive path on land.

198. What are the wind speeds of a tornado? Scientists estimate that the maximum speeds of winds rotating within a tornado reach more than 400 or 500 miles an hour near the center of the vortex. These are maximum speeds—most tornadoes probably have wind speeds somewhat lower than 300 miles an hour.

199. What methods are used to determine maximum wind speeds? Few instruments can stand up to the violent impact of tornado winds. Even though tough instruments such as air-speed indicators of jet aircraft can withstand forces of fast violent winds, such instruments usually become damaged by flying debris when subjected to winds of tornadoes.

Some estimates of tornadic winds are based on analyses of the effects produced on structures whose mechanical properties are known. For instance, an idea of the tremendous wind velocities can be gained by inspecting various wheat straws and frail grass stems that have been propelled as much as an inch deep into the trunks of trees, telephone poles, fence posts, and sides of buildings. During the May 27, 1896, tornado that struck St. Louis, winds were so strong that a solid ⅝-inch-thick iron sheet was pierced by a two-by-four-inch pine plank.

200. What other system are scientists hoping to use to record tornado speeds? One potential method of determining the high wind speeds of a tornado might be a Doppler radar that can measure the speed of individual particles such as water drops or bits of dust and debris in the rotating winds by the changes in frequency of returning radio waves.

The Doppler effect is a phenomenon based on the frequency changes of waves, as the source of these waves approaches or retreats from a reference point. These waves may be light waves, radio waves, radar waves, or sound waves. The change of the frequency depends on the speed of movement of the object toward or away from the reference point.

One aspect of the Doppler effect is shown by the change in sound of a train's whistle as the train approaches or departs. As the train approaches, a greater number of sound waves per second is received by the observer's ear, and therefore the pitch seems higher. As the train passes the observer and retreats in the distance, relatively fewer sound waves reach the ear per second, and the pitch appears lower. The Doppler effect is also apparent in the change of frequency in light waves radiating from distant stars approaching the earth, or traveling away from it.

201. In which direction do tornadoes usually spin? Tornadoes spin the way they do because the funnel winds are set in motion by masses of air whose movements depend upon the rotation of the earth.

In the Northern Hemisphere, the warm moist air moves northward from the equator. Colder air moves from the west or northwest. When these two currents of air meet, the resulting wind currents blow in a counterclockwise rotation.

In the Southern Hemisphere the winds creating tornadoes blow in opposite directions from those of the Northern Hemisphere, and hence form a clockwise rotating funnel.

202. Do tornadoes always spin in the same direction? In the Northern Hemisphere, tornadoes usually spin in a counterclockwise motion—but not always. In a study of 550 tornadoes, one scientist counted 29 that may have rotated clockwise. Reports from England have also mentioned a few clockwise-spinning tornadoes.

203. How fast does a tornado travel across land? The forward speed of a single tornado varies greatly, and speeds of different tornadoes vary. Some tornadoes have traveled as slowly as 5 miles an hour, and some have ripped across the land at nearly 70 miles an hour. The average speed is about 25 to 40 miles an hour.

204. What was the fastest speed at which a tornado has traveled forward? One of the fastest tornadoes on record, the Tri-State tornado of 1925, reached a peak of 73 miles per hour in the last stage of its rampage. (See Question 176.) The fastest average speed on record for a tornado is 65 miles per hour, held by a tornado passing through southeast Kansas on May 25, 1917.

205. What was the slowest speed of a tornado? Possibly the slowest traveling tornado may have been the one that remained in nearly the same place for about half an hour on June 2, 1929, at Hardtner, Kansas.

Another tornado, on May 24, 1930, traveled only 5 miles an hour as it spun near Pratt, Kansas.

206. What is the earliest record of a tornado in the New World? The earliest account of an American tornado was given by a colonist who described a whirlwind striking a meetinghouse in Massachusetts in 1643, killing an Indian. The wind was regarded more as a curiosity than as a significant event. A tornado was reported in New Haven in 1682, and another in Charleston in 1762. Another tornado struck Charleston on May 4, 1761, sinking five warships offshore and breaking the masts of several more. On shore the tornado tore up trees and destroyed houses.

By the time the early pioneers journeyed to the Central Plains and saw the violent behavior of these tornadoes, they began to regard these storms with awe. Astounding reports have been given of tornadoes sweeping across closely packed ranks of buffalo herds, carrying individual animals weighing several tons through the air, and breaking nearly every bone in the animals' bodies when they were crashed to the ground.

207. What was the earliest written account of a tornado? Possibly the earliest reference to a tornado in writing may have been about 600 B.C. in the Old Testament, when Elijah was snatched up to heaven in a lightning-filled tornado. A description appears in Ezekiel, Chapter 1, verse 4:

And I looked, and, behold, a whirlwind came out of the north, a great cloud, and a fire infolding itself, and a brightness was about it, and out of the midst thereof as the colour of amber, out of the midst of the fire.

208. What is a prester? A prester is what the ancient Greeks and Romans called a whirlwind or waterspout in the Mediterranean area when it was accompanied by lightning. The word comes from the Greek word meaning venomous snake, a scorching whirlwind, or a neck swollen with anger.

209. In what general direction do tornadoes usually travel? Tornadoes in the United States usually travel from the southwest to the northeast, in front of an advancing turbulent weather front. As warm moist air usually travels northward and dry cold air pushes eastward, when the 2 masses meet, the resulting front moves northeastward forming tornadoes as it advances.

Reports show that nearly 90 percent of U.S. tornadoes come from a westerly direction, and 4 percent come from the east. However, tornadoes have been known to come from almost any direction.

210. How long does a tornado's destructive path extend? A tornado usually travels along the earth's surface for a distance of from 10 to 40 miles. At least 7 tornadoes have traveled more than 200 miles, and a few have plowed destructive trails for nearly 300 miles. The average length of a path is about 16 miles.

Many tornadoes east of the Mississippi River travel 150 to 200 miles before they die out.

The path of a tornado is usually clearly marked and well defined. Sometimes violent side winds obscure the sharp edges of the path, but only rarely.

211. What was the longest tornado path? The longest continuous path on record for a tornado is 293 miles—forged by an energetic tornado on May 26, 1917. This twister traveled from the town of Louisiana, Missouri, to the eastern boundary of Jennings County, Indiana.

212. Where are tornado paths shortest? Tornado paths in the Midwest tend to be shorter than those anywhere else in the nation— possibly because there is not much moisture in the air to sustain the energy of the winds. In Kansas and Oklahoma, for instance, the average length of several hundred tornadoes was found to be only 10 miles.

213. How wide are these paths? The width of the tornado's path varies as the storm journeys across country. Sometimes tornadoes shrink to less than 100 feet in diameter. One was recorded as only 9 feet across. Other tornadoes have expanded to cover an area 2 miles wide.

Various averages have been calculated on different studies: the average path was 396 yards wide for a thousand tornado paths; 440 yards for 924 tornadoes; and 432 yards for 209 tornadoes that occurred in 1950.

From many statistics, however, the average width has often been given as 250 yards.

214. What is the general appearance of a path left by a tornado? A tornado can leave a path of destruction similar in many ways to the devastation caused by a heavy demolition squad.

If the tornado has passed over a city or town, the wreckage of houses, furniture, and vehicles is strewn throughout the path, sometimes piled several feet deep, while only 10 or 15 feet away, outside the path, objects have been undisturbed by the destructive winds.

Buildings have been exploded and demolished by tornadoes, leaving strewn wreckage of broken boards, shattered glass, cracked concrete, and twisted pipes. Large trees have been torn up by their roots, twisted and shattered. Automobiles have been rolled over and crushed like toys. Trains have been pushed over and the track torn up. Few people survive if they are caught in the open by a tornado. Sometimes freak incidents save their lives, but usually anyone in the direct path of the storm is killed.

The path of a tornado can be easily traced through the countryside by twisted, shattered trees, stripped of foliage and limbs and often uprooted. Fields of corn or meadows look as if a bulldozer has passed across them. Nothing is left in peace when these vicious storms touch earth.

215. What are some odd paths of tornadoes? Tornado funnels have been known to make U-turns and even complete circles. Sometimes they stop moving forward altogether. A tornado in Sioux City, Iowa, on June 16, 1944, spun in one place for 10 to 20 minutes, then made a U-turn, traveled toward the southeast about 3 miles, then moved south, then east, then north, and finally headed east again. The tornado that passed through Buchanan County, Iowa, in 1942, moved in a crescent-shaped path, traveling toward the northwest, then north, and then to the northeast.

216. What are some freakish tricks of tornadoes? The intensity of tornado winds and the funnel's erratic movement across the earth's surface have caused many astonishing and freakish effects, some so unusual they seem unbelievable.

One tornado of April 16, 1880, carried heavy timbers of an entire house a distance of 12 miles near Marshall, Missouri, and an 800-pound ice chest for 3 miles.

Tornadoes have been known to snatch up blankets and mattresses from beds while leaving the surprised occupants shaken but unharmed.

The tornado of June 23, 1944, passed over the West Fork River, West Virginia, and left it dry for a few moments. A 1912 tornado lifted a telephone pole straight up from its site and carried it down the street thumping it along in a vertical position.

217. When have tornadoes lifted trains? Tornadoes have picked up and knocked down many freight and passenger trains.

On May 27, 1931, a tornado struck the train Empire Builder as it was traveling 60 miles an hour near Moorhead, Minnesota. The powerful storm picked up five coaches weighing about 70 tons each and lifted them off the tracks. One coach, weighing about 83 tons and containing 117 passengers, was lifted by the tornado, carried 80 feet, and then dropped in a ditch.

Another tornado struck a moving train on June 22, 1919, near Fergus Falls, Minnesota. Seven heavy coaches were picked up and thrown from the tracks. The baggage car was lifted up and set down about 30 feet away from the rest of the train, at right angles to the rails.

One tornado picked up a locomotive from one track, spun it around in mid-air, and set it down on a parallel track facing the opposite direction.

218. Have human beings been lifted up by tornadoes and set down alive? Human beings have been carried aloft by tornadoes and set down alive to tell the story. During a tornado at El Dorado, Kansas, on June 10, 1958, a woman was plucked through a window and carried 60 feet away, landing beside a phonograph record titled "Stormy Weather." On April 9, 1947, a man opened his front door near Higgins, Texas, and was scooped up by a tornado and carried 200 feet over the treetops.

In Topeka, Kansas, a small tornado picked up a man, carried him 100 feet through the air, and let him down unharmed but covered with mud.

219. What other fragile objects have tornadoes carried? One tornado was reported to have moved a crate of eggs 500 yards without cracking a single shell.

Another story of the gentleness of a tornado is the saga of a jar of pickles carried 25 miles through the air and lowered unharmed into a ditch. Mirrors have traveled air-borne for miles and been set down undamaged.

The reason fragile things can be deposited so gently is that objects frequently descend through an ascending air current which allows them to be lowered to earth at a comparatively slow rate of speed.

220. Do tornadoes occur more frequently singly or in groups? Tornadoes are generally "loners"—that is, they occur singly.

Yet when weather conditions are right, tornadoes occur in families of 2 or more which are associated with a parent thunderstorm or squall line extending over a distance of several hundred miles. As the storm front moves, tornadoes may drop at intervals along the path, travel for a few miles, then dissipate. A short time later, another tornado may form, travel for a while, and then disappear in the same manner.

221. What was the largest number counted in a series of tornadoes? Sometimes 2 or 3 tornadoes are generated side by side and travel in parallel paths covering an area several miles wide. A dozen tornadoes can form and disappear in a day. Probably the largest number in a tornado series counted in one day was 60—reported on February 19, 1884. (See Question 181.)

222. Are tornadoes affected by the terrain or obstacles they pass over? Hills or cliffs, trees or buildings, lakes or rivers seem to have little or possibly no effect on the speed and power of tornadoes, which appear to move across all obstacles with equal ease. Scientists are accumulating data to determine the behavior of tornadoes as they pass over various objects.

A series of tornadoes in 1944 rode over mountains in West Virginia and Pennsylvania and dropped into nearby valleys with such energy that creeks were sucked dry.

223. What happens when tornadoes move out over water? Tornadoes can move from land to water and back again without any apparent diminishing of energy or any change. They often are called waterspouts if they remain over water for any noticeable length of time, enough to slow them down, for instance. (See Questions 251 and 252.)

One tornado of June 28, 1924, traveled across Lake Erie for a distance of 25 miles, crossed the shoreline, and struck Lorain, Ohio, with undiminished force.

224. How long do tornadoes last? Tornadoes usually spin across a given point on earth in a minute or two.

The whole funnel structure may have a life history of one hour between birth and the time it dissipates. Some tornadoes have lived only a few minutes, and other series of tornadoes have lasted 9 or more hours.

225. How many tornadoes occur each year in the United States?
No completely accurate count has been made on yearly numbers of tornadoes.

From 1916 through 1952, fewer than 300 tornadoes a year were reported in the United States.

In 1953, the Weather Bureau initiated its tornado forecasting program, and tornado watchers and reporters became more numerous and more precise. During the period 1953 through 1966, the yearly average of tornadoes reported rose to about 670. The increasingly large numbers of tornadoes spotted in recent years underscore the lack of significance of such statistics of yearly averages.

226. Why are more tornadoes being reported in recent years?
As warning systems are perfected and more citizens keep watch for tornadoes and report all they see, the number of these storms on record increases, and the yearly average keeps rising. A sharp increase in the number of reported tornadoes began in 1953, when the Weather Bureau's tornado forecasting program stirred up interest in these storms. For instance, during the 192-year span from 1682 to 1874, a total of only 150 tornadoes were reported in the New World. In the 51 years from 1916 through 1967, however, there have been nearly 16,000 tornadoes reported.

● **227. What was the highest number of tornadoes reported in one year?** In the year 1967, a total of 912 tornadoes was recorded by the Weather Bureau, including 115 tornadoes observed spinning off hurricane Beulah. This is the highest figure on record through 1968. Each successive year may bring even higher numbers as more people and instruments are on watch.

Other years with high numbers of tornadoes include 1965, with 898 tornadoes reported; and 1957, with 864.

● **228. What states have been hardest hit each year, on an average?**
During a 13-year period, 1953 through 1965, Texas had the highest annual average number of tornadoes—109. Oklahoma was next with about 77 tornadoes per year, and Kansas averaged 72.

Other hard-hit states were Nebraska with 39 a year, Florida with 32, Missouri with 27, Illinois and Indiana with 24, Iowa with 22, and Arkansas and Alabama each with 18.

● **229. What states have been hit least, on an average?** Alaska, Hawaii, Rhode Island, and the District of Columbia were four areas not hit by any tornadoes during the years 1953 through 1965.

Oregon and Nevada each averaged less than one tornado a year. (This means each had from 1 to 12 during that 13-year period.)

Washington, Utah, West Virginia, Delaware, New Jersey, Connecticut, and Vermont each averaged about 1 tornado a year.

● **230. What are the averages for the other states?** Other states averaged the following yearly numbers during the years 1953 through 1965.

South Dakota	20	Pennsylvania	6
Georgia	17	Wyoming	6
Louisiana	16	Kentucky	5
Minnesota	15	Massachusetts	5
Wisconsin	15	Maine	4
Colorado	15	Montana	4
Michigan	15	Virginia	4
Mississippi	14	Maryland	3
North Dakota	14	California	3
Ohio	11	Arizona	3
South Carolina	10	Idaho	2
New Mexico	9	New York	2
North Carolina	8	New Hampshire	2
Tennessee	7		

231. Can tornadoes strike 2 or more times in the same place? Like lightning, tornadoes sometimes can strike more than once in the same place. Even though chances are low for the narrow short-lived storms to hit the same place twice, many instances have been reported where tornadoes have hit several times during the years, particularly in the regions where tornadoes are frequent. Reports on specific cities being hit repeatedly by tornadoes are increasing, but this may occur because the cities are continually growing and expanding their boundaries. Thus a city might be hit several times by tornadoes, but not in the same spot.

232. What cities have been struck more than once by tornadoes? Oklahoma City can easily claim the title as the most tornado-battered city, with a record 26 tornadoes since 1892. Chicago has had 5 tornadoes strike within its boundaries. Four tornadoes have been known to strike Gainesville, Georgia; and Minneapolis, Minnesota, has also been struck four times. Codell, Kansas, has been hit three times on May 20—in 1916, 1917, and 1918.

The city of Austin, Texas, was hit twice by tornadoes within half an hour, on May 4, 1922. Baldwyn, Mississippi, was hit twice on March 16, 1942—25 minutes apart. Ellis County in Oklahoma was hit once on April 9, 1947, and again on May 31 of the same year. Louisville, Kentucky, and Great Bend, Kansas, have each been hit twice by tornadoes.

233. What is "tornado weather"? "Tornado weather" is a distinctive, ominous atmospheric condition that breeds tornadoes. It can threaten an area for hours or even days before a tornado forms. And sometimes no tornadoes appear.

At this time, the air seems very oppressive and sultry, uncomfortably warm and humid. People often describe the weather as being "sticky" or "threatening." The relative humidity is unusually high, as well as the temperature.

There are often dark thunderclouds in the sky, especially in the western and northern part. The winds are generally from the south or southwest, flowing into the darkening and lowering sky. As the atmosphere becomes more agitated, the bottoms of the clouds may become bulbous in appearance and seem to hang downward like giant balloons. Some observers have described the cloud masses bulging downward as like a huge quilt fastened to the sky by hundreds of invisible pins, between which the quilt droops in loose, sharply curved folds. These strange clouds can become quite dark and menacing. They seem frequently to be almost black, often with a greenish tint.

Heavy hail and downpours of rain often precede a tornado, and lightning flashes frequently with strong intensity.

From out of such a sky descends the dreaded tornado funnel.

234. What cloud formations look like tornadoes but are not? Several weather conditions and clouds can look startlingly like a tornado, but the Weather Bureau points out they can be quite different and comparatively harmless.

Seen at a distance, for instance, a column of rain water falling from a thundercloud can look deceptively like the funnel of a tornado.

Another deceptive cloud is an approaching roll-type squall line that is slowly revolving about its horizontal axis. As the squall cloud passes overhead, its ragged, windtorn edges can be seen moving to the right or left and up and down. To the untrained eye, this movement is much like that of a forming tornado funnel.

Sometimes the base of a turbulent storm cloud consists of bulbous hanging pouches that look like the beginning of tornado funnels—but these bulges do not spin as tornadoes do.

235. What electrical discharge has been observed with tornadoes?
Some of the most spectacular phenomena associated with tornadoes are discharges of electricity, suggested by some scientists as the driving force behind tornadoes. (See Question 175.) Eyewitness reports of lightning have described the unusual brightness of the flashes discharged in connection with tornadoes. The flashes seem "brighter, bluer, and more vicious" than those from any other type of storm. One person described the funnel as being "continuously illuminated with a blue-green light flashing like an electric welding torch."

Strokes of lightning sometimes occur as often as 10 or 20 per second, and a steady electrical discharge has been observed in connection with tornado funnels. One particular photograph, taken on the evening of April 11, 1965, showed two diffuse, almost vertical pillars of light—possibly the electrical glow from two funnels of tornadoes that cut a swath of destruction in and around the city of Toledo, Ohio.

236. Does hail accompany a tornado? Heavy showers of hail often occur before or during the passage of a tornado. Often the hailstones are of large size. Disk-shaped hailstones measuring 6 to 10 inches in diameter and 2 to 3 inches thick were reported falling from a tornado traveling near Topeka, Kansas, on June 5, 1917.

237. What does a tornado sound like? When a tornado is high in the air, it often has a peculiar hissing, whistling noise that can be heard for several miles over the countryside. This sound has been described by observers as "a screaming, hissing sound," or "like the buzzing of a million bees."

The shrill high-pitch shriek may change to a terrible deafening roar when the cloud dips to the ground and starts churning up the earth.

238. What are some other descriptions of the sounds of a tornado? The sounds of an oncoming, earth-rending tornado have been described in various ways. Some say there is a roaring rushing noise that might be produced by thousands of trains speeding through tunnels or over trestles. One observer said an approaching tornado sounded like a "giant blowtorch." Others likened the deafening noise to the "bellowing of a million mad bulls," the "roar of ten thousand freight trains," the "blast of a million cannons," or the "thunder of a squadron of jet airliners."

239. Why do tornadoes produce these strange sounds? Scientists do not yet know what causes the high shrieks, hissings, and roars of tornadoes. Some have suggested that the tornadic winds may reach the speed of sound and produce shock waves of small amplitude. Others suggest the electrical disturbances as producers of the sounds, particularly the buzzing and hissing.

240. Where do tornadoes strike? Tornadoes occur on all continents of the world, but they strike most often and most violently in the great Central Plains of North America, from the Gulf of Mexico to Canada and from the Rocky Mountains to the Atlantic seaboard.

Tornadoes are moderately frequent around the Great Lakes, the Ohio River region, and the southern United States. They rarely hit the northeastern United States, are most unusual over the Rocky Mountains, and are virtually nonexistent from the Rocky Mountains westward to the Pacific Ocean. Yet every state, including even Alaska and Hawaii, has experienced at least one tornado in its history.

241. Why do so many violent tornadoes form in the U.S. Central Plains? The topography of the Central Plains of the United States is such that more tornadoes form there than in any other part of the world. Over these plains masses of warm moist air, called tropical maritime air, sweep up from the Gulf of Mexico or the Caribbean Sea unhindered by mountain barriers. In spring, the warming earth of the Southern states adds to the layer of warm moist air traveling relatively close to the ground.

At the same time there are frequent invasions of cool dry air masses riding high from the north or northwest. This is the maritime polar air from the northern Pacific Ocean which is blown eastward across the Rocky Mountains. Most of the moisture is dropped on the western slopes of the mountains, and the air that flows across the mountains becomes drier when it descends the eastern slopes. A tongue or wedge of cool dry air spreads out over the Great Plains states more than 10,000 feet above the earth's surface, well ahead of the main block of cold air which drags behind, blocked by the Rocky Mountains. The tongue of cool air rides over the top of warm humid air traveling close to the warming earth surface in a northerly direction from the Gulf.

Riding in from the west comes another ingredient—the powerful jet stream, helping to stir up the already unstable, turbulent atmosphere.

Air masses forming tornadoes over U.S. Central Plains

242. What is Tornado Alley? So many tornadoes travel across the 4 states of Texas, Oklahoma, Kansas, and Missouri that the region has sometimes been called Tornado Alley.

Central Oklahoma has been found to be more tornado-prone than any other area of comparable size in the world.

243. What is the cyclone state? Kansas has sometimes been nicknamed the cyclone state because it has been struck by a higher number of tornadoes in proportion to its size than any other state. Tornadoes in the Midwest are often erroneously called cyclones.

244. In what other countries do tornadoes hit? Although the most frequent and most violent of tornadoes occur in the United States, these storms also strike in many other parts of the world, especially in the north and south temperate zones where there are rapid changes in temperatures and air masses.

Tornadoes frequently occur in Australia, especially in the southern part. Although they are quite severe, they are not so violent as those found in the United States. Since there is a sparse population spread over large tracts of land, the damage is not very great. Strong tornadoes form in the southern and middle part of Russia. These storms are also known in India, China, Japan, Canada, England, France, Holland, Germany, Hungary, Italy, the Bermuda Islands, and the Fiji Islands. The so-called tornado of west Africa is actually a thunderstorm.

245. Where do tornadoes occur in England? A study by the Royal Meteorological Society of England lists a total of 50 tornadoes occurring over an 82-year period in England. Tornadoes there have occurred mostly over the lowlands and river basins in every month of the year except December. Most occur in October. Three of the most destructive tornadoes on record in England occurred on October 19, 1870; October 27, 1913; and May 21, 1950.

246. What were the dates of 2 early tornadoes in England? One of the earliest tornadoes on record in the British Isles struck London on October 17, 1091, blowing down 600 houses and many churches. One vivid account of a disaster attributed to the devil took place on Sunday, October 21, 1638, in a church at Widecombe-in-the-Moor in Dartmoor, England. Just as the Vicar began the service, "everything went black" with a violent storm of wind and lightning that wrecked the roof and tower and killed about 60 people.

247. What tornadoes occurred in France? Tornadoes have struck in France at relatively rare intervals. One tornado hit Monville on August 10, 1845, and completely destroyed several buildings. A tornado in 1950 killed several persons at Cambrai, causing damages of over $3 million.

248. When do tornadoes occur? Although tornadoes can be generated at any month of the year, about 70 percent of them occur in the United States from March through June, with the peak number occurring in May. The season of their least activity is during the winter months—about 8 percent of the total yearly crop form in December and January. About 20 percent of all tornadoes break out during late summer and early autumn.

249. Why does the area where tornadoes strike most frequently shift northward as the tornado season wears on? In February, at the beginning of the annual tornado season, the center of their maximum activity lies over the warming central Gulf states. In March, this center shifts eastward slightly to the southeast Atlantic states where the greatest number of tornadoes form in April. In May, the frequency center moves to the Southern Plains states, and in June it moves north to the Northern Plains and the Great Lakes area, and as far east as western New York State.

In essence, this is a shifting to the north of the storm track. The reason for this drift is the increasing penetration northward of warm moist air from the south as the sun moves north, and spring journeys up the country. Meanwhile the cool dry air from the north and northwest continues to surge southward and southeastward. Most tornadoes are generated where these two air masses clash. Thus, early in the year, most tornadoes form in the Gulf states, when the warm spring air starts to move northward. After May, when there is little cold air to give turbulence to the sky, the number of tornadoes drops. There are few tornadoes in winter because the air is generally cool over large land areas, and there are fewer encounters between air masses of the widely differing temperatures found in spring.

250. What time of day do tornadoes usually appear? Tornadoes may occur at any hour of the day or night but most of them form during the warmest hours of the day, usually in late afternoon. About 82 percent of them occur between noon and midnight, while the greatest number—23 percent of the total tornado activity—strike between 4 P.M. and 6 P.M.

251. What are waterspouts? Waterspouts are divided into two classes: true waterspouts, or tornadoes over water, and fair-weather waterspouts.

They frequently appear in groups. As many as 30 in one day have been spotted at sea.

The shape of a waterspout varies greatly in different conditions—from a straight vertical column to a very long, slender, and twisting coil.

As the rotating moist air in the vortex rises and cools by expansion, it condenses and the resulting water droplets make the whirling wind visible. Contrary to public opinion, they do not suck up much water, although ships have reported being sprayed with fresh water as they pass through a waterspout.

Waterspouts are accompanied with sounds of hissing, sighing, and sucking. Hail and chunks of ice have sometimes fallen from them, as well as live fish and other small water creatures.

252. What are true waterspouts? Most waterspouts are tornadoes at sea produced in essentially the same way as tornadoes with cold air aloft and warm air beneath. They have the same characteristics as tornadoes—rotating winds and funnel-shaped clouds dipping from overhanging clouds. The lower portion of the funnel stirs up spray and water rather than dust and debris, sometimes forming 20-foot fountains and cascades of heavy spray from the base of the funnel.

Waterspouts may be tornadoes that have developed over land and traveled to sea; or they may have developed over water. A land tornado crossing a river or small stretch of water and continuing on land on the other side is still called a tornado.

Waterspouts, or tornadoes over sea, move slower than tornadoes over land and last a shorter time—rarely for more than an hour. The majority of sea tornadoes form and disperse in half an hour. Some last only 2 to 5 minutes. Their winds spin at lower speeds than those of tornadoes. All these conditions exist especially in tropical regions because upper winds are weaker, warmer air masses move slower, and there are few strong steering winds.

253. What causes waterspouts? Waterspouts, like tornadoes, are created in unstable air conditions, when cool layers of air are blowing across the water from above at high levels and warm moist air is sweeping upward from below. Waterspouts have been generated in both high- and low-pressure areas, in calm weather and in gales, in warm and cold weather, at night and during the day.

254. Do ships encounter waterspouts? There have been several reports from ships encountering waterspouts. A number of waterspouts bore down on the ship *Hestia* in 1902 off Cape Hatteras. One large spout hit the ship with a deafening roar and shock of winds, but little damage was inflicted.

255. What waterspout appeared off Martha's Vineyard? A waterspout was observed by many people only 6 miles from the beach at Martha's Vineyard on August 19, 1896.

The spout was estimated to have been 3,600 feet high, its column 840 feet wide at the crest, 140 feet at the center, and 240 at its base. The cascade surrounding it at the surface was 720 feet broad and 420 feet high. The waterspout lasted for 35 minutes, during which time it dissolved and reappeared 3 times.

According to eyewitness reports, some of the more impressionable members of the crowd watching were stricken with fear, and many women and children were thrown into hysterics by the awful sight. As the local Weather Bureau observer wrote in his official report, "there was so much confusion, women and children crying, that I was not very observant until it was all over."[*]

256. How large are waterspouts? Heights and diameters of waterspouts vary widely. They may range in height from only a few feet to close to a mile high. A waterspout off Rabat on the Moroccan coast was reported nearly 1,000 feet high.

[*] Slater Brown, *World of the Wind* (Indianapolis: The Bobbs-Merrill Company, Inc., 1961), p. 167.

Waterspouts may be a yard or less thick in diameter—or several hundred feet.

257. How fast do waterspout winds rotate? The speed of the rotating winds of a waterspout may reach 60 to 120 miles an hour. Because they have such relatively weak rotating winds, waterspouts often disappear when they cross over to land and are slowed by friction.

258. How fast do waterspouts move? Waterspouts may move slowly, at the rate of a mile or 2 an hour, or they may progress forward at higher speeds, perhaps 10 or more miles an hour.

259. When do waterspouts occur? These columns of water can occur in any month, although they form most frequently from May to September over ocean areas that are quite warm.

260. Where do waterspouts occur? Waterspouts can occur on all oceans, from the polar regions to the equator. They are also found spinning across lakes and rivers.

The most favorable regions for formation of these sea tornadoes include the stretch of the North Atlantic Ocean from the Gulf of Mexico, and Caribbean Sea, up along the Antilles and the northern course of the Gulf Stream. They also have risen off the West Coast of Central America and south Mexico. They have been sighted in the San Diego Bay when cold winds aloft blow over the warm water. They have also been observed on the southern parts of the Great Lakes and on Lake Geneva in Switzerland, in the autumn, when the lakes are still warm and a cold atmospheric low exists.

They have been observed off the coasts of China and Japan, and in the conflicting winds off the western coast of Africa.

261. What are fair-weather waterspouts? Fair-weather waterspouts are whirling columns of water droplets that rise in an upward spiral from a lake or sea. They differ from true waterspouts in that they start from the surface and develop upwards, while a true spout forms in the clouds and develops downwards.

Fair-weather spouts are usually solitary. They frequently develop under a clear sky, when no clouds are present.

Because there is so much moisture present, however, a small cloud sometimes forms over this kind of waterspout, and occasionally a thunderstorm develops.

They usually are small and do not develop into dangerous storms. If they pass onto land, they may collapse after only 10 or 20 seconds.

262. What causes a fair-weather waterspout? These fair-weather waterspouts are caused when a quiet, super-heated layer of humid air remains over water, producing convection currents and rising rotating columns at different spots. Occasionally one develops higher into waterspout size.

263. Is a dust devil like a tornado? Dust devils may look like tornadoes, but they behave differently and never attain the violence of the larger storms. The funnels of dust devils do not drop from clouds, as do those of tornadoes. They develop on the earth's surface and rise.

Sometimes rapidly whirling air on hot summer afternoons rises to form a relatively high column of rotating dust, small twigs, leaves, and other lightweight debris. On dusty roads and dry plowed fields in temperate zones, these dust eddies are sometimes only a few inches across and a few feet high and last only a short while.

In subtropical deserts, under a blazing sun with dry winds blowing, dust whirls grow to imposing heights of 2,500 feet and diameters of 280 feet. Some can last several hours. One dust devil near Cairo, Egypt, lasted 5½ hours.

264. What causes a dust devil? Two conditions are needed to form a dust devil—a flat hot surface, such as a dust road or a desert, and a gentle breeze.

As the sun beats down on a bare flat surface and warms it, the air above the surface becomes heated. As it heats, it expands and rises. Surrounding air currents blow inward to replace this rising hot air. They become heated and also rise. A gentle breeze, blowing 2 to 5 miles an hour, helps give this rising air current a twist, and suddenly an eddy is formed, ascending rapidly in a spiral. The larger the mass of superheated air, the larger these dust devils grow and the longer they last.

265. Do dust devils cause any damage? Dust devils usually advance at such slow speeds that people can easily avoid them, merely by moving rapidly away. Even if a dust devil hits a person, no severe damage is done beyond a noseful, mouthful, earful, and eyeful of dust.

266. Where are dust devils being made in the laboratory? Miniature dust devils have been created in a Harvard University laboratory on a heated circular platform with a 10-foot high cylinder of screening that revolves slowly around the platform. Instead of using desert dust, scientists use dense white smoke, usually produced by heating ammonium chloride. The smoke slowly whirls and curls upward and soon becomes a tight, rapidly twisting wind. Results of the model studies may shed some light not only on dust devils but on larger scale twisting phenomena, such as tornadoes and hurricanes.

267. What are fire tornadoes? People have often observed and reported on tornadoes seen spinning out of intense heat from great city fires, forest fires, or from volcanoes. These are not true tornadoes, but are large whirlwinds set in motion by the violent convection currents as hot air rushes upward.

268. What severe fire whirlwinds have occurred in burning cities? Dragon twists, as the fire-induced whirlwinds were called by the Japanese, caused terror and damage as they spun out of the huge fires of Tokyo started by the 1923 earthquake.

A bombing raid on Hamburg, Germany, on July 27, 1943, during World War II, started huge fires that converged and sent many fire whirlwinds gyrating through the city. It was a hot sultry night and the heat was so great that a strong updraft formed, violently sucking in air from all sides. A firestorm developed that devastated more than 8 square miles of the city.

269. What tornadic winds have been observed with volcanoes? Volcanoes sometimes generate whirlwinds by the convection currents from the heat of their eruptions. A heat whirlwind, called a miniature tornado, some 20 feet in diameter, was created in the erupting volcano Paricutin in Mexico, 1943.

Eyewitnesses of the 1815 eruption of the Tambora volcano in

Indonesia reported that several violent whirlwinds were formed in the suffocating heat.

● **270. How are tornadoes being spotted and studied?** Many varied kinds of instruments have been developed for studying the formation and behavior of tornadoes.

Some of the oldest methods of gaining information about these wind systems include sending balloons aloft with radio-transmitting units and weather-recording devices. Radio signals from these balloons are tracked on the ground to help define the air circulation in tornado areas, as well as to keep track of temperatures and moisture conditions in the atmosphere.

Weather stations on the ground constantly record temperatures, rain, winds, and humidity to give a continuous report on changing atmospheric conditions.

Radio, radar, light, photographic, and other kinds of equipment are mounted on aircraft and satellites to keep constant observations on possible tornado-generating cloud systems.

271. How are satellites used in watching for tornadoes? Constantly circling satellites, such as TIROS, Nimbus, and others, record patterns and formations of clouds. Even though tornadoes form too fast for such satellite reports to be effective, satellites can spot high-reaching cloud systems, from 2 to 10 miles in diameter, some of which show a very brilliant reflection of sunlight. These dense clouds invariably contain severe storms and often tornadoes. When these clouds are recorded, weather observers on earth are alerted to watch for actual tornadoes in these areas.

When potential tornado weather is observed, National Aeronautics and Space Administration's (NASA) Applications Technology Satellites are programmed to take photographs of the changing cloud systems in specific areas, to be used in future research.

272. What is the tornado's radar fingerprint? Sometimes, but not always, tornadoes have a radar fingerprint that shows up readily on the radar screen. This is a distinctive hook-like or figure "6" pattern generally found in the southwest corner of a strong thunderstorm echo on radar. The tornado is usually associated with the revolving ball of the hook. By tracking this radar image, the speed and direction of movement of the tornado itself may be determined.

273. How are laser beams used in tornado studies? Temperatures, pressures, and humidity can affect the intense beam of amplified light produced by a laser. When laser equipment is set up in tornado generating areas, scientists measure minute changes in the beam that indicate weather changes for possible formation of tornadoes.

274. Can anything be done to stop or control a tornado? No known force or device can yet prevent tornado formation or can artificially dissipate these storms after they are formed.

People have sometimes tried to stop a tornado by firing pistols or cannons into the funnel, hoping to disrupt the air currents and collapse the rotating winds. But the total masses of air are so vast that nothing short of a nuclear explosion could change the course of the winds— and this probably would do more harm than good.

Scientists constantly discuss possible control measures in terms of electrical mechanisms or lasers that might restrict the electrical energy exchange in a tornado. The "defusing" by electrical discharges is based on the as yet entirely unproved hypothesis that tornadoes are caused or promoted by electric forces. (See Question 175.)

275. What seeding experiments are being considered for tornadoes? Meteorologists experimenting with seeding hurricanes to disperse wind energy (see Questions 150 through 152) are also considering the effects of seeding on tornadoes. By seeding thunderclouds that might produce tornadoes with chemicals such as dry ice, meteorologists hope to dissipate their energy source and prevent tornadoes from forming.

276. What is the National Severe Storms Forecast Center? The National Severe Storms Forecast Center, part of the Environmental Science Services Administration, located in Kansas City, Missouri, maintains constant analysis of atmospheric conditions over the 48 conterminous states in order to identify general areas where tornadoes may develop and send out forecasts and warnings as far ahead as possible. Tornado watches are prepared and issued by the Severe Local Storms (SELS) Forecast Center, also in Kansas City.

Because a tornado occurs so suddenly and is concentrated in a small area, much of the burden of spotting, warning, and evacuation falls on individual communities and citizens. Tornado detection requires a network of many alert reporters and a swift reporting procedure within each tornado watch area. Nearly 500 local networks supply information to the Weather Bureau.

277. What is NADWARN? In 1966, after the disaster of the 1965 Palm Sunday tornadoes, the Environmental Science Services Administration put into effect a single system for swift public warnings of many environmental hazards, including tornadoes, hurricanes, blizzards, floods, seismic sea waves, and storm surges. This nationwide Natural Disaster Warning (NADWARN) system is designed to improve detection of natural hazards, methods of rapid communication and warnings, and projects for community preparedness. Each year new and improved techniques are developed and incorporated into the system.

278. What is SKYWARN? SKYWARN is the name of a concentrated program of tornado preparedness set up in 1969 by ESSA's nationwide Natural Disaster Warning system (NADWARN) for tornado seasons. This program expands existing networks of volunteer tornado spotters and involves more people in the responsibility of watching for tornadoes, spreading warnings, and knowing exactly what to do. By improving everyone's awareness on national, state, and local government levels, as well as in private organizations, schools, and hospitals, weathermen hope to further reduce the number of lives lost by these violent storms.

279. What is a tornado watch? A tornado watch is the first alerting message from the National Severe Storms Forecast Center to areas where tornadoes may occur during the next several hours. The size of the area covered by the average watch is about 100 miles wide and 250 miles long.

The object of a watch is to alert people in the potentially threatened area to the possibility of tornadoes and advise them to get ready for immediate action if a tornado is actually sighted.

Watches are teletyped directly to local offices of the Weather Bureau and broadcast to the public by radio and television. Police officers, emergency forces, volunteer reporters, and others are all alerted, but do not interrupt their normal routines except to sharpen their watch for the increasing threat of tornadoes.

280. What is a tornado warning? A tornado warning is issued when an actual tornado has been sighted by human eyes or detected by radar. Reports of an actual tornado often come from responsible persons who immediately notify the nearest Weather Bureau office or community warning center.

Warnings are sent out swiftly over television and radio, indicating time of detection, the area through which it is expected to move, and the time when it is likely to strike.

Persons in the path of the tornado should immediately take safety precautions and run for shelter.

281. What are some safety rules for tornadoes? Several safety rules have helped people escape the deadly impact of tornadoes:

As soon as a tornado warning is given, run to the nearest tornado cellar if possible. If you are in open country, move away from the tornado's path at right angles. If you are in an automobile, you can easily outdistance a tornado if there is a clear road ahead. Remember, tornadoes move forward at speeds of about 25 miles an hour.

If you are on foot, and there is no time to escape, lie flat in the nearest ditch or ravine. Almost anything is better than remaining erect in the open.

If you are in a city, seek shelter inside, preferably in steel-reinforced buildings. Avoid auditoriums, gymnasiums, or other large halls with large, poorly supported roofs.

282. What safety rules should be observed in a house? If you are in a house, stand in an interior hallway on a lower floor, or climb under heavy furniture in the center of the house. The safest spot is the corner of the basement toward the direction from which the tornado is approaching. Keep some windows open to avoid destructive pressure changes (see Questions 189 and 190), but stay away from them—people have been literally sucked through them. Electricity and fuel lines should be shut off.

Keep tuned to the radio or television for latest news and instructions. Telephones should not be used, in order to keep emergency lines open.

283. Why are tornado cellars so important? In parts of the country where tornadoes are frequent, a tornado cellar is one of the safest protections from the destructive storms, well worth many times the effort and cost of preparing it.

Such a cellar should be located outside, and near the southwest corner of the main house, but not too close to possible falling walls. The entrance should face northeast, with the doors opening inward, in case debris blocks the entrance. The cellar should not be connected in any way with house drains, gas pipes, sewers, and should have the floor sloped to a drainage outlet. A vertical ventilating shaft extending from near the floor through the ceiling is recommended. Emergency supplies should include water, canned food, and first aid kits. Equipment should include such things as a pick, shovel, lantern, and a hammer.

284. What material is best for a tornado cellar? Reinforced concrete is the best material for a tornado cellar, although split log planks, cinder blocks, and bricks are also good. The roof should be covered with a three-foot mound of well-pounded soil.

A tornado shelter 8 feet long, 6 feet wide, and 7 feet high offers adequate protection for eight people.

285. What buildings are considered safe against tornadoes? Engineers and architects have drawn various plans for tornado-proof houses, built with various testing materials. So far the modern steel-reinforced concrete building is proving the safest.

Certain specifications for withstanding violent tornadoes include the following:

Houses must be strong enough to resist the impact of direct tornado winds and of any heavy object carried by the winds. They must also be able to withstand suction on the lee side, and pressure differential between the exterior and interior. For instance, vents help insure a rapid flow of air and pressure as the funnel of the tornado passes over the building.

286. How can a community set up a warning center? In tornado country, every community should have an active warning center, with an office open at all times during tornado seasons—like the police station or telephone exchange—to receive Weather Bureau tornado watches and warnings. This community center should be set up to receive reports of approaching tornadoes from local observers and issue local warnings. When a tornado is sighted, the center should notify nearby towns in the path of the storm, and telephone the nearest Weather Bureau office to warn other areas.

A community warning network relies heavily on alert volunteers assigned responsibilities of observing, watching, reporting, and helping. Usually trained observers are stationed in quadrants about 2 miles apart around the population center, with the heaviest concentration of observers to the southwest. In other, less-populated areas everyone can be part of the network.

Prearranged systems should be set up, whereby radio and television stations issue alarms, the city fire alarms or Civil Defense sirens sound, and farmers in threatened country areas are warned by telephone.

287. How effective are tornado warnings? The following story shows the value of tornado warnings. On May 31, 1947, a tornado funnel was sighted in the distance by a telephone operator in Leedey, Oklahoma. He immediately sounded the fire alarm and gave warning through a high-powered public address system. Almost the entire population had moved into their storm cellars by the time the tornado arrived, only half an hour later. As a result, there were only 6 fatalities, although two thirds of the town was completely demolished.

Another warning sent out to citizens of Haynesville, Louisiana, half an hour before a tornado struck on February 13, 1950, enabled everyone to be saved. When the tornado struck, streets and houses were deserted, as people had sought shelter.

288. Are tornadoes causing more destruction? Tornadoes are causing more property damage each year as population in tornado regions increases and cities grow larger and more numerous.

However, people are becoming more "tornado conscious" as special pamphlets and programs on radio and television are helping them understand what tornadoes are, how they start, and when to expect them. With more attention paid to tornado weather conditions, more accurate spotting methods, faster communications, and more informed persons, the death rate will continue to decrease.

Also, more tornado cellars or structures are being built, and houses and buildings are being designed to resist tornado destruction in an effort to save not only lives but also the buildings.

289. Where are special studies on tornadoes under way? The National Severe Storms Laboratory in Norman, Oklahoma, is the headquarters for continual research on the origin, life, and behavior of tornadoes. A network of many weather stations is coordinated with aircraft and satellite reconnaissance to accumulate data on all severe storms in that storm-ridden area.

Tornadoes are also being studied at various universities and colleges such as Oklahoma, Texas, Chicago, Harvard, and Michigan State.

The Disaster Research Group is an activity of the Division of Anthropology and Psychology of the National Academy of Sciences/ National Research Council in Washington, D.C. It is concerned with problems of disseminating warnings and of proper interpretation of warning communications, as well as the reaction of people when disasters strike.

Operation Rough Rider is the name of an investigation program that is put into effect during tornado seasons. With headquarters at the National Severe Storms Laboratory in Norman, Oklahoma, research planes from the Air Force and ESSA's Research Flight Facility fly over the storm area, taking photographs of cloud formations and patterns. Various instruments collect data on wind turbulence, temperatures, air flow patterns, and other phenomena at many different altitudes. When a tornado is spotted, instruments record the size, intensity, rate of growth, and direction.

III. THUNDERSTORMS

Introduction. One of the most common and dramatic storms on earth is the thunderstorm. Towering several miles high into the spring or summer sky, these giant clouds generate huge amounts of energy which is released or transformed throughout the atmosphere in spectacular and destructive forms.

These often localized and short-lived storms occur in nearly every part of the world in endless procession. Several million thunderstorms are formed each year throughout the world. Scientists estimate that, at any given instant, nearly 2,000 storms are in the process of brewing or dying.

The number and frequency of these storms, and the violence and variety of forms in which they release their energy, make them one of nature's most consistent destroyers of life and property. In the United States, for instance, lightning alone may kill as many as 600 people each year—more people, on the average, than are killed by hurricanes or tornadoes. Pelting hailstorms cause millions of dollars of damage each year by destroying crops, smashing buildings and vehicles, and killing and injuring animals, birds, and livestock. Cloudbursts bring flash floods that wash away houses, sections of highways or railroad tracks, and other objects in their paths, and sudden gusty downdrafts and squalls have capsized boats, flattened trees, and knocked over buildings.

Yet these destructive storms are also benefactors, helping to keep a balance of electrical discharge between earth and sky and bringing essential rains to parched fields and forests. Even the great killer lightning constantly sustains life on earth by sending down tons of nitrogen from the atmosphere into the soil, a vital element in the food chain for plants, animals, and men.

290. What is a thunderstorm? A thunderstorm is a towering cloud system with violent upward and downward air currents, containing enormous amounts of energy which is expended in various forms of lightning, thunder, wind, rain, and hail.

Thunderstorms may be frontal and move across the earth's surface with the prevailing winds, or they may be local heat storms that grow and diminish above one particular area. They consist of one or more storm cells, each one of which is a separate wind system containing strong vertical air currents and precipitation, and following a special pattern of growth, maturity, and dissipation. Powered by vast amounts of moisture and warm air in an unstable atmosphere, these thunderclouds sometimes loom more than 12 miles high in the sky. The upper regions, often containing ice crystals, may spread out in the characteristic shape of a blacksmith's traditional anvil. The broad, relatively flat base of a thunderstorm system may extend over distances of hundreds of miles.

291. What causes a thunderstorm? Thunderstorms are formed under unstable atmospheric conditions when warm moist air is lifted vertically into cooler air at great heights by special processes of vertical circulation called either thermal or mechanical convection. (See Questions 293 and 294.)

Two important factors are involved in the making of these immense cloud structures: strong rising currents of heated air and a lot of water vapor. As the warm moist air rises, it expands and cools. The water vapor in this air condenses into tiny airborne droplets that are large enough to be visible yet too small to drop to earth. These form the towering clouds known as cumulonimbus or thunderclouds.

As the warm air rises and cools, the cooler air descends, and a violent mixing of layers of air occurs, engendering vigorous vertical convection and the turmoil known as thunderstorms.

292. What is convection? Technically, convection is a process of heat or energy transfer in gas (or liquid) produced by moving masses of matter. In the atmosphere, convection most frequently consists of vertical air movements.

Examples of convection can easily be seen in everyday events. The rising of smoke from a house chimney or of smoke blown over a warm radiator or light bulb indicates the convection of hot air with visible flecks of ash. The accompanying downdraft of cooler air in these examples is somewhat more difficult to detect, for there may be no visible particles to indicate the downward path.

Convection, the most important method of heat transfer in weather phenomena, is an essential factor in the production of thunderstorms.

293. What is thermal convection? Thermal or heat convection occurs when air is warmed and its molecules become agitated and move away from one another. Thus the air expands, decreases in density, and rises in an upward current through layers of relatively cooler air.

In the formation of thunderstorms, this occurs when the sun has been shining on a relatively quiet location such as a field or forest and warming the air over the ground. The heated air rises vertically over the sun-warmed areas.

294. What is mechanical convection? Mechanical convection of air occurs when air is forced upward by mechanical or physical means. For instance, warm air may be pushed upward as a wedge of an advancing cold-air front moves beneath it. Warm air may also be deflected upward as winds blow across mountain slopes or high hills, or over warmed sloping beaches and seashores.

When any of these conditions is sustained, the resulting turbulence of air can contribute to the formation of a cumulus cloud that grows into a cumulonimbus thundercloud.

295. What does a thunderstorm look like? From a distance, a thunderstorm seems to rise like a majestic snow-covered mountain. The clearly defined globes of white clouds have been described as looking like cauliflower heads or like the domes and turrets of a castle stacked one above another. The billowing high clouds often reflect colors from the sun—yellow, pink, or mauve—and they cast dark purple or black shadows. When they rise high into the air, ice crystals form at the top and the clouds sometimes flatten out into a distinctive anvil shape.

As this remarkable phenomenon increases in size, its beauty becomes awesome and terrifying as the dark underside of the cloud blocks out the horizon, thunder becomes louder, and lightning flashes become brighter and more frequent. Turbulent patches of clouds rise and fall, churn and roll. The gentle breezes that had been wafting toward the storm center disappear, and violent cold winds rush out from the base of the dark squall cloud. Large splats of raindrops start to fall and soon drop in torrents. For 15 to 20 minutes the raging storm seems to fill the sky with rain, wind, thunder, and lightning, shutting out everything else.

Then the wind decreases, the rains slow down, lightning flashes less often, and the rumble of thunder retreats in the distance. Soon the rain ceases completely, the clouds break up, the air feels cool and refreshing. The sun may shine through, perhaps lighting up raindrops like diamonds dripping from bushes and trees, perhaps creating a rainbow in the sky.

The storm has passed.

296. What damage do thunderstorms cause? Thunderstorms produce flashes of lightning that may sear, burn, and kill people and destroy property. Sudden torrents of rain produced by the storms can wash away houses, fields, and roads in a matter of minutes.

All thunderstorms produce violent vertical winds that can shake up and damage any object flying through them, and some are part of the system that generates the most violent of windstorms yet known, the tornado. (See Chapter II.) Some of these storms, but not all, produce tons of hailstones that can flatten crops and buildings, causing enormous amounts of damage, especially in the Midwestern United States. (See Chapter IV.)

Although no one thunderstorm causes any spectacular cataclysm by itself, the constant toll of death and damage from thousands of these storms each year adds up to considerable numbers.

297. How did a thunderstorm defeat Napoleon? Torrential rains from a thunderstorm on June 17 and 18, 1815, created such mud on the battlefield of Waterloo, south of Brussels, that Napoleon Bonaparte delayed attacking the British army for about 6 hours. By this time Prussian reinforcements had arrived to help the British defeat the French and end Napoleon's dream of an empire.

Another instance of storms molding the course of history occurred during the retreat from Gettysburg in the U.S. Civil War, when the pursuit of Confederate troops was thwarted by storm-swollen streams.

298. What was the worst disaster caused by a single thunderstorm? About one hundred years ago, in 1856, lightning flashed from a thunderstorm and struck a supply of gunpowder stored beneath the church of St. Jean on the Mediterranean island of Rhodes, causing an explosion that killed more than 4,000 villagers.

299. What was another disaster? Another notable disaster took place in the little town of Brescia at the foot of the Italian Alps in

July, 1769. A powerful bolt of lightning struck the tower of the church, traveled to the vaults beneath, and set off 100 tons of gunpowder in one huge explosion that destroyed one sixth of the city and killed 3,000 persons.

300. What caused the tragedy of the German airship *Hindenburg*? On May 6, 1937, the huge, hydrogen-filled dirigible *Hindenburg* burst into flames over Lakehurst, New Jersey, killing 36 people. Many indications showed that the explosion was caused when a so-called harmless electrical flickering from a thundercloud ignited a leaking mixture of air and hydrogen.

301. How did the *Hindenburg* tragedy occur? Having completed one successful year of service, the luxurious 803-foot German passenger airship *Hindenburg*, some 100 feet longer than the largest battleship then afloat, was completing its first flight of the 1937 season across the Atlantic Ocean. Thousands of spectators, news photographers, and reporters were on hand to welcome its arrival at Lakehurst, New Jersey. Since there were thundershowers present in the area, the captain cruised the airship around until the storms passed. Then as the ship hovered 200 feet above the landing site, the crew began to lower handling lines to men on the ground. Suddenly the stern of the ship burst into flames that spread rapidly through the gas-filled balloon and the passenger gondola beneath. The flaming ship sank as it burned.

Some of the 97 passengers and crewmen jumped clear, but the tragedy engulfing the others has remained such a terrible memory that no dirigible has carried paying passengers since that date.

302. What was the *Shenandoah* tragedy? The first large dirigible of the United States, the U.S. Navy's *Shenandoah*, took 4 years to build and a few minutes to destroy. After her first flight in 1923 and 56 other successful flights, the huge airship was caught in a thunder-squall over Noble County, Ohio, on September 2, 1925. A sudden updraft carried the ship from its 2,500-foot cruise altitude to over 6,000 feet. Then it was plunged downward at 25 feet per second. A few minutes later it was carried upward and broke apart as it lurched violently, separating into 3 pieces that fell to earth in a plunge, killing 14 of 43 men aboard.

303. What other dirigibles have crashed in storms? Another U.S. dirigible, the *Akron*, 785 feet long, encountered a storm off the New Jersey coast in April, 1933. The ship was struck by a sharp gust of wind that caused it to drop its stern into the ocean, bringing death to 73 men.

The British airship *R-101*, 777 feet long, crashed in a thunderstorm while cruising over France in 1930, killing 48 people.

304. What happens when a plane flies into a thunderstorm? As a plane flies into a thundercloud, sudden gusts of violently rising winds or downdrafts can cause it to roll and twist, pitch forward or backward, or flip over. A powerful gust can even rip off a plane's wings.

Sometimes planes have been carried 2,000 or 3,000 feet up in a few minutes in a strong updraft. Powerful downdrafts have plummeted planes for distances of as much as 1,500 to 2,000 feet at speeds of about 20 feet a second.

Within these clouds, aircraft have been severely buffeted by blinding snow flurries, torrents of rain, and chunks of hail. Ice often forms on the plane, and flashes of lightning have momentarily blinded pilots and knocked radio and other instruments out of order. Claps of thunder are so powerful they have caused temporary deafness.

The action of snow, ice, and rain rubbing and bouncing over the plane sometimes charges it with so much electricity that streamers of blue light some 10 or 15 feet long have been seen shooting from the propellers, wings, nose, and other protuberances.

305. What was the tragedy of five German glider pilots? In Britain and Europe sailplaning or gliding in engineless planes has been a great sport, and gala exhibitions are held from time to time to set records for gliding height, length of time aloft, and other feats. Often glider pilots would deliberately fly into thunderclouds in hopes that the powerful updrafts would boost them higher in the atmosphere.

Just before World War II, a German glider society held an exhibition in the Rhön mountains. Five pilots steered their crafts into a towering mature thundercloud—and encountered violent destructive winds that threatened to tear their planes apart. All the men bailed out and were caught in updrafts that carried them to cold regions of hail, snow, and ice. As the young men were sustained aloft by the wind currents, they became coated with ice and all but one froze to death by the time they finally tumbled down to earth.

306. Who fell through a thundercloud and lived? Several people have flown planes into a thundercloud, encountered devastating winds and precipitation, parachuted, and never survived.

But in August, 1959, Lt. Col. William Rankin of the U.S. Marine Corps bailed out of his single seater jet fighter, an F-8U Crusader, when the engine failed at 46,000 feet above the Carolina coast—into the boiling pot of a thunderstorm. His preset parachute did not open until he had fallen to 10,000 feet, where he began being tossed about in violent updrafts and downwinds of the storm, much like being thrown into an elevator shaft and shooting up and down at speeds of a mile a minute. He was pelted with hail, snow, and rain.

For a trip that in normal weather would have landed him on the ground in 13 minutes, the journey through the storm cloud took him nearly 45 minutes.

He finally dropped through the base of the cloud, about 550 feet from the ground, onto a field. He made his way to a nearby road where a passing motorist picked him up and took him to a hospital. Fortunately Lt. Col. Rankin suffered little worse than temporary shock and frostbite from the experience few have ever lived through. He left a remarkable record of his encounter.

307. What does it feel like to fall through a thunderstorm? The harrowing journey of Lt. Col. Rankin as he fell through a thundercloud has been vividly described in his book:

After the first few shocks of turbulence, virtually straight up-and-down actions, I felt a queasiness in my stomach. . . .

. . . My mind changed rapidly after I was hit by the first real shock from nature's "heat engine." It came with incredible suddenness—and fury. It hit me like a tidal wave of air, a massive blast, as though forged under tremendous compression, aimed and fired at me with the savagery of a cannon. I was jarred from head to toe. Every bone in my body must have rattled, and I went soaring up and up and up as though there would be no end to its force. As I came down again, I saw that I was in an angry ocean of boiling clouds, blacks and grays and whites, spilling over each other, into each other, digesting each other.

I became a veritable molecule trapped in the thermal pattern of a heat engine. I was buffeted in all directions—up, down, sideways, clockwise, counterclockwise, over and over; I tumbled, spun, and zoomed, straight up, straight down and I was rattled violently, as though a monstrous cat had caught me by the neck, and was determined to shake me until I had gasped my last breath. I felt all the painful and weird sensations of the forces—positive, negative and zero. I was pushed up, pushed down, stretched, slammed, and pounded.*

308. What kinds of thunderstorms are there? Thunderstorms are usually classified into two basic categories: frontal or general thunderstorms found in groups or lines along cyclonic weather fronts sweeping across wide areas of the country; and local, air-mass thunderstorms that are often isolated convective storms, sometimes called heat thunderstorms. The thunderstorms that consistently rise over particular areas of mountains or seacoasts are air-mass storms.

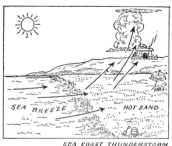

SEA COAST THUNDERSTORM

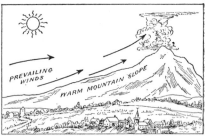

MOUNTAIN THUNDERSTORM

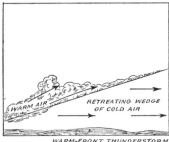

WARM-FRONT THUNDERSTORM

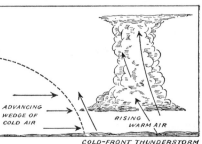

COLD-FRONT THUNDERSTORM

Different types of thunderstorms

309. What are frontal thunderstorms? There are two kinds of frontal thunderstorms—those formed in a cold weather front and those in a warm front.

A weather front is the area where two air masses, cold and warm, meet and often clash, as the warm tries to rise and the cold tries to

* Lt. Col. William H. Rankin, U.S.M.C., *The Man Who Rode the Thunder* (Eng'ewood Cliffs, N.J.: Prentice-Hall, Inc., 1960), pp. 166, 167.

sink. If the cold mass advances and wedges beneath the warm, the front is called cold. If the warm air advances into or over the cold, the front is called a warm front. In either case, frontal weather is unsettled, often stormy, and brings bad weather.

310. What is a cold-front thunderstorm? A cold-front thunderstorm is formed when a vigorous wedge of cold air pushes into and under a mass of warm air, forcing the warm air aloft where it condenses into huge cumulus clouds that grow into thunderstorms. A cold front moves at speeds of about 20 to 30 miles an hour in summer and about 30 to 40 miles an hour in winter when air is colder and exerts greater pressure. This fast-moving wedge creates a steep advancing edge and limits the area of turbulent air to a very narrow band. Thus cold-front thunderstorms are usually brief but violent. A line of such thunderstorms may extend for hundreds of miles along the advancing cold-front line, strung out like beads on a string. These storms can arrive at any time of day or night, and at any season.

311. What is a warm-front thunderstorm? Warm-front thunderstorms occur less often than cold-front thunderstorms since there is less storm-producing turbulence as higher warm air advances over a retreating cold wedge. Warm fronts move at about 15 miles an hour, and the vertical slope is less steep, so the uplift of air is slow and gradual and has little turbulence. These warm-front storms are less severe, and much of their action takes place far above an obscuring layer of clouds.

312. What is an air-mass thunderstorm? An air-mass thunderstorm is one that occurs within one relatively homogeneous air mass and does not appear to be associated with any frontal activity.

There are generally three kinds of air-mass thunderstorm: the local or heat thunderstorm, the mountain thunderstorm, and the seacoast thunderstorm.

313. What is a local thunderstorm? This type of air-mass storm is the most common of all thunderstorms. Literally hundreds of these billowing clouds may form over the land in a single hot summer day. They rise over the warm parts of the earth, in the warm seasons of the year, and in the warm part of the day.

Local heat thunderstorms need a deep layer—some 10,000 feet

or more—of moist air, as well as unstable atmospheric conditions and warm sun rays.

314. How does a local thunderstorm form? A local thunderstorm usually begins to form in the morning, when the sun shines down undisturbed upon a field, plain, or city for several hours, warming the air above the earth, and causing it to rise in distinct convection columns. Around noon, small puffy clouds begin to form. Some of these evaporate in only a few minutes as they rise into the dry upper air, but gradually they become more numerous and larger, accumulating into a mountain of white cumulus clouds.

Many such thunderclouds are small and ephemeral, although some can grow into large turbulent storms in the afternoon. They usually die down at evening, when the sun's heat wanes and the cooling ground ceases to supply the necessary energy.

Since they may form over many kinds of terrain and can travel in almost any direction, it is difficult to forecast the time or place they will appear.

315. What are mountain thunderstorms? Thunderstorms often rumble throughout mountain ranges or hills that lie across paths of prevailing winds. These mountains act as permanent barriers, mechanically pushing up winds as they blow across the land. Aided by updrafts rising from sun-warmed slopes, the moist air rises and invisible water vapor condenses into visible tiny droplets that form clouds and then larger raindrops that fall in torrents on the slopes and valleys.

Thunderstorms formed in the mountains are called orographic storms, from the Greek word *oros* which means mountain.

316. How can mountain storms be predicted? The particular shape of the slopes and valleys of a mountain governs the formation of storms. By studying these slopes and weather conditions, weather researchers can define quite accurately the time and place the storms will occur, and how intense they will be.

317. What is a seacoast thunderstorm? On summer afternoons, great thunderstorms build up over heated seashores and beaches, the products of basic laws of temperatures and winds.

Land and sea absorb and reflect solar radiation in different ways. The land tends to absorb more heat during the daytime and at night gives off more radiation than the oceans, which are slower to warm up and to cool down, and hence have more stable temperatures.

Thus, during the course of the day, the land becomes more heated from the sun's rays than the sea. As the warm air over the land rises, moist air moves in from the sea to take its place. This sea air blows over the warm and slightly elevated land, heats up, rises, and forms thunderclouds. The same phenomenon occurs over lake shores and beaches.

318. Why do sea thunderstorms form at night? At night, the movement of air over the seacoast is reversed, since the land cools faster than the sea, and the cool denser air flows toward the relatively warm ocean. This is a gentle slow movement, sometimes hardly perceptible. Rarely is the sea warm enough at night to cause air to rise. Sometimes, however, if the coast is curved to form a bay, the converging winds from the land may crowd together and rise, forming billowing clouds and sometimes sea thunderstorms.

Sometimes convection currents occur over warm ocean currents or lakes at night, and cumulus clouds are formed. These rarely have much strength to generate storms, since there is not enough heat for growth.

319. What is a thunderstorm cell? A thunderstorm cell is part of the complete storm system, functioning much like a chimney of vigorously rising warm updrafts, with cooler downdrafts that form later.

Each cell has a fairly well-defined life cycle of three general stages: 1) the beginning or cumulus stage when a strong updraft of warm moist air occupies the whole cell; 2) the mature stage of greatest activity when downdrafts form, and rains and cold winds fall; and 3) the final or dying stage when weak downdrafts prevail throughout the cell. During its existence, each cell produces its own assortment of rain, snow, ice, electrical discharges, and sometimes other phenomena such as hail.

320. How many cells does a thunderstorm have? A thunderstorm may be composed of only one convective cell or a series of many cells. Most thunderstorms have, at maturity, from 3 to 5 cells, each of which is living its own independent cycle.

321. What is the beginning stage of a thunderstorm? A thundercloud starts to form when moist warm air begins gently rising in vertical updrafts as in a chimney. This updraft develops in a region where surface winds are gently converging and atmospheric pressure is slightly lower than in the surrounding areas. The air expands as it rises, the cooling water vapor condenses into visible particles, and a cumulus cloud develops which continues to grow as more heated air rises into it—somewhat like a balloon being blown up in a vertical position.

As the air continues to rise in the updraft, it is replaced by air flowing into the base, like the inward currents of air sucked into the draft of a fireplace. Air also flows in through the sides of the cloud, mixing and rising with the updraft. These rising air currents may extend for several thousand feet throughout the entire cloud structure, from below the base to the top.

322. What happens as warm air rises? As expanding warm moist air rises through regions of lowering temperatures at increasing heights, it gradually cools. This rising air is also cooled by a process called adiabatic change. (See Question 324.) As it cools, water vapor begins to condense, forming tiny droplets of water too small and light to fall out as rain. The process of water condensation releases heat energy into the cloud system, which further helps the ascent of the updraft. The rate at which the energy is released is directly related to the amount of water vapor condensed into liquid water.

As the updraft reaches the upper parts of the thundercloud, small particles of snow and ice crystals begin to form, and fuse together into larger particles, releasing heat energy. The latent heat of fusion contributes to sustain the updraft of air currents.

At this stage the cloud is swelling, new cloud turrets and domes bulge out of the top and along the sides, and the water content builds up within the cloud system.

323. How cold does it get at higher altitudes? The higher the altitude from the earth's surface, the lower the temperature of the atmosphere up to the troposphere. (See Question 991.) The temperatures themselves vary from hour to hour, day to day, and season to season. Their rate of decrease is also variable.

The following chart shows a somewhat typical example of the drop in temperature on a mild summer day:

If the air temperature at the earth's surface is 82 degrees Fahrenheit,

    5,000 feet up, temperature may drop to 62° F.
    15,000 feet up, temperature may be freezing
    25,000 feet up, temperature is about 3° F.
    35,000 feet up, temperature is minus 36° F.
    40,000 feet up, temperature is minus 60° F.

The rate of decrease of temperature with height is called the lapse rate. In general, a steep lapse rate, or a rapid decrease of temperature with height, is most favorable for the growth of thunderstorms.

324. What is adiabatic cooling and heating? An adiabatic change— either cooling or heating—in an air mass means a change in temperature brought about by processes in which heat is neither received from nor given to the surrounding air or other outside source.

It is a natural law of gases that when air, which is a mixture of gases, rises, it expands because there is less weight or pressure of air around it at higher altitudes. Energy is needed to bring about this expansion. This energy is taken in the form of heat from the rising air mass and brings about a cooling and reduction of temperature within the rising air mass itself. In the atmosphere, a rising parcel of air is cooled at a rate of about 5.5 degrees Fahrenheit for each 1,000 feet of ascent.

Conversely, when air descends from higher altitudes, it becomes compressed at lower levels, and its temperature rises at about the same rate—5.5 degrees Fahrenheit for every 1,000 feet of descent.

Such adiabatic changes in temperature, which take place during the lifting and sinking of large masses of air, occur within the masses, with no addition or subtraction of heat from the outside.

325. What is the mature stage of a thunderstorm? The mature stage of a thunderstorm is its period of greatest activity and violence, when both upward and downward air currents are at their peak. At this stage, tiny water droplets have grown in size by colliding and coalescing with other droplets until they are large enough to fall down against the rising updrafts. Solid and liquid particles of water in the form of rain, ice, and snow begin to fall within the cloud, creating the first downdrafts.

At this stage the top of the ascending air has reached heights of 35,000 to 40,000 feet.

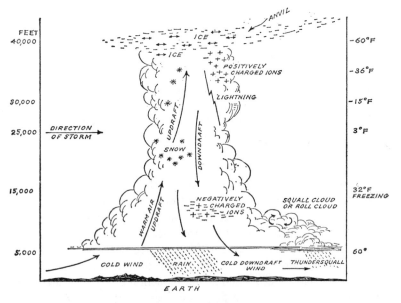

Anatomy of a thunderstorm

326. How does a downdraft develop? As more and larger water droplets form and develop, they become heavy enough to fall through the updraft of the rising cell and a downdraft is generated. At first this downdraft of cold rain and air takes place in only a small area of the thunderstorm cell, on one side of the updraft. As more particles evaporate and cool, more winds start shifting downward. At this time some of the water evaporates, absorbing heat in the process, and cooling the system. This feeds and strengthens the downdraft. As the amount of falling rain, ice crystals, and sometimes hailstones increases, the downdraft area gradually expands, joins other downdrafts, and begins to cut off the updraft. Soon the storm system in the thunderstorm cell begins to abate.

American Red Cross

A man braces himself against towering waves of the September, 1947, hurricane pounding the roadway and Baker's Haulover bridge, north of Miami Beach, Florida.

TOS DATA OVER U.S. AND SURROUNDING OCEANS

NASA (National Aeronautics Space Administration)

Spirals of turbulent clouds and rainbands of Hurricane Gladys over the Gulf of Mexico, October 16, 1968, photographed by crew of Apollo 7.

Opposite: Five hurricanes in motion at the same time, photographed from a weather satellite on September 14, 1967.

Hurricane Camille battered three large sea-going vessels against the shore near Gulfport, Mississippi, on August 17, 1969.

Shattered church, houses, and trees of the town of Vialet in Haiti lie in the wake of Hurricane Flora in October, 1963.

U.S. Weather Bureau

The threatening tornado of July 8, 1927, as it approached the town of Vulcan, in Alberta, Canada.

ESSA: Eric Lantz, Walnut Grove (Minn.) Tribune

A long tornado funnel reaches out of the sky near Tracy, Minnesota, on June 13, 1968.

Aftermath of a tornado that cut like a knife through an apartment building in St. Louis, Missouri, in February, 1959.

A woman surveys the rubble of her house, destroyed in a few minutes as tornadoes swept through Texas, Oklahoma, and Arkansas in April, 1957.

A spectacular flash of lightning streaks across the sky. Lightning causes hundreds of deaths each year and destruction by fire in cities and forests.

Hailstones develop within violent thunderclouds thousands of feet high and crash on earth in many shapes and sizes, causing millions of dollars of damage to crops each year.

Cornstalks were shattered and stripped by a heavy hailstorm which ruined many crops in Boone County, Iowa.

U.S. Weather Bureau

Hazardous fog banks rolling over the sea nearly obscure U.S. naval ships. The battleship U.S.S. *Missouri* is in foreground.

After an ice storm, many tons of ice bend and break telephone wires and trees throughout New England.

Man is pouring disastrous industrial and urban wastes into the atmosphere, bringing death and illness, and causing changes in our climate.

Right below: Majestic but treacherous icebergs threaten shipping lanes in the North Atlantic Ocean during the spring and summer months.

Above: The luxury liner *Titanic* as she started her maiden voyage in April, 1912, **hours** before she struck an iceberg and sank with more than 1,500 people.

A snow-blinded calf near Rapid City, South Dakota, stands stunned and lost during the February, 1966, blizzard, one of the worst storms in recorded history.

Crystals of rime ice can grow to huge proportions and weird shapes in windy cold areas such as Mount Washington, Wyoming.

ESSA

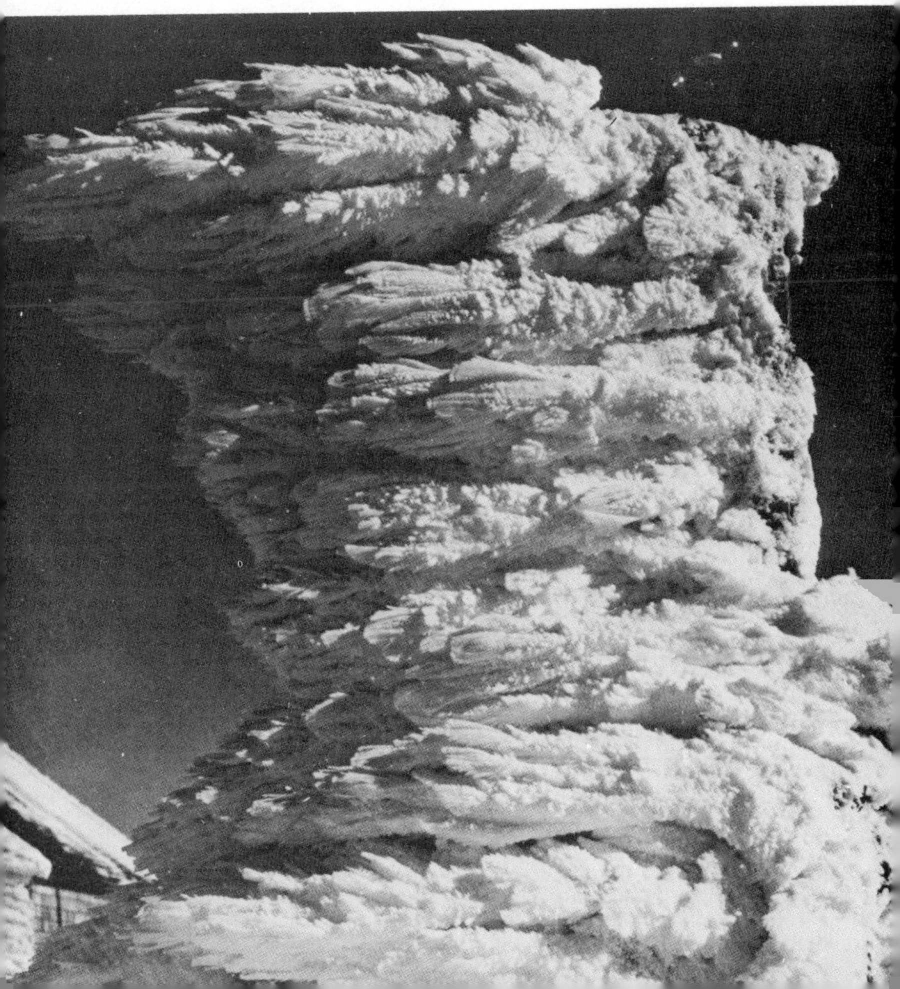

Dramatic flickering colors of aurora borealis occur when high concentrations of solar radiations penetrate the earth's atmosphere near the poles.

327. What happens at the top of a mature thundercloud? At maturity, the thunderstorm cloud may tower some 40,000 feet high or more. Ice crystals have begun to form at high altitudes, and fibrous streamers of frozen precipitation elements appear, blown by the winds. These streamers may spread out in straight lines, forming a flat top called an anvil because it resembles the blacksmith's tool. This frozen cap, visible in dry regions for hundreds of miles, is technically an incus.

328. What is the final stage of a storm? In its final or dissipating stage, a storm may have such large downdraft areas throughout its cells that the updraft winds are choked off and the whole cloud becomes composed of weak, sinking air and light precipitation.

Since no more warm, moist air is rising and cooling, little or no rain may be formed. The large drops have fallen out of the cloud, and only small light drizzles remain. Finally, even this ceases.

Now the towering cloud has degenerated into broken stratus clouds, with ragged edges that are blown away and dispersed by outside winds.

329. What is a thunderhead? Thunderhead is a popular term for a towering cumulonimbus cloud with shining white edges, forerunner of a thunderstorm. The word was used in early technical works of meteorology, but now is seldom used by meteorologists.

330. How large is a thundercloud? In its early stages a moderately sized, expanding cumulus cloud can extend about a mile in diameter and stretch some 15,000 feet high to near the freezing zone in the atmosphere.

In 10 minutes, the cloud may swell to a diameter of 5 to 10 miles, towering 25,000 to 40,000 feet high. This is the average size of thunderclouds.

A few great thunderclouds may reach 60,000 feet high, and the dark canopy of its underside may be 14 miles in diameter, covering an area of about 200 square miles.

331. How high is the cloud base? The base of a thundercloud may be some 500 to 13,000 feet above the earth's surface. Sometimes it is only a few feet above the surface. In the Midwest of the United

States in summer, the bases of many of the clouds are about 5,000 feet above ground.

332. How long do thunderstorms last? The duration of a thunderstorm varies, depending on different conditions of location, season, number of storm cells involved, and temperatures of the air masses.

The life span of a single thunderstorm cell can be about one hour or less.

Since thunderstorms are often composed of more than one cell at different stages of development, the whole storm may last for several hours. With a series of storm cells, one cell may be dissipating, while another is at its height of activity and still another cell is just beginning to form.

333. How many thunderstorms occur on earth? Throughout the world each year about 16 million thunderstorms are formed. At any given moment, about 2,000 such storms are brewing, growing, or dying at some place on earth.

334. How much energy is contained in a thunderstorm? The energy contained in a thunderstorm has yet to be calculated. Some scientists estimate that an average thunderstorm on a summer day, about 25,000 to 40,000 feet high, can release more energy in one minute than a 120-kiloton nuclear bomb.

335. Where does a cloud get its energy? A thundercloud obtains much of its energy from the updraft of warm moist air and the condensation of water. When water evaporates, the heat of vaporization is stored as latent energy in the water vapor. When the vapor condenses into droplets, this latent heat is released. Further heat energy is released when the snow and ice particles high in the cloud fuse to form larger particles. This heat is called heat of fusion.

336. Where do thunderstorms occur? Thunderstorms can appear nearly everywhere on earth. Local convective storms form almost every day in the warm humid regions of the tropics. They occur in the summer season in humid continents, over heated plains, in mountain regions, or along coasts.

Frontal thunderstorms are found in the middle latitudes, associated with squall lines, or along windward mountain slopes. They develop in the temperate zone in advance of cold fronts, usually forming more frequently at night than during the day.

337. Where do thunderstorms occur most often? Probably the location with the greatest number of thunderstorm days is Kampala, Uganda, which has about 242 thunderstorm days each year. The island of Java, in Indonesia, averages about 223 thunderstorm days a year.

The doldrum belt along the equator, with its constantly high temperatures and moist air, averages about 75 to 150 thunderstorm days a year.

338. Where are thunderstorms rare? In regions above the 60-degree latitudes of the Northern and Southern Hemispheres, thunderstorms seldom occur because the air is too cold and generally too stable to foster these storms. In arctic regions, where the air is cold and quite dry, thunderstorms occur only about once every few years.

In desert areas in the low latitudes, where there is enough heat but not enough moisture for formation of thunderstorms, these storms occur only about 5 days of the year.

339. Where do most thunderstorms occur in the United States? In the United States, thunderstorms occur most frequently in Florida and in states along the Gulf of Mexico, along the southern Atlantic coast, and in the southwestern mountainous areas.

The southeastern United States, with its high heat, humidity, and unstable atmosphere, is the most stormy area outside the tropics, with about 70 to 90 thunderstorm days in the year. The Rocky Mountain area, high barrier of the moist eastward blowing winds, has 50 to 70 thunderstorm days. Numerous storms drift across the Central Plains, which have about 45 thunderstorm days.

340. In what particular areas of the United States do mountain storms occur? Orographic storms are common in the summertime over the Rocky Mountains, the Sierra Nevada Mountains, and the Allegheny Mountains.

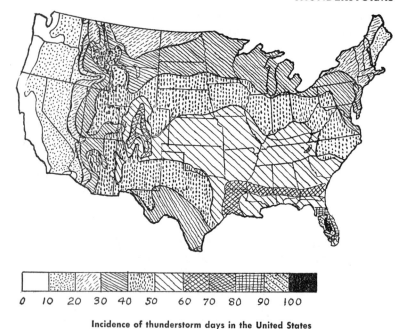

0   10   20   30   40   50   60   70   80   90   100

Incidence of thunderstorm days in the United States

341. Where are thunderstorms rare in the United States? Thunderstorms are relatively rare in the northern states of New England, North Dakota, Montana, and other northern states where the air is often too cold for storm generation. They are also rare along the Pacific coast where summers are too dry for thunderstorms to form.

342. When do thunderstorms occur? On land, in the middle latitudes, air-mass thunderstorms usually occur in late spring and summer when large amounts of tropical maritime air move across the continent. They usually build up in the middle of a sunny afternoon, about 2 to 4 P.M., when the surface air is most heated from the sun.

343. What is the thunderstorm season in the United States? In the United States, thunderstorms occur mainly in the summertime, becoming most severe from May through August.

344. When do thunderstorms occur most often in the tropics? In the tropics, thunderstorms occur most frequently during the rainy seasons or monsoons when there are profuse masses of warm, moist, and unstable air. They often occur each day in the afternoon.

345. When do frontal thunderstorms occur? Frontal thunderstorms may occur at any time of the day or night as the storm line travels across the country. Yet even these storms are most likely to break out in the mid-afternoon.

346. What are some of the dramatic, destructive phenomena accompanying these storms? The tremendous amounts of energy generated in towering thunderstorms can take many spectacular and destructive forms, such as lightning (see Questions 347 through 404), thunder (see Questions 405 through 409), winds (see Questions 410 through 414), rains (see Questions 415 through 417), snow and ice, and sometimes hailstones (see Question 418).

347. What is lightning? A flash of lightning is the electrical discharge from one region of charged particles to another region of oppositely charged particles. The lightning process is actually built from countless small charges of electricity. The light is the glow of atoms and air molecules that are energized and agitated by the electrical discharge. This discharge can take place within a single cloud, from cloud to ground, or from cloud to cloud.

348. What causes lightning? Lightning occurs when positive electric charges and negative electric charges separated from each other are built up in large enough quantities to start a stream of electrons flowing through the intervening space. The electric stream continues until the electrical charges on gases, water molecules, dust, and other particles become neutralized again.

349. What damage does lightning cause? Of all the phenomena created within a thunderstorm, lightning has caused the most consistent damage throughout the world. Occurring frequently, these electrical discharges slowly stack up a toll each year of hundreds of people killed and thousands wounded. Lightning is one major cause of fires in cities, rural areas, and forests. It has been a considerable problem in sophisticated societies, where it damages power lines, disrupts communications, and sometimes blacks out whole cities.

● **350. How many people are killed each year by lightning?** Determining the statistics of how many deaths are caused by lightning is difficult, and results may often be erroneous. For instance, if a person dies of a heart attack when lightning flashes nearby, is the death caused by lightning or by heart trouble?

Figures given for annual deaths in the United States range from 130 to 500. Some reports state that as many as 600 people die each year from lightning. One or two thousand are injured each year.

● **351. When are people hit most often by lightning?** About 25 percent of all lightning fatalities occur when people or animals seek shelter beneath tall trees. These trees are high, easy targets for lightning strokes.

352. Where are people most often killed by lightning? Insurance companies report that more people are killed by lightning in the country than in urban regions. Nearly 5 times as many males are killed as females—possibly because men and boys participate more often than women in outdoor activities such as fishing, golfing, camping, and swimming.

353. Why are animals struck more often by lightning than human beings? Some say that animals are more affected by lightning strokes than human beings because their four legs, in contact with the current running through the ground, can receive a greater shock than a person's two legs.

Lightning causes severe losses among domesticated animals such as sheep and cattle grazing in open fields during a thunderstorm. Often these social animals huddle so close to one another during a storm that a lightning stroke passes through all of them at one time.

354. What was a record number of animals killed? On June 22, 1918, lightning struck a flock of sheep bunched together in an open field in the Wasatch National Forest, Utah, and killed 504 of them in one stroke.

355. What was one of the most expensive lightning strokes? On July 10, 1926, a bolt of lightning struck a small brick building in the U.S. Navy's largest ammunition depot at Lake Denmark, New Jersey.

Even though the building was heavily supplied with lightning rods, the lightning struck depth charges and TNT bombs stored within, causing explosions that demolished all buildings within 2,700 feet and scattered debris as far as 22 miles away. Sixteen people were killed, and property worth $70 million was destroyed by this flash, the most expensive lightning stroke in U.S. history to date.

356. How is electricity generated within a thundercloud? No complete explanation has yet been given as to the nature of the electric field found in a cloud and the earth beneath it or as to how the lightning really is set off.

Some scientists believe that electrical energy is created within a cloud as billions of ice and snow crystals, particles of dust, and raindrops collide, shatter, and rub against one another in the turmoil of the cloud's violent winds.

In this restless motion, particles are broken, fractured, and splintered, each piece acquiring or losing a tiny charge of electricity.

357. Where are the electrical charges built up? Within most tumultuous thunderclouds, some scientists believe, many large particles of snow and ice crystals and raindrops, each with a negative charge of electricity, drop to the lower levels of the cloud, which therefore acquires a negative charge. The smaller, positively charged particles tend to remain in the middle layers of the cloud or may be carried upward in rapidly ascending air currents to the freezing temperatures of the upper cloud layers where they tend to produce a positive field.

Hence, a thundercloud is usually negatively charged at the lower levels and positively charged at the top. Positive charges are also built up on the ground surface, directly beneath the cloud, in a phenomenon sometimes called a mirror image. (See Question 358.)

358. What is the mirror image of electrical charges? As a storm cloud moves over the land, a strong positive charge of electricity can be generated on the surface of the earth, which at other times is usually negative. This phenomenon, called mirror image, is actually a reverse image of the cloud's negative charges. It is a condition wherein positive charges race along the earth's surface like a shadow in an area beneath the cloud. As the cloud moves, billions of these electrical charges race across the fields, forests, and earth beneath, swarming up any prominences they encounter in an effort to get nearer to the cloud. Electrical charges form on steeples, tall buildings, chimneys, telephone poles, even animals and people.

359. How does the air act as an insulator? Between the negatively and positively charged regions of a thundercloud, and those of a thundercloud base and the ground, lies a generally neutral area of air with no predominance of negative or positive charges. This band of air may sometimes be hundreds or thousands of feet thick. Since air is not a good conductor of electricity, it acts as a thick insulator.

As the electrical potentials build up, small charges leak out from the negative areas and flow swiftly and invisibly through this insulating air, forming conductive channels through which the current flows.

360. What happens as lightning flashes? As charges within the negative and positive areas of the cloud or clouds or between earth and cloud build up, the negative charges begin to seep out in an invisible streak of free electrons, called a pilot leader. This in essence is a tunnel or ionized path about 50 to 75 feet long, as small in diameter as a piece of string or as thick as a man's wrist. Down this ionized tunnel surges a current, called a step leader. Immediately after this, another step leader plunges down the ionized tunnel, lengthening it and bringing it nearer to earth. Other leaders follow, each elongating the path until the track consists of about 40 or 50 darts, giving lightning its characteristic zig-zag shape.

Meanwhile in the area on earth directly beneath the cloud, positive charges are swarming upward to meet the downward surging negative charges. At the time the downward leader is close enough for the upward positive streamer to meet it, the flash of lightning occurs, actually a leap of positive charges from the earth into the sky. As the main flash ascends, it energizes, agitates, and lights up the downward thrusting branches made in the atmosphere by the leaders.

The whole thing happens so fast that an entire series of discharges may take place in less than ten millionths of a second.

361. Why does it seem as if lightning comes from the sky? Our eyes play tricks on us in making us think the flash comes down from the sky, instead of reaching up from the ground.

As leaders of negative electrons push down from the cloud, we cannot see them because they give almost no light. But when the

Buildup of electricity and a lightning flash

positive charges travel up from earth in ionized tunnels, they create a brilliant light along all the various prongs and ramifications of the tunnels down which the downward electrons have been boring. Since the flash of lightning travels so fast, our eyes cannot tell from which direction it comes, and it seems to have come from the sky.

Also as the main flash ascends, it lights up the branches made by the leader track from the sky to earth. These branches point downward, so when they all are lit up it seems the flash is directed toward earth.

362. How do thunderstorms and lightning maintain electrical balance between earth and sky? Thunderstorms and their flashes of lightning involve many complex electrical imbalances and exchanges still not understood by scientists. Some meteorologists believe that these storms may help restore the prevailing charge distribution between earth and the atmosphere.

Normally, as in fair weather, authorities say, the earth quietly is "leaking" negatively charged ions into the atmosphere. Other authorities say the atmosphere conducts or "leaks" a current from the positively charged ionosphere to the negatively charged earth.

During thunderstorms, the negatively charged ions are fed back to earth, either in lightning strokes or by point discharge, which is the electron discharge from the sky into earth through natural points, such as blades of grass, tips of leaves, small trees, ship's masts, or other pointed protrusions.

During thunderstorms, scientists have calculated, the flow of negative electricity to earth is considerably greater than the flow from earth in fair weather.

363. How long does a cloud take to recharge? The electrical charges of the cloud and earth are renewed in about 20 seconds after the lightning flashes, and the 2 poles are ready for another discharge.

364. What are some benefits of lightning? One of the most positive benefits of lightning is the production of millions of tons of valuable nitrogen fertilizer each year, produced through a process called nitrogen fixation.

The earth's atmosphere consists primarily of about four fifths nitrogen and one fifth oxygen, plus small amounts of carbon dioxide and rare elements. When lightning flashes through this atmospheric mixture, its heat combines the nitrogen and oxygen chemically into nitric oxide gas, which is brought down to the soil by rains. Nearly 100 million tons of nitrogen are washed into the soil each year in this manner. From the soil, the nitrogen is taken up through the roots into trees, crops, and other plants, where it is assimilated into grains and other food parts to be eaten by animals and men.

So far, scientists have not been able to duplicate nature's nitrogen-fixation process to obtain nitrogen in such large quantities.

365. How could lightning have started life on earth? Scientists believe that an electric spark of lightning may have helped create life on this planet some 3 billion years ago.

By sending electrical charges through a simulated primeval atmosphere of the 4 basic gases—ammonia, methane, hydrogen, and water vapor—scientists found that amino acids were formed. These amino acids are some of the primary building blocks of all proteins, indispensable for all life forms.

366. What kinds of lightning are there? The electrical discharge from a thundercloud may take various distinct forms and shapes, depending on many factors such as wind, clouds, and other atmospheric conditions. Lightning has appeared in the shape of streaks, sheets, balls, and other forms. (See Questions 367 through 375 for more details.)

367. What is streak lightning? Streak lightning is the most common type of lightning, occurring as one or more vivid lines or streaks, seen in a zig-zag shape, moving from cloud to earth or from cloud to cloud.

368. What is forked lightning? Forked lightning is streak lightning which forms two or more branches simultaneously.

369. What is ribbon lightning? Ribbon lightning is streak lightning blown sideways by a heavy wind, moving the ionized path so that individual strokes are seen separately and parallel as a series of bands or ribbons.

370. What is bead lightning? Bead, chain, or pearl lightning is a form of streak lightning that has been interrupted or broken into more or less evenly spaced segments or beads.

371. What is sheet lightning? Sheet lightning is usually the reflection of a hidden flash of streak lightning, obscured or diffused by clouds or distance. A whole wide expanse of a cloud appears to be lit up, giving the impression of a sheet of light.

372. What is heat lightning? Heat lightning is seen flickering along the horizon on hot summer evenings when no thunderclouds are visible. This may be caused by reflection of lightning occurring below the horizon. Heat lightning is usually silent merely because it is so far away.

373. What is ball lightning? Ball lightning is a peculiar, spectacular kind of lightning in which the electrical discharge appears as a round moving blob of light, sometimes about 4 or 5 inches in diameter, sometimes larger, traveling at several hundred feet per second and moving in erratic paths.

These balls have been reported reddish or yellow, white or blue. Some seem to decay or dwindle slowly and others disappear rapidly with a sharp explosive pop.

374. What are some strange stories about ball lightning? Balls of lightning have been reported following air currents, sometimes against the wind, drifting through the air, gliding down a tree, or hovering near the ground for several minutes before suddenly starting out again or disappearing. They have been reported traveling along wire fences, following terrified people around the kitchen, falling into a bucket of water and causing it to boil, or disappearing through the keyhole in a door.

375. Are reports of ball lightning considered authentic? Considerable controversy surrounds the existence of ball lightning and the many weird reports on its behavior. It has been called an optical illusion and sheer imagination. Some people even claim this lightning is the basis for mysterious "flying saucers."

Growing evidence of its scientific existence has been shown in laboratories of Westinghouse Electric Corporation, where an electronic computer in a simulated model predicts many strange properties of ball lightning. Scientists describe it as a luminous, hot region of air having high electrical conductivity, with temperatures ranging from 6,300 to 10,800 degrees Fahrenheit.

376. What is St. Elmo's fire? St. Elmo's fire is the luminous activity of weak static electricity that sometimes glows on pointed objects with bluish or greenish colors.

The weird fire has been seen lighting up exposed points, such as masts of sailing ships, mountain peaks, steeples, lightning rods, ears of horses, horns of cattle, and even the hairs of people's heads. It is commonly seen on the wing tips or noses of airplanes.

Scientists explain this fire as a phenomenon of point discharge, formed when an electrified cloud rubs against or passes a point such as a ship's mast. At this time molecules of gas in the air around the point become ionized and glow.

377. Where did St. Elmo's fire get its name? This strange fire received its name from an Italian corruption of the name Saint Ermo of St. Erasmus, who was an Italian bishop and patron saint of Mediterranean sailors. This bishop, living around A.D. 300, was famous for his piety and goodness. Sailors regarded the glow from a ship's mast as a visible sign of his guardianship. It never seemed to destroy anything and was considered beneficial and a good omen.

378. When did St. Elmo's fire encourage an expedition? There are many tales of events that may have hindered expeditions or caused them to continue—no one really knows for certain. At least one account gives full credit to the comforting glow of St. Elmo's fire for encouraging one of the most important expeditions in history—that of Ferdinand Magellan around the world. On October 5, 1519, as his ships sailed south of Cape Verde into unknown waters, the crew became more apprehensive of the future as they were physically and psychologically shaken by the storms. Suddenly someone noticed the glow of St. Elmo's fire around the masts. For several hours the light flickered around their ships, and the seas became more calm. The superstitious sailors considered the fire to be the holy body of St. Elmo himself and were consoled by this auspicious omen enough to continue the voyage.

379. How fast does lightning travel? As the first jab of a leader pokes down from the cloud, it may travel at speeds of some 100 miles a second. The next leader may go at faster rates—about 1,000 miles a second. When the return stroke or streamer shoots up from earth, it can travel about 80,000 miles per second. This is less than half the speed of light, which is 186,000 miles per second.

380. How long is a stroke of lightning? A stroke of lightning may travel for various distances. Strokes have been measured from a few hundred feet long to a few miles long. From cloud to earth, they are rarely longer than a mile, although flashes 4 miles long have been reported.

In mountainous regions, flashes may be 100 yards long—or less when the clouds are low.

381. How often does lightning strike? Definitive figures on how often lightning occurs are extremely difficult to obtain, but scientists calculate that every 24 hours, the earth is struck 8½ million times by lightning.

382. Is it true that "lightning never strikes in the same place twice"? No. Lightning does strike twice and even more often in the same place. Many lightning flashes can strike one place in multiple succession in a few weeks. The Empire State Building in New York city has been struck as many as 12 times during one storm.

383. How powerful is a lightning stroke? Individual lightning strokes vary in the amounts of energy they carry.

A single discharge may have from 10 million to 100 million volts. The amount of electricity involved in an average lightning stroke is 30,000 amperes.

384. What was an extremely powerful lightning stroke? One of the most powerful strokes yet measured and recorded in the United States took place in Pittsburgh, Pennsylvania, in July, 1947. This great flash of lightning yielded 345,000 amperes—enough power to light more than 600,000 60-watt light bulbs for as long as the flash lasted.

385. What is a "bolt from the blue"? Long strokes have been known to travel from one part of a cloud to another part, or from one cloud to another. One bolt was recorded as 20 miles long. Some researchers believe these long bolts travel great distances from their parent cloud through clear skies and suddenly turn to hit earth as a "bolt from the blue."

386. What color is lightning? Lightning is most often seen as whitish or as white-yellow. Under certain conditions, such as in humid moist air, it appears pink. Against yellowish artificial lights, lightning may appear blue.

387. Is lightning hot? About three quarters of the energy of a lightning bolt is expended in heat. In only a fraction of a second, the temperature around a bolt rises to about 27,000 degrees Fahrenheit. One report listed temperatures as high as 36,000 degrees Fahrenheit. For comparison, the visible surface of the sun is considered to be 10,000 degrees Fahrenheit.

388. What is cold lightning? Cold lightning is a flash of lightning that lasts a very short time but which has extremely high currents. Wherever it strikes, it has explosive effects.

Hot lightning, in comparison, lasts longer, but has less current. Hot lightning is more apt to set objects it strikes on fire.

389. When is danger from lightning greatest? Scientists are constantly studying the effects and patterns of lightning. According to some observations, the danger from lightning may be greatest at the time the thunderstorm is approaching, just before the rain begins.

390. How is a person affected when hit by lightning? A stroke of lightning can electrocute or kill a person when a strong electric current passes through his body.

The electric shock of a lightning stroke can cause paralysis of the muscles and nerves that control the movement of a person's diaphragm. Hence the lightning victim cannot draw breath. When the victim regains consciousness, he often is numb and unable to move or speak for a while. Even after he recovers, he can have muscular pain and stiffness for a long time. Often severe or minor burns may result from lightning strokes. In some cases, permanent injury or death is caused when the victim has been roughly tossed about by the blast of the strike.

391. Do people's bodies retain an electric charge? A person or animal struck by lightning does not carry any electric charge on his body and is quite safe to touch, contrary to many popular opinions.

392. Do most lightning victims recover? Most people struck by lightning recover completely. Many more would recover or survive if more people nearby came to their assistance instead of assuming them to be dead, since breathing and pulse may have stopped. Sometimes essential aid is not given because persons near the scene of a lightning strike may themselves have become hit or disoriented. Or they may have the fallacious idea that the victim's body still retains electricity.

393. What first aid is given for lightning strokes? Since lightning can paralyze the diaphragm or lung muscles of a person, artificial respiration should be given to someone knocked down by lightning, just as if he had drowned. Once a person's breathing is started again, he should be carefully examined and treated for burns.

Sometimes the heart stops beating after a lightning stroke. The victims can often be revived by external heart massage.

394. Why do objects explode when hit by lightning? Trees literally burst open when hit because the intense heat volatilizes the sap, which explodes in a fraction of a second.

A rock can explode in a similar manner when the high heat of lightning turns the moisture in a crack into superheated steam which explodes like a charge of dynamite.

Chimneys have been shattered when the moisture of the bricks is suddenly expanded.

395. What are fulgurities? Along ocean beaches, sandy hills, or dunes, people sometimes come across strange hollow tubes of cemented sand, perhaps in the shape of a tree branch, perhaps like a small rough rod.

This is a piece of petrified lightning—created in a fraction of a second when lightning struck the earth and fused sand particles into silica. Scientists call them fulgerites, from the Latin word *fulgur*, meaning lightning.

When a bolt of lightning strikes an area of dry sand, the intense heat melts the sand surrounding the stroke into a rough glassy tube, forming a fused record of its path.

Fulgurites sometimes have diameters from half an inch to 2 inches, and extend 10 or more feet long. They are nearly always found broken in pieces, for the brittle cast breaks easily if jarred, dropped, or stepped upon. The inside walls of the tube are glassy and lustrous, and the outside is rough with sand particles adhering to it. Fulgurites are usually tan or black, but one found in Florida was almost translucent white.

Lightning also leaves its mark on rocks, usually on mountain summits. Pieces of rock may be fused together or the surface of a boulder may be fired to a lustrous glassy material. One scientist described a rock fulgurite as a "white incrustation," as if white paint had been splattered about or spread in a rough, branching, straggling line.

396. Where does the word electricity come from? The word electricity was first used to express the "power of attraction" by William Gilbert, personal physician of Queen Elizabeth I of England, about 1575.

The yellowish clear substance amber, which is fossilized sap or resin from pine trees some 50 million years old, had long been known to have powers of attracting small light objects such as lint and bits of dry paper—static electricity, or electricity at rest. The ancient Greeks thought perhaps gods resided in the material.

Gilbert noted this power and called it electricity from the Greek word for amber, which is *elektron*.

The word electricity remains as the word for the energy that exists in all matter in the universe.

397. When did men begin to experiment with electricity? In the mid-seventeenth century, scientists throughout Europe were beginning to experiment with electricity, particularly with static electricity generated by rubbing pieces of amber or other kinds of materials.

Friction machines were built to generate electricity, which then could be stored for long periods of time in the curious Leyden jars invented in 1745 by Pieter van Musschenbroek at the University of Leyden, in the Netherlands.

At this time, electrical experiments and tricks were performed as a matter of curiosity and amusement rather than as a science. No practical applications for the "electrical fluid" had been discovered.

398. What experiments showed the electricity of the sky to be the same as that of earth? The idea that lightning might be electricity was first discussed in 1708 by William Wall, an English scientist who suggested that sparks and electrical discharge from a piece of amber were the same as those of lightning.

In 1749, Benjamin Franklin set forth his belief that thunderclouds were charged and that lightning was an electrical phenomenon. He proposed a specific experiment with a sentry box and a 30- or 40-foot pointed iron rod to draw down the electricity from a thundercloud. In May, 1752, a passing thundercloud caused sparks to leap from such a rod erected by a French electrician, Thomas-Francois D'Alibard. This was the first proof that thunderclouds held electricity.

399. What was Benjamin Franklin's kite experiment? In June, 1752, during a thunderstorm, Benjamin Franklin and his young son flew a kite. He attached the kite to a long line of twine, holding it with a silk handkerchief to insulate the kite and string from himself and the ground. Just above the knot between the silk and linen twine, Franklin tied an iron key, to "attract the electric fluid from the sky."

The first thing he noticed was that loose fibers of the linen twine bristled outward—evidence of an electrical charge. Franklin touched his knuckle to the key and received a shock.

Actually the electricity of that shock was created merely by the ionized atmosphere. If lightning had struck at that instant and traveled down the twine, Franklin undoubtedly would have been killed.

From this experiment Franklin surmised that the electricity of thunderstorms and the electricity of earth were the same thing.

400. Have other people tried Franklin's experiment? Other people have tried the kite experiment, but no one has yet been able to duplicate the series of events that Franklin experienced. The experiment is considered extremely dangerous, since a shaft of actual lightning traveling along the twine could kill a person touching the key.

401. Why cannot lightning be harnessed? Lightning's tremendous energies have little or no practical use for man as yet, for they last for such a small fraction of time.

402. Can man make lightning? Scientists have made synthetic lightning in high-voltage laboratories of organizations such as the General Electric Company and the Westinghouse Electric Corporation and in research institutions and universities.

Two generators are usually used in the experiments—one manufacturing electricity with a positive charge, the other with a negative charge. These charges are led off to condensers or storage systems standing about 50 feet apart. When the electric tension between them grows strong enough, a brilliant crackling stream of lightning leaps from one condenser to the other. With this man-made lightning, studies are made on protection devices for buildings, telephone poles, transmitters, and other equipment.

403. How does a lightning rod work? Traditionally, a lightning rod is a metal rod placed above a house or structure and connected to the ground. Since electrical charges are seeking the shortest path to each other, the positive particles from earth climb up the high pointed rod, the negative particles from the sky reach down toward it—and when the bolt occurs, it is conducted through the metal into the atmosphere, rather than through the house.

Metal is used because it conducts electricity readily, whereas other materials, such as wood, brick, or concrete, do not. Electricity, like water, tends to take the path of least resistance. Major damage from lightning results from the heat and explosive forces generated in non-conductive materials.

Another form of lightning rod is built into house walls themselves. These metal frames run throughout the height of the building and conduct the lightning from the base of the house to the outside atmosphere.

Power lines are protected by shielding wires erected parallel to the lines and grounded into the earth. These wires are called arresters and range from small matchbox sizes that protect railway signal lines and fire alarm systems, to huge 30-foot-high arresters beside generators in vast hydroelectric plants.

404. Can lightning be predicted? A new radar device that pinpoints lightning areas within storms, developed by the University of Miami Radar Meteorological Laboratory, will enable airline pilots to guide their planes through non-violent storm areas, greatly reducing lightning risks and saving wide detours.

The new device, called a sferics-to-radar converter, is housed in a compact box and can be readily adapted to any existing radar set. It requires 100 transistors and takes coded radio noises emanating from electrical storms and converts them into information on a radar screen. Arrow-like markings on the screen point to the most intense zones of electrical storm activity and also indicate the intensity of the lightning produced.

405. What is thunder? Thunder is the result of sound waves set in motion by the powerful compression and sudden violent expansion of air heated by the lightning stroke. Great claps of thunder are produced when the intense heat and ionizing effect of lightning along a highly heated path cause the air in the atmosphere suddenly to expand, sending a vibrating pressure or shock wave outward at the speed of sound.

406. How is thunder heard? Thunder is heard as a single sharp crack when a person is near a stroke of lightning.

Farther away, thunder can be heard as a continuous rumble as the sound waves, which travel relatively slowly, move away from various points along the lightning stroke—somewhat like the sound of cloth ripping along a seam. Since sound travels at speeds of about a thousand feet per second, and since the stroke may be several miles long, the thunder is usually heard as a drawn-out rumbling.

Thunder is heard as a reverberating sound that echoes and re-echoes as sound waves are reflected and bent in different patterns around objects such as hills, mountains, and forests.

407. What is a thunderbolt? When a flash of lightning is accompanied by a clap of thunder, the combination of the two is called a thunderbolt. This term was often used in ancient legends and the bolt was supposed to cause much of the damage actually caused by lightning.

408. How far away can thunder be heard? Ordinarily, thunder can be easily heard for a distance of 6 or 7 miles, though on a quiet summer day, rumbles can be heard for 20 miles. It is rarely heard at greater distances.

Actually the distance the pressure waves of thunder travel depends on temperature and wind distribution in the atmosphere. Meteorologists are finding that, because of the decreasing temperatures at increasing height in the atmosphere, sound waves bend upward, away from the earth, and thunder cannot be heard far. This is particularly true when the sound waves of thunder are moving against the wind.

409. How does one judge the distance of a thunderstorm? A person can judge how far away a thunderstorm is by timing how long it takes the thunder to reach him after he sees the flash of lightning. By counting seconds between the flash and the thunder he can obtain an approximate number. He then divides this number of seconds by 5. The resulting number is the approximate number of miles away the thunderclap occurred.

This simple calculation is based on the fact that light travels about 186,000 miles a second, and sound travels about a thousand feet per second—or one mile in a little less than 5 seconds. The lightning flash and thunder start from the same place at the same time. A person sees the lightning at essentially the same time it occurs, but the sound waves roll over the land at 5 seconds a mile.

410. What winds blow within a thunderstorm? As the thunderstorm cell develops, the column of ascending warm air blows upward at speeds of about 20 miles an hour. In a mature cloud, these speeds increase to 60 or 70 miles an hour. The most violent updrafts can blow 150 miles an hour, usually in the right front part of the cell. The wind columns tend to slope somewhat toward the rear, carrying raindrops with them as they go.

Just behind this main violent updraft is a zone where the rising air slows down and the region of cold, downward-rushing air begins. This is the area of heaviest rainfall, bringing with it downdrafts of cold air from upper regions at speeds of 20 to 30 miles per hour. These downdrafts crash into earth and blow outward from the storm in heavy gusts, often forming a squall or roll cloud. (See Questions 411 through 414.)

The violent updrafts and their relatively more gentle downdrafts may extend from 1,000 to 25,000 feet in length within the cloud, constantly changing speed and position. They create all sorts of strong eddies and whirlwinds as they move and clash, causing the violent turbulence that has sent airplanes shooting up or down or ripped their wings apart.

411. What is a thundersquall? A thundersquall is a strong cool wind that blows out from the lower front part of an approaching thundercloud. These gusty winds can blow across the land at speeds of 60 to 80 miles an hour, sometimes even reaching speeds of over 100 miles an hour and causing much damage.

The wind is caused by strong downdrafts of cool air that fall from the thundercloud along with falling rain. Since the draft is cold, it is heavy and dense and spreads out along the ground in front of the storm, beneath the warm air. It brings welcome cool relief on a hot afternoon.

412. What is a squall cloud? A thundersquall wind helps form the squall cloud, which is a dark-gray boiling roll of cloud rotating on a horizontal axis, pushed ahead of the oncoming thunderstorm. This cloud is formed by the downward cold currents blowing out of the storm, and the upward current of warm air blowing into the storm.

From the ground, a squall cloud line sometimes may look like a long, low wave of billowing smoke rolling horizontally above the ground. It sometimes appears light colored or even white in contrast to the dark storm clouds behind it.

413. What is a squall line? A squall line is a line of gusty thunderstorms advancing well ahead of a cold front pushing into warm, muggy air. Extending in a line sometimes hundreds of miles long and a few miles wide, the squall line moves somewhat like a shock wave, bringing strong winds, rain, and violent electric activity. From the ground it may look like a wave of rolling turbulent black clouds. The squall line advances at night as well as in the day, since it journeys ahead of a large storm system that may be moving across the country for several days and nights.

414. How do squall lines progress? In the Northern Hemisphere, this line moves generally from the southwest toward the northeast some 50 to 150 miles ahead of a fast-moving cold front—at speeds of about 20 to 30 miles an hour in summer, and 30 to 40 miles an hour in winter. They may last only a brief 20 minutes or for several hours.

415. What is a cloudburst? A cloudburst is a sudden torrential downpour of rain, when rainwater seems to fall in continuous streams rather than in separate drops. To a person standing beneath, it seems as if a cloud literally has burst wide open and dumped all its contents at one time in sheets of solid water.

A cloudburst occurs when raindrops within a cloud are no longer held up by strong updrafts of wind. The supporting air column suddenly gives way, and rain comes tumbling down in concentrated amounts. The rain does not actually burst out of a cloud, but only sweeps in torrents across the earth.

Rain may fall from these cloudbursts at the rate of 3 inches or more an hour—at times even as high as one inch per minute.

416. Where are cloudbursts particularly destructive? Sudden cloud bursts are particularly destructive in dry or mountainous regions where steep bare slopes converge into deep valleys. Within a few minutes these dry valleys and ravines may be filled with one or more waves of water 10 to 20 feet high.

417. How much water may a thundercloud hold? A large thundercloud may hold as much as 150,000 tons of water in the form of raindrops and ice crystals. This water would be enough to fill a pond a mile long, 300 feet wide, and a little over 5 feet deep.

418. What devastation does hail bring? Major thunderstorms may generate severe hailstorms that drop tons of ice pellets within a few minutes over certain sections of the land. These hailstones can cause millions of dollars worth of damage to crops, as well as to buildings, cars, and other objects. Hail has killed much livestock, as well as wild animals and birds, and has brought death to a few people. (See Chapter V.)

419. What are some ancient myths about storms? Nearly every ancient civilization had stories of their thunder gods—powerful masters who rode the spirited storms and hurled thunderbolts and lightning shafts at their enemies.

Thor was the mighty god of the Norsemen, a fierce deity with flaming red hair, drawn in a chariot pulled by two goats. Thunder rumbled as his chariot wheels struck the tops of storm clouds. Thunderbolts erupted whenever he threw his hammer, Mjollner, which always returned to his hand.

Lei Chen Tzu, ancient Chinese god of storm, wore a halo of fire and lightning and produced thunder by beating on his drums.

Zeus was the king of the Greek gods and god of the thunderstorms, hurling down lightning, sheets of rain, and wind to route his enemies.

Jupiter or Jove was the Roman king of gods and god of the storm, hurler of thunderbolts.

Laurel, a shrub sacred to the god Apollo, was thought to be immune to lightning. The Roman emperor Tiberius, reigning during the time of Christ, was so frightened of thunderstorms that he kept laurel wreaths within reach and put one on his head during storms.

420. What was the thunderbird? The American Indians of the plains, in particular, told beautiful but fearful stories about the great thunderbird whose shadow was the cloud, whose great flapping wings created thunder, and who shot out flashes of lightning whenever he opened and shut his eyes.

Several tribes of Africa today still believe in the thunderbird as a bird of destruction. Witch doctors of the Bantu tribe of southern Rhodesia search beneath a tree shattered by lightning for the eggs or droppings of the fearful bird.

421. Why did people ring church bells during a storm? Since storms and lightning were regarded as acts of God, people would hasten to church to pray when a thunderstorm approached.

Bell ringing was considered one way to divert evils of the storm. Many church bells dating from the eleventh, twelfth, and thirteenth centuries have a Latin inscription cast on their rims: *Fulgura frango*, which means I break lightning. To superstitious people, ringing bells tended to drive away evil spirits riding the storm. To others, the ringing was supposed to send up sound waves in the air to interrupt the path of the thunderbolt.

However, bell ringers often were in susceptible spots for lightning to strike, and many were killed. As early as the eighth century, Emperor Charlemagne reportedly issued a decree forbidding the ringing of church bells during thunderstorms. Nevertheless, the custom remained and continued to be observed, despite many edicts. In 1784 a book reported that in Germany lightning had damaged or destroyed 386 church steeples and killed 103 bell ringers during the course of 33 years.

In England, the church of St. Pauls in London has been struck by lightning many times. In Venice, Italy, the bell tower or campanile of St. Mark's was struck and damaged many times before a lightning rod was set up in 1766 to keep it safe.

422. What are thunderstones? In Scandinavia, bits of iron or stone in the shape of eggs are sometimes plowed up by farmers in their fields. The farmers call them thunderstones, from the old myths saying they are broken pieces of the thunder hammer of Thor, the Norse god of storms. Some of these stones are actually small meteorites. Others have been identified as tools of the Stone Age.

423. How valid are some still prevailing proverbs? "Flies bite more fiercely before a storm." Untrue. Flies can bite just as fiercely before, during, and after a storm.

"Thunder and lightning can turn milk sour." Untrue. Thunder and lightning cannot turn milk sour. However, if milk is left out during the hot moist weather in which thunderstorms are produced, the growth of bacteria is favored, and the milk can turn sour.

"Red at night, sailor's delight. Red in the morning, sailor take warning." Possible. A red glowing sun seen in the morning could mean that storm elements such as rain, clouds, and wind are present and could brew into a storm. The red sun at night is the reflection of the setting sun on the evening clouds . . . a normal phenomenon.

"A halo around the moon brings rain." Possible. The presence of a halo or bright ring around the moon or sun indicates the presence of high ice-crystal clouds called cirriform clouds. These clouds often indicate an approaching warm front which may or may not bring rains or storms.

424. Can people feel when a storm is coming? Many people claim they can tell when a thunderstorm is brewing by specific signs they feel or see. Some say they can feel a storm in their bones. Arthritic people complain of more painful, more swollen joints. High-pressured people feel their hearts beat faster, and others say their old scars start hurting when a thunderstorm is approaching.

Dogs have been reported to become more nervous and edgy before a storm, and farm animals such as cows and horses may become unruly.

The various complaints and feelings have a certain foundation, for changes in humidity and pressure of the air can bring about changes in pulse, respiration, and blood pressure in human beings.

There is also some basis for the claim of people who say they can smell a storm coming. When the atmospheric pressure starts to decline before a storm, various decaying odors from ditches, marshes, or newly cleared land are released and spread to other regions.

425. What warning system is used for thunderstorms? The Weather Bureau maintains a Severe Local Storm Forecast Center at Kansas City which gives a report each morning on the most likely areas of severe thunderstorm activity during the next 24-hour period.

During the day, a continuous watch is kept on possible storm development. An IBM computer is used to speed the processing of hourly reports and make computations on weather changes. Using this data when a storm develops, weathermen issue severe-weather forecasts, sometimes 6 hours in advance, but usually 1½ hours ahead of the storm. The areas covered by the forecasts are usually about 20,000 to 30,000 square miles.

426. What precautions should be taken while indoors during thunderstorms? If you are in a house when an electrical storm hits, keep calm and don't be afraid. Remember, thunderstorms are usually of short duration; even squall-line storms pass in a few hours.

Be sure to stay away from all electrical appliances, such as radios, television sets, telephones. It is best also to stay away from walls, fireplaces, radiators, and any large metal objects that might serve as conductors of electricity.

The center of the room is probably the safest place to stay. The cellar is also quite safe.

427. What are some thunderstorm safety rules for outdoors? If you are outdoors, avoid high ground such as tops of hills or cliffs. Avoid open spaces and lone trees. If you are caught in a field with a few trees, lie down in the open at distances at least as far away as twice the height of the nearest tree. The best thing to stand under is a small tree in a dense wood with taller trees. You are also relatively safe in a deep valley or canyon or at the foot of a cliff. However, remember, that thunderstorms may produce flash floods. So stay out of dry creek or stream beds.

If you are out hunting, fishing, or playing golf, drop all metal tools and gear. If you are swimming or in a small rowboat or sailboat, head for shore. All-metal ships are rarely damaged, but small boats have been sunk when lightning came down the mast and struck a hole in the bottom of the hull.

Keep more than 100 feet away from wire fences, and do not touch one after the storm is over—it may still be storing a lethal charge of electricity.

428. What was the Thunderstorm Project? The Thunderstorm Project was an intensive study undertaken by the U.S. Weather Bureau, the Army Air Force, the Navy, commercial airlines, the Civil Aeronautics Board, the Massachusetts Institute of Technology, the University of Chicago, and the Carnegie Institute of Washington. For two years, 1945 and 1946, flights were made by experienced pilots into thunderstorms to study the nature of the storms. This was the most comprehensive program of thunderstorm investigation ever undertaken. The 1,363 airplane penetrations into the storms, along with 1,375 balloon soundings, yielded information on major structural features and gave new understanding about the storms.

429. What research projects are under way with thunderstorms? Since the end of World War II, thunderstorms and lightning have been intensely investigated by scientific methods involving radar, balloons, airplanes, rockets, and satellites. They have also been studied in laboratories and with computer systems.

The National Severe Storms Laboratory at Norman, Oklahoma, part of the Environmental Science Services Administration, has conducted research programs to study the causes and effects of thunderstorms. In cooperation with the Army, Air Force, and Federal Aviation Administration, planes are sent into Midwestern storms to investigate their wind flow, direction, and other phenomena. Cumulus-cloud research programs are also under way in the Caribbean Sea and nearby land areas, in direct cooperation with hurricane research.

Lightning-modification projects are being carried out jointly with the Environmental Science Services Administration and the U.S. Army at Flagstaff, Arizona. Field experiments and aircraft flights indicate that it may be possible to short-circuit the charged field in which lightning develops and thus reduce the danger.

Year round theoretical studies of field and laboratory lightning are being conducted at the University of Arizona, Tucson. With a 162-

foot tower topped with a 35-foot needle, scientists are accumulating some of the most complete data on single strokes of lightning ever gathered, including statistics on duration of strokes, channel size, current, luminosity, ionization, and the nature of the cloud's electrical fields.

Other research is being carried on in the New Mexico Institute of Mining and Technology; the U.S. Forest Service in the Department of Agriculture; the Lightning and Transients Institute in Minneapolis, Minnesota; and the Cambridge Research Center in Bedford, Massachusetts.

IV. HAILSTORMS

Introduction. Suddenly, on a warm oppressive summer afternoon, tons of ice pellets shoot down from a towering thundercloud and in a matter of minutes flatten a ripening field of grain as if an army had marched across the land. The hailstorm lasts less than 10 minutes, yet millions of dollars worth of crops are completely destroyed.

This is the scourge of hail—the summer ice called White Plague by farmers who watch in helpless anger, unable to protect their crops.

Hail particles are small lumps of ice, each one of which is composed of concentric layers of clear and opaque ice and snow. They are formed in conjunction with severe thunderstorms, thousands of meters high in freezing temperatures of the atmosphere. Caught within violent updrafts and downdrafts of wind in the storm cloud, hailstones are formed layer by layer as frozen water accumulates around a nucleus of dust or an ice particle. Riding through cold and freezing layers of air, the stone grows in size until it becomes too heavy for the wind currents to sustain it and falls to earth.

Strangely enough, true hail forms in the warm summer months, when ground temperatures are above freezing. At this time of year, sheets of hail fall most often in the interior of continents of the middle latitudes. There are other kinds of hail that fall in the winter months—graupel and soft hail—but these are not so destructive as true hail.

Hailstorms have caused more property damage than tornadoes in the United States. A deluge of hail can strip leaves and ripening ears off cornstalks, leaving a naked forest of sticks. It can puncture melons and tomatoes, bruise and shred tobacco leaves, knock down grape arbors. A particularly destructive aspect of hailstorms is the fact that they form in many locations just at the peak of the local growing season of crops.

Hail can also cause extensive damage to houses—smashing through roofs, shattering windows, or knocking off chimneys. Hail can batter, dent, and tear a vehicle to irreparable junk. It can force airplanes into emergency landings, sometimes damaging them permanently.

Hail has also taken toll of human life. In the densely populated countries of India and China, hail has killed thousands of people.

430. What is hail? Hail is precipitation in the form of small concentric-layered ice balls or irregular lumps of ice, snow, and sometimes rime. A hailstone is a single individual unit of hail, sometimes as small as a pea, sometimes as large as a tennis ball or a grapefruit. On rare occasions these ice balls grow even larger.

Hail is the heaviest and largest unit of precipitation.

Hailstones usually consist of onion-like layers of clear ice alternating with opaque, partially melted and refrozen snow, all structured around a tiny central core or nucleus. Sometimes there are 3 or 4 concentric layers, sometimes more. As many as 25 layers have been counted.

There are several kinds of hail, including graupel and small hail, but true hail is a peculiar product of violent thunderstorms, causing much damage as it pelts down from summer skies.

431. What causes hail? Most researchers agree that hail is formed in conjunction with cumulonimbus or thunderclouds. Hail appears to be associated with only certain cells in these thunderstorms rather than with the entire storm.

A parcel of hailstones usually starts to form in the front part of a large thunderstorm, when freezing water and ice begin to cling to small nuclei in the air, such as dust, smoke, chemical particles, or fragments of ice and snow. This initial formation may take place one or more miles above the ground, sometimes higher. Caught in the powerful winds of the storm cell, these icing particles are blown through zones of freezing temperatures and those of warmer temperatures.

In the turbulent journey through the storm cloud, hailstones accumulate additional layers of ice, melting snow, freezing rain, and frost and increase in size.

432. What air movements help generate hailstones? Intensive hailstorms with their tons of ice pellets appear to be created by a combination of three kinds of air movements—1) a well-defined vertical movement of air that helps supply energy to drive and sustain the thundercloud, 2) a horizontal flow of air in from the sides that helps determine whether many small stones form or a fewer number of large ones, and 3) the presence of strong upper winds that blow across the top of the thundercloud and carry smaller-sized water and ice particles away from the storm system. The large heavy particles that fall back into the supercooled moisture of the storm grow into hailstones.

433. How does hail develop? There are various and diverging theories concerning the development of hailstones in the parent cloud before they shoot down to earth.

Some scientists believe each hailstone accumulates additional layers by rising and falling through updrafts and downdrafts within the storm cell.

This theory states that a developing hailstone may move into a weak updraft and tumble down into a region of supercooled water

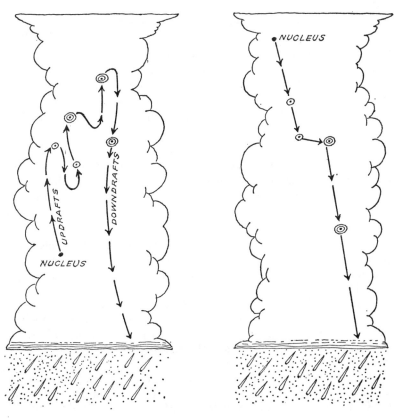

Two theories of development of hailstones

droplets (tiny drops of water with temperatures below the freezing point of water but which are still unfrozen because of many reasons, one of which is lack of nuclei). Here the hailstone may gather a film of supercooled water that immediately turns to ice. According to some scientists, the pellet may fall for a considerable distance, then meet an upward blast of air and be propelled into freezing temperatures again where it gains another layer of ice.

With each upward or downward journey through the turbulent thundercloud, the grain of hail gains other layers of ice, snow, or frost and increases with size. Finally the pellet becomes so large and heavy, or the updrafts become less strong, and it plunges to earth as hail.

434. What is another theory about hail development? Other scientists believe that instead of riding up and down the drafts of a thunderstorm, the hailstone develops from a small nucleus into a larger pellet of ice in one long, continuous descent. In the long slow fall, sometimes several miles in distance, the ice stone may remain almost stationary at times, held up by powerful updrafts. As it floats through layers of clouds the hail pellet becomes coated with layer after layer of ice and freezing rain.

435. How does it feel to be in a heavy hailstorm? An account in the newspaper *The Ottawa Citizen* gives some idea of the hailstorm that struck Kemptville, about 30 miles south of Ottawa, Ontario, on June 26, 1952:

Stones pounded down as big as 4½ inches in diameter. This hailstorm, the Ottawa Valley's biggest in 50 years, bombarded Kemptville on the afternoon of June 26 and missed Ottawa by a thunderclap, Rockliff meteorologists said. . . .

During the 15 minute deluge there, an estimated 5,000 windows were broken, 500 dogs, cats and fowl were killed, 200 automobiles severely damaged, almost 100 metal roofs punctured and 15 acres of demonstration and experimental crops ruined. Damage estimated in the district ranged up to $500,000.

Only three persons were injured, none seriously. The storm began at Renfrew and turned to hail as it swooped down from the northwest to Kemptville. The hail extended to Mountain, five miles east of

Kemptville, and then turned to rain again as high winds whipped it down to Prescott and across the St. Lawrence River to Ogdensburg.

The hailstorm at Kemptville was followed by a torrential rain. Hail smashed 1,700 panes of glass in buildings and greenhouses at the Kemptville Agricultural School. Officials said apple, potato, cabbage and tomato crops were smashed to pulp, destroying a year of experimental work by government experts.

Windows on the north and west sides of almost all buildings were shattered in an instant. Broken branches crashed to streets to mix with thousands of glass slivers and the bouncing stones of ice. Windshields were smashed and bodies badly dented on 200 parked cars along the main street at the time of the storm. On convertible cars canvas roofing was ripped to shreds.*

436. How do hailstorms kill people? Hailstorms can cause deaths by physically pounding people—and animals—to death. Sometimes people may be knocked unconscious and buried under drifts of several feet of hailstones, where they may die of cold and exposure. Hailstones can inflict severe head concussions on people before they reach shelter.

437. What was the worst hailstorm disaster on record? The hailstorm considered as the most disastrous to human life on record occurred during a violent storm over the Moradabad and Beheri districts of India on April 30, 1888. At this time, 246 people were reported killed by hailstones "as large as cricket balls." Some people were hit and battered by the stones, and others were buried under thick cold drifts of hail and died of exposure to cold. More than 1,600 cattle, sheep, and goats were killed.

438. What was another disastrous hailstorm? Another severe hailstorm occurred over the western part of Honan Province in China on Sunday, June 19, 1932, killing some 200 persons and injuring thousands of others. Large stones were reported falling for 2 hours, destroying houses, crops, animals, and trees over an area encompassing 400 villages.

* *The Ottawa Citizen* (Ottawa, Ontario, Canada: Southam Press, Ltd., June, 1952).

439. What was another destructive hailstorm in India? On October 30, 1961, an intensive hailstorm fell over 45 villages in Madhya Pradesh, India, and in one hour killed 12 persons, seriously injured 97 more, killed 700 to 1,000 head of cattle, and caused extensive damage to crops over an area of about 4,000 acres.

440. Where else have deaths been caused by hail? In Pakistan, in 1956, more than 30 people died in a severe hailstorm.

At Rostov, Russia, 23 people died trying to save their cattle during a hailstorm on July 10, 1923.

In the Siatista district of Greek Macedonia, a hailstorm killed 22 people on June 13, 1930.

In 1936, 19 Bushmen in South Africa were reported killed by a hailstorm that deposited large stones over the area, covering their bodies 3 feet deep.

In Klausenberg, Rumania, 6 children were killed by hail and 10 others injured on May 1, 1928.

441. Has anyone been killed in the United States by hail? Even though hailstorms have killed and wounded hundreds of people in other lands, they have hurt few people in the United States and claimed possibly only one person's life. This figure may be so low because of the relative scarcity of population in the hail belts, and also because there is always some sort of shelter nearby.

The one person reported killed was a 39-year-old farmer who was caught in a storm when he was in an open field near Lubbock, Texas, in 1931. He tried to run for shelter, but hailstones pelted him so roughly that he was struck down and died several hours later.

442. How does hail harm livestock and wildlife? Hailstones have caused much damage to livestock, killing animals as large as horses, cows, and water buffalo. In Canada, a 600-pound hog was pounded to death by hail. Livestock have been blinded by hailstones, and have suffered large welts on their necks and backs. Sometimes their legs have been broken, or other parts of their bodies have become injured so badly that they had to be shot.

Smaller animals such as chickens, dogs, and cats have been knocked down and killed by hailstones, and wild animals, such as lizards, rabbits, and squirrels, have been found literally flattened out on rocks or logs.

443. What storm was particularly devastating to wildlife? The hailstorm over the Alberta Province of Canada on July 14, 1953, destroyed an estimated 36,000 ducks and thousands of other birds such as owls, songbirds, and other wildfowl. Another devastating hailstorm only 4 days later killed another 27,000 ducks in the same area.

444. What does the land look like after a hailstorm? The damage inflicted on land, plants, and wildlife can be great. One storm created a path of destruction some 140 miles long and 5 miles wide over the Alberta Province of Canada, on July 14, 1953. Hailstones were as large as golfballs, driven by winds whipped up to 75 miles an hour. Here Allen G. Smith, a biologist of the U.S. Fish and Wildlife Service, gives an eyewitness account of the aftermath:

A close investigation of the hail damage on the ground presented a picture of unbelievable devastation. Grasses and herbs were shredded beyond recognition and beaten into the earth. Trees and shrubs were stripped of all leaves and small branches and the bark on one side of the larger trees had been torn away or deeply gouged by hailstones. Plants growing in waters of the potholes and lakes were reduced to nondescript pulp. Emergent vegetation had disappeared, destroyed and beaten under the water's surface by the weight of the hail. Ponds that had been choked with grasses, sedges, cattails and bulrush since June were stripped of all evidence of former plant growth.*

445. What crops are damaged by hail? Because of the size, weight, and speed of impact, the hard bullet-like pieces of ice hurtling from the sky can literally strip an orchard of leaves and fruit, puncture watermelons or cantelopes in a melon patch, or flatten a field of wheat in a matter of minutes.

The extent of damage to crops depends upon the severity and duration of the storm and the type of crop.

* Allen G. Smith, "Hail, Great Destroyer of Wildlife," *Audubon Magazine*, 1960, Vol. 62, No. 4, p. 170.

Tobacco plants are very susceptible to hail damage. A promising crop has been utterly destroyed by a short but heavy fall of hailstones—mainly because the broad tender leaves are punctured and torn and often break off at the stalk. In North Carolina, the great tobacco state, annual losses caused by hail exceed $3 million.

Melon patches in Georgia have also suffered irreparable harm under hailstorms, as vines are broken and torn, and fruits punctured. Tomato plants also suffer from hail.

Grapevines, citrus trees, and any kind of fruit trees such as apple, peach, or plum rarely can withstand the battering of a hailstorm. The fruit states of California, Washington, Michigan, New York, Pennsylvania, and West Virginia are hard hit by hail.

446. What U.S. crop suffers most damage from hail? Of all the crops grown in the United States, wheat and corn seem to suffer the most damage each year from hail. This damage occurs partly because the most plentiful fields are in the same areas that are most often frequented by hailstorms.

447. What was a record U.S. crop disaster? One of the worst hailstorms to hit an agricultural crop lashed an area in southeastern Iowa on August 18, 1925, flattening large fields of corn to such an extent that many tenant farmers were forced to leave their farms and find other work. Losses from this one storm were estimated at $5 million.

448. How much money do hailstorms cost? In the United States, hail destroys from $200 million to $300 million worth of crops each year. It causes about $25 million worth of damage to property each year.

Single hailstorms frequently cause crop damages of a million dollars or more in such states as Montana and Kansas. A single hailstorm flattened and ruined wheat in Nebraska that would have yielded an estimated 3 million bushels. Hail damage in Kansas alone cost more than $100 million during a 10-year period, 1944 through 1953.

449. How does hail damage buildings? A hailstorm is particularly damaging to buildings when windows and skylights are shattered. Wooden siding of houses has been ripped off by hail and heavy winds, paint has been chipped and scratched, and wooden knots have been completely pushed out of pine paneling. House chimneys have been knocked over and roofs have been ripped up and cracked by hailstones which sometimes have hit so hard they break through the roof and crash into the room beneath. Greenhouses in particular suffer severe damage from hailstorms, ruining not only the glass or plastic roofs and windows but also valuable plants and seedlings.

450. What roofs are most damaged by hail? Researchers have found that hail causes most damage to roofing when the roof is only slightly slanted. Steep-pitched roofs are less vulnerable because hailstones strike them at a sharp angle and bounce off.

Asbestos shingles fare better than wooden ones, and tile roofs are considered even safer. Yet even tile has been cracked or broken by falling hailstones.

Metal roofs of heavy tin or sheet iron offer still better protection, especially if they are laid upon a base of heavy boards. Softer metals such as aluminum are often dented or punctured.

Statistics show that concrete slab roofs offer the best resistance against hail damage.

451. What other kinds of damage do hailstorms bring? Within a few minutes, a severe hailstorm can cause damages of thousands of dollars to cars in a crowded city, denting and puncturing the tops, hoods, and fenders; shattering windows and headlights. Hailstorms also can create extensive traffic jams and accidents as piles of icy stones build up along the streets and highways.

452. Which hailstorm was most disastrous to property? The most damaging hailstorm on official record in the United States took place on June 23, 1951, when a 200-mile long path was plowed from Kingman and Sumner counties of Kansas into Missouri. Wichita and the surrounding area received the worst damage from the storm, which left thick blankets of hailstones 12 inches deep, broke windows and roofs, and dented automobiles—all costing over $14 million.

453. How does hail damage aircraft? Airplanes suddenly caught in a bombardment of hail can be seriously damaged, whether flying through a storm or stationary on the ground at an airstrip. When flying, the impact of the hail is many times increased by the forward motion of fast-moving planes. Parts of nose sections and leading edges of wings and tail may be severely damaged. Landing lights and windows have been shattered by ice chunks and the metal of the frame severely ripped and dented. Controls have been damaged, and pilots hurt by shattering glass.

So far as is known, however, no plane has crashed because of hail damage.

Today's modern airplanes are built to withstand the impact of hailstorms, and pilots avoid severe storm regions as much as possible. A minor nuisance is caused when the rattle of hail hitting the plane's hull is so loud that communication is impossible between people in the cabin.

Gliders have been extensively damaged by hail. Wings have been torn off and holes blasted in the rudder and body fabric.

454. What severe plane damage has been caused by hail? On May 25, 1949, a Douglas DC-6 was hit by hail at 14,000 feet over Texas and suffered damage to its nose, windshield, cowlings, and all leading edges, causing a $25,000 repair bill.

A severe hailstorm near Cheyenne, Wyoming, on May 20, 1948, broke the windshields of a DC-3 aircraft, shredded the ailerons and elevators, and created such major structural damage to the fuselage and wings that an emergency landing was ordered.

A Viscount airplane flew through a hailstorm over Delhi, India, on May 27, 1959, and on later inspection at the landing field a hole was found measuring 5 inches in diameter, along with several dents measuring 10 by 15 inches.

455. What kinds of hail are there? There are generally three kinds of hail: graupel or soft hail; small hail; and true hail.

456. What is graupel? Pellets of graupel or soft hail are white and opaque, usually round, and about the size of small peas. These tiny snowballs are seldom larger than a quarter inch in diameter. They are crisp and easily break apart as they rebound after striking hard ground. This kind of hail falls mostly inland, before or along with snow, occurring mainly at temperatures above freezing at ground level.

457. What is small hail? Small hail contains more ice in its structure than graupel and is somewhat transparent. This kind usually has a center of soft hail with a very thin layer of ice around the outside which makes it seem glazed. These balls are generally circular, sometimes conical, and they hold their shape even when they hit hard ground. Small hail falls at temperatures above freezing and often accompanies rain.

Sometimes small hail is formed by raindrops falling from a warm layer through a thick layer of cold freezing air. More accurately, this is called a freezing rain.

Both graupel and small hail are not large enough to be very destructive, and both may form in winter without thunderstorms.

458. How does true hail differ from graupel and small hail? True hail can always be identified as different from the other kinds of hail because the stones have typical layered coatings of ice and snow around the nucleus.

459. What shapes are hailstones? Hailstones occur usually in the shape of rounded balls, yet they have been found in other shapes such as oblong, pear, egg, pyramid, or hedgehog with icy spikes as long as an inch sticking out. Some stones occur in the shape of a lens or a disk.

These unusual shapes are caused by varying conditions of moisture, temperature, and air currents molding the hailstones as they are being formed within a thundercloud.

460. How large are hailstones? Many hailstones are reported as large as a golf ball, or if ovoid, as a hen's egg, about 1½ to 2 inches long. Others are reported as large as a baseball, about 3 inches in diameter, and the largest is the size of a small grapefruit, about 4 inches in diameter. Meteorologists report the average hailstone to be about a quarter-inch in diameter.

Even larger sizes have been reported, but these may be conglomerations of several hailstones that partly melted and stuck together as they fell.

461. How can you tell if you have one single huge hailstone or a conglomerate? Proof that a hailstone is a single stone and not a collection of several pellets stuck together can be made simply by cutting it in half and inspecting the cleft. If the pellet consists of concentric opaque and clear layers of snow and ice around a central nucleus, it can be considered a single hailstone. If there are several nuclei, each with their concentric circles, it obviously is a cluster.

462. What determines the size of hailstones? The size to which hailstones can grow seems to depend upon the speed and strength of the updrafts in the thundercloud. If the updrafts are powerful, they can keep the icy pellets suspended in the clouds for a relatively long time, circulating them around inside the cloud until they grow quite large.

If the updrafts are weak, only small hailstones will form, with only one or two concentric layers.

● **463. What was the largest U.S. hailstone found?** The largest hailstone officially recorded in the United States weighed a pound and a half, and measured 5½ inches in diameter, according to a Weather Bureau official. It crashed to earth in a hailstorm of large stones near Potter, Nebraska, on July 6, 1928. When cut open, it had the typical concentric layer formation. In order to validate the size of the stone, J. J. Norcross, who lived in Potter, drew up the following statement and had it signed by a notary public:

To whom it may concern:

I, J. J. Norcross, proprietor of the Potter Drug Co., do affirm the following facts:

That on July 6, 1928, following the hailstorm in Potter, Nebr., I gathered several hailstones and measured and weighed them on standard store scales, and that one stone measured 17 inches in circumference and that it weighed one and a half pounds. . . . [It] was round and hard, with a smooth surface, and upon breaking it open I found it was composed of concentric layers built around one centre. . . .*

464. What other large stones have been found? Hailstones weighing up to 7½ pounds were reported falling in Hyderabad state in India on March 17, 1939. Stones with somewhat similar weights have been reported from other parts of India and from Africa—although

* George H. Kimble, *Our American Weather* (New York: McGraw-Hill Book Company, 1955), p. 109.

scientists say these huge stones may actually be composites of several stones frozen together.

465. Which size stones cause the most damage? Most damage to crops and property is caused by hailstones the size of walnuts or marbles. They are particularly destructive when they fall thick and fast in large showers, driven by powerful gusts of wind.

466. Why don't larger hailstones cause much damage? Hailstones the size of grapefruit do not cause such extensive damage as the smaller sized hail—possibly because fewer of them fall.

467. What odd objects have been found in hailstones? Sometimes small objects such as leaves, twigs, or nuts have been found embedded in hailstones. These pieces of debris evidently had been carried aloft by strong undrafts of a storm and became encased by ice and snow in the process of hailmaking.

Some people have even found larger hailstones with fish, turtles, or frogs still alive inside them. Some think there is enough entrapped air within a hailstone to enable small creatures to continue breathing for a while.

468. What larger objects and men have been caught by hailstones? Ducks and other birds have become so encased with ice from hail that they have fallen to earth, unable to continue flying. If they cannot break through the ice coating, they are unable to move and soon freeze to death.

The 5 German glider pilots that soared into a thunderstorm over the Rhön Mountains (see Question 305) were subjected to low temperatures and became so covered with layers of ice after they had bailed out that 4 froze to death.

469. How fast do hailstones fall? A 3-inch hailstone has been timed as falling at a speed of about 60 miles per hour, at the end of its fall. The rate of fall of a 1-inch hailstone has been theoretically calculated as 40 miles per hour. Stones with diameters of 4 inches or more may fall as fast as 100 miles an hour, while stones with about 5-inch diameters may hit earth at speeds of nearly 140 miles per hour.

470. How much hail can fall in a hailstorm? Although exact amounts of hail have not been determined, researchers estimate that in a single storm, thousands of tons of hailstones can drop to earth.

471. How thick do these stones pile up? People who have never seen a severe hailstorm scarcely believe the enormous amounts of pellets that accumulate in only a few minutes—sometimes a few inches thick, sometimes several feet.

472. What were some record accumulations of hail? In Chad, Africa, on May 9, 1935, hailstones piled 19 inches deep over an area of about one square mile.

Hailstones piled up to a depth of 12 inches on June 23, 1951, in an area in the northern part of El Dorado, Kansas.

There were drifts 6 feet deep after a storm in Washington County, Iowa, on September 1, 1897.

The 6-foot hailstone drifts that accumulated in parts of Iowa in August, 1890, lasted 26 days.

473. How long and wide are hailstone paths? Batches of hailstones tend to fall in relatively small and irregularly shaped paths. On the average, hail paths measure about one or two miles wide, but they do not form a continuous path. Sometimes the stones can fall over an area 10 miles wide, and as many long. Yet some storms have been known to drop hail over a path several hundred miles long—as many as 500 miles. This occurs when a series of thunderstorm cells drop their load of hail in a continuous path as the thundercloud moves across the country.

474. What are hail streets? Hail streets are the narrow paths of hailstorms that exist near mountainous regions. Along these streets, which are channeled by the configuration of the hills, hailstorms form and travel the same paths again and again.

475. Why is hail called freakish? Hailstones often fall in groups or batches that give freakish results. Hail may flatten a field of wheat on one side of a road, while on the other side no hail falls and the crops remain undamaged. A hailstorm may break all the windows in one building and leave all intact in a building next to it.

476. Why do hailstones fall in batches? Hail damage is always spotty throughout an area. This happens because the hailstones are formed within individual thunderstorm cells that make up a large thunderstorm. Each of these cells develops in its own life cycle. At its mature stage it can drop torrents of rain or tons of hail and then dissipate. Another cell of the storm may be just forming or just maturing and dump its load of hail in another location.

The result is that bursts of hail can pelt the earth in several places miles apart, while areas in between remain untouched.

477. What happens after a cloud cell dumps its load of hail? Some storms are known as single hail producers. This means they dump only a single load of hail, then continue as rainstorms. Other storms may have many hail cells.

Each cell in a large many-celled storm may have a load of hail. After a cell drops the load of hail, it continues as a rain cell and never develops hail again. But another cell is growing and may dump another load.

478. How do weak hailstorms suppress their own hail formation? Paradoxically enough, the mechanism of some hailstorms works to suppress the formation of hail within their own storm systems.

When the upper-level winds are weak, they do not blow away the many small ice and snow particles at the top of the thundercloud. These particles fall back into the updraft of the storm cell, where they all compete for the water within the cloud. Since there are so many, they all remain small and melt before they drop to earth.

479. How long does a hailstorm last? Hailstones dropping from the front of a thunderstorm rarely fall for more than 15 minutes. Sometimes one hailstorm follows another in quick succession. The hail is then dumped from successive storm cells, so that it seems as if there is one continuous fall of hailstones.

Some hailstorms have been recorded as lasting half an hour. Others occurring in Kansas, Nebraska, and Montana have been known to last for more than an hour.

480. What was the longest local hailstorm on record? The longest local hailstorm anyone has yet timed occurred at Sheldon, Kansas, on

June 3, 1959, when hailstones pelted down continuously for 85 minutes.

481. Where do most hailstorms occur? Hailstorms occur in many parts of the world, but the most violent and frequent storms generally fall in the warm areas outside the tropics, in the interior of continents, where convection currents on hot days form particularly violent storms. Statistics show that areas between 30 and 50 degrees north latitude are most affected by hail.

482. Where do hailstorms fall in the United States? In the United States, the most heavily hit region is the high-altitude area east of the Rocky Mountains and in the Great Plains where many important crops are grown—"the breadbasket of the nation."

The hail belt includes Wyoming, Colorado, Kansas, and Missouri. The valuable wheat state, Kansas, has more hail damage than any other state.

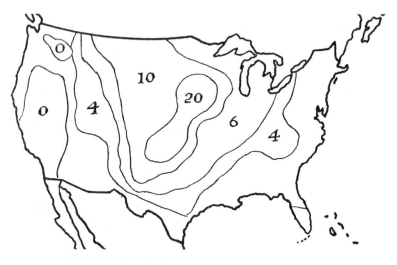

Average number of hailstorms in regions per growing season

483. Why is hail so frequent in central United States? As formerly described (see Question 431), hailstorms form in conjunction with thunderclouds during warm seasons of the year. This situation often occurs over the Great Plains of North America in spring and summer, when warm moist air is blowing up from the Gulf of Mexico, moving beneath masses of cooler, drier air that are blowing eastward from over the Rocky Mountains. Under certain conditions on warm sunny days, convections of warm air penetrate through the overlying cool air masses, and severe thunderstorms and hailstorms are generated.

484. What city is hardest hit by hail? Cheyenne, Wyoming, the most hail-ridden city in America, is lashed by 9 or 10 hailstorms a year. Southeast Wyoming has more hailstorms than any other area of comparable size.

485. Where else in North America do hailstorms occur? Canada receives several hailstorms a year, usually in the interior. Severe storms have caused extensive damage to tobacco crops in southern Ontario and to wheat in the prairie provinces. The storm of June 26, 1933, caused heavy damage to 300,000 acres of crops in eastern Saskatchewan. In 1966, hail damage in Alberta alone was $3 million.

Mexico City averages 5 or 6 hailstorms each year, as do other areas on the Mexican plateau. The lowlands of that country, however, receive little or no hail.

486. Where does hail fall in Europe? Severe hailstorms sometimes occur in England and other parts of Europe. Württemberg, Germany, records about 13 days of hail a year, and Paris gets about 10.

The Mediterranean area receives most hailstorms during the winter months.

487. Where else in the world have severe hailstorms occurred? South Africa receives many severe hailstorms that result in damages costing thousands of dollars. Countries such as Guinea, Ghana, and Liberia along the west African coast have had violent hailstorms.

India reports high death tolls—more than any other nation.

China also has reported severe losses in life and property, and northern Japan is hit by many heavy hailstorms.

488. Do hailstorms always occur with thunderstorms? Most hailstorms occur in conjunction with thunderstorms, but not all thunderstorms contain hail. In the United States, for instance, hail occurs most frequently over the Great Plains, whereas thunderstorms occur most frequently over the southeastern states. Florida has the highest number of thunderstorm days in the nation, yet hail rarely falls there. Along the West Coast, hail falls more frequently than thunderstorms occur—but this is graupel or small hail, not true hail.

489. Why are hailstorms rare in the tropics? Hailstorms form in the tropics, but the hailstones seldom reach the earth's surface—they melt as they fall through the warm atmosphere. These stones form in the cooler layers of air, often several miles high in the towering thunderclouds. By the time they travel the long distance to the earth's surface, through warmer layers of air, they usually melt and fall to earth as drops of rain.

For the same reasons, hailstorms are extremely rare over tropical oceans, coastal regions, or low-lying lands.

490. Why don't hailstorms occur in cold weather or in polar regions? In the colder climates of the world, toward the polar regions, and in the winter months, hailstorms do not occur because the atmosphere is so cold that convection currents seldom are formed to generate thunderstorms and their accompanying hailstorms.

491. At what time of day do hailstorms usually occur? Hail may drop at any time of the day, but most true hailstorms occur in the afternoon and evening, when the updrafts of warm air and the formation of cumulonimbus clouds are greatest.

492. When do most hailstorms appear in the United States? In the United States, 73 percent of all hailstorms are created between 2 P.M. and 9 P.M. They occur most frequently between 4 P.M. and 7 P.M.

493. Does it ever hail in the morning? Only 10 percent of hailstorms occur in the morning, in the hours between midnight and noon.

494. Why does hail sometimes fall from a calm blue sky? Hail has suddenly surprised people by falling from a clear blue sky. Actually this occurs either in front of a thunderstorm or behind one.

Meteorologists believe such hail was formed in a thundercloud but was thrown from the top of the air current rising in the cloud and hurled some distance away by strong upper winds.

495. What time of year do hailstorms occur? True hailstorms occur during the spring and early summer, in conjunction with thunderstorms. Graupel and small hail can form without thunderstorms in the winter seasons.

496. Which is the month of most hail? June is the month when most hailstorms occur in the United States.

497. What is the seasonal path of the U.S. hailstorm zone? In general, the zone of greatest hail frequency tends to "follow the sun" on its apparent northward journey in spring and summer.

Hailstorms begin to arrive in the southern Mississippi Valley in March. The zone extends in an arc from central Texas through southeastern Oklahoma to Arkansas and southern Missouri.

In April, the zone broadens out and lengthens, extending from central Texas to Colorado, southern Illinois, and northern Alabama through Arkansas.

In May, the hail zone stretches from the Rio Grande in Texas, through Idaho and Indiana to Montana and the Canadian border.

498. What is the track of hailstorms after May? More than three quarters of all Kansas storms occur in May, June, and July, when the wheat crop is ripening.

Hailstorms arrive in the Corn Belt around Iowa, Minnesota, and Nebraska in June and July.

In June and July hailstorms reach into Wyoming, Montana, southern Alberta, and Saskatchewan.

By July the zone of greatest hail frequency centers in Cheyenne, Wyoming, and Butte, Montana.

In August, hail is more frequent in Montana than in any other month, just when the last summer wheat crop is maturing.

Hailstorms in the United States are less frequent after August.

499. Who calls hail the White Plague? White Plague is the nickname many United States farmers of the Midwest give hail which has devastated their growing crops with a lethal blanket of white ice.

500. What hailstorm has changed history? One hailstorm has been called responsible for calling off a war and contributing toward the signing of a treaty. In May, 1360, Edward III, King of England, was regrouping his army, which had been repulsed in their invasion of France. On the eighth of May, south of Paris, a violent hailstorm burst over the British invaders and in the resulting pandemonium, hundreds of men and thousands of horses were reported dead. It created so much terror and confusion that the King vowed to the Virgin Mary that he would sign a peace treaty. The Treaty of Brétigny was the result.

501. What other hailstorm helped shape history? Another blast of hailstorms has been credited for changing French history by helping precipitate the French Revolution. A large storm on July 13, 1788, ruined so many crops that the resulting famines contributed to the misery and unrest of the lower classes and helped stir them to revolt against the government.

502. What monument was raised to a hailstorm? A brick monument was built in Barton, England, with these words inscribed:

In memory of the great hailstorm at Barton, July 3, 1883, 10:30 to 11 P.M. Ice 5 inches long, three inches wide—15 tons of glass broken—ice weighed 2½ ounces.

503. What were some myths and superstitions about hail protection? In ancient Greece, men believed hail could be warded off by sacrificing a lamb or chicken to the storm god.

In the Middle Ages, peasants nailed scraps of paper with incantations against the hail god onto tall poles stuck in their fields.

British gardens were believed to be protected from hail when the skin of a seal, hyena, or crocodile was hung out, according to *The Profitable Arte of Gardening* by Thomas Hill.

Italian farmers used to ring silver bells, hang amulets on trees, and sprinkle pieces of burned Yule log over their fields.

Throughout Europe for a while it was standard practice to ring church bells to ward off hail, as well as lightning. Kites and balloons were sometimes flown into clouds in an effort to draw off electricity and hail.

In the late nineteenth century it was believed that loud noises from cannon fire would prevent hail, and canonnades were set up when dark thunderclouds appeared. This practice was common in the Alpine regions of France, Switzerland, Italy, and Austria.

504. What hailstorms are mentioned in the Bible? The Bible mentions hail some 29 times.

One hailstorm was sent by the Lord to aid Joshua as he and the Israelites sought to evict the Canaanites from their promised land (Joshua, Chapter 10, Verse 11):

And it came to pass, as they fled from before Israel, and were in the going down to Beth-horon, that the Lord cast down great stones from heaven upon them unto Azekah, and they died; they were more which died with hailstones than they whom the children of Israel slew with the sword.

505. Is there any protection against hail? Buildings can be constructed of sturdy material to resist damage in hailstorm-prone areas. Metal roofs and concrete slab roofs can be installed on houses in the hail belts, and wire-mesh hailguards can protect windows from breaking during the storms.

As for protection for people, the best method is to run for cover and stay there until the hailstorm is over.

506. Can agricultural crops be protected? Farm crops seem to be completely at the mercy of hail. Until more research can be undertaken to understand the structure of the storm, there is little that can be done scientifically to protect crops from the physical impact of hail.

Until that time, probably in the late 1970's, the only sure financial protection for farmers against the ice pellets is hail insurance.

507. How does crop hail insurance work? Crop hail insurance is the main form of protection against this kind of natural disaster. There are many forms of insurance against hail, some separate and others combined with insurance against other risks.

Crop hail insurance was first organized in the United States in 1880 and today is more than a billion dollar a year business. The Crop-Hail Insurance Actuarial Association of Chicago, Illinois, is an effective group of more than a hundred insurance companies that have pooled their information and analyzed hail risks by areas. The Association is supporting research to find more information about the nature and distribution of hailstorms.

Coverage against hail damage can be bought on any crop of commercial value—alfalfa, tomatoes, cucumbers, watermelons, and other crops. Most insurance is written on wheat, tobacco, cotton, and corn, in that order. All crop-hail policies are written on a per acre basis which means crops are insured for a certain amount per acre and losses are adjusted on that basis.

For most hail insurance, there is a 24-hour waiting period between the time of application and the time of effect. This is to prevent farmers from running to insurance agents when a hailstorm is forecast for the next few hours.

508. Along what lines has hail research progressed? In the United States, the general plan for probing the systems of hail has proceeded along two fronts—theoretical research and field experiments.

Theoretical work has resulted in more comprehensive knowledge of the dynamics and physics of hailstorms. This basic understanding is necessary for any specific program to investigate the storms with practical experiments in the field—the second aspect of hail research.

509. How well does radar indicate hail? Ground radar can sometimes give warning of approaching hail in a storm.

Radar mounted on airplanes can detect definite shafts or edges of hail in clouds for a radius of about 20 miles. The latest technique utilizes a Doppler radar-navigation system so precise that airborne scientists can measure the velocity and direction of the winds as they fly through them.

510. What are anti-hail guns? Around the turn of the century, an Italian grape farmer vented his anger against hail by loading an unused cannon with gunpowder and firing into a storm cloud. Surprisingly enough, that year his vineyard was damaged far less by hail than those of his neighbors. The next year other farmers tried firing cannons and were delighted at the success. The news soon spread into other countries, and soon thousands of anti-hail guns were being designed and built.

Shaped like megaphones, these anti-hail guns were designed to shoot great blank charges of doughnut-shaped hot air up into the clouds, where the air was thought to mingle with the clouds and disperse the hail.

An international hail-shooting congress was held to exchange ideas and experiments. But by the time a second congress was called, there was a marked decline of optimism as people realized that much of the first successes had been a matter of luck, not science. It was pointed out to them, for instance, that the hot air rarely reached the altitude of the clouds, and the guns had been working at times when there naturally was less than the usual amount of hail.

511. What was the electric niagara? Another method devised to combat hailstorms was called the electric niagara. This was a group of high masts or lightning rods set on a hill in order to dissipate the difference of electrical potential between earth and clouds and thus reduce the hailstone formation. Meteorologists have not considered the results from such experiments conclusive.

512. How are hailstorms forecast? By using radar and high-floating radiosonde (see Question 148), meteorologists can detect the presence of hail in thunderclouds and issue warnings with regular weather storm forecasts. The growth and action of storm cells are so complex, however, that it is almost impossible to predict exactly where or when one of the cells will dump its load of hail.

Hailstorms were first forecast in 1950, when military aircraft operated from the Tinker Air Force Base in Oklahoma. Shortly afterward the U.S. Weather Bureau began to forecast hail for civilian fliers and for the general public in certain areas.

Today the Severe Local Storm Warning Center at Kansas City, Missouri, carries out regular hail-warning forecasts stating the areas where hail is likely to fall, the probable size of the stones, and other information.

513. What are some more recent research techniques? One recent and powerful technique being used in hail-research programs is the infrared mapping of the surface of earth where a hailstorm has already dumped its load of hailstones. This is made by an airplane equipped with infrared instrumentation flying low over terrain where a hailstorm has just passed. The exact path of hailstones can be determined by the low-temperature measurements, distinguished from the warmer areas where rain has fallen. These infrared temperature maps can encompass several miles in width.

514. Can hail be controlled or prevented? Scientists have long debated the possibility of controlling or preventing hail by modifying hail-producing clouds. Some believe that of all the disastrous storms—

including hurricanes, tornadoes, and snowstorms—hailstorms seem to be the most likely to be understood and controlled.

Most promising experiments have been in seeding the parent thunderclouds with dry ice, or silver or lead iodide. (See Question 150.) By introducing a large number of artificial nuclei upon which form raindrops or ice crystals, scientists hope to precipitate rain at a stage before hailstones have a chance to form in the rapidly developing clouds. Another approach is to precipitate hailstones while they are still tiny and have not grown into larger hailstones, and hence can melt harmlessly before striking the earth.

515. What are the results of seeding potential hail-bearing clouds?
Results of seeding hail-generating thunderclouds have often been inconclusive. Some tests have been ineffective because once clouds have reached the stage when hailstones are formed, it is too late to prevent them from developing. The one positive result occurs when isolated clouds are seeded while they are still in the process of forming and developing. Seeding at this time may cause clouds to pass quickly through the hail stage before hailstones of any considerable size form. Unfortunately, hailstorms occur within a complex system of many severe local storms in the form of squall lines. They can be so embedded in these frontal systems that it is difficult to separate the hailstorm elements from accompanying thunderheads. Researchers have been conducting many experiments in the lee of the Rocky Mountains, on the high plains of Colorado, where hailstorms are often isolated and well defined.

516. What is the National Hail Modification Project? In 1968 the National Science Foundation, the National Center for Atmospheric Research, the Environmental Science Services Administration, and the Department of Agriculture formulated a 5-year pilot project for the purpose of studying and eventually modifying hail formation.

The current project combines instrumented research aircraft, ground-based radar, and data collection systems.

517. What was Project Hailswath? In the summer of 1966 a team of scientists representing some 23 organizations, in conjunction with the National Science Foundation, conducted a 30-day hailstone survey in western South Dakota. This was the largest study of its kind to date.

In the program, called Project Hailswath, clouds were seeded with silver iodide by various methods and compared with other clouds left unseeded. Less hail fell from the seeded areas, and scientists agreed that under certain conditions hail possibly can be reduced. Since different factors control the weather in different parts of the country, however, it is debatable whether operations effective in one area would be effective in other regions.

It will be several years before further field experiments will provide any firm conclusions of the control of hail.

518. Where are other experiments and research under way in the United States on hail? Hail and hailstorms are being studied at the National Center for Atmospheric Research, in Boulder, Colorado. Here researchers are working particularly with dropsonde and other instruments in an effort to gain more information on hailstorm dynamics and structure.

The Severe Local Storm Warning Center in Kansas City, Missouri, is conducting research and experiments to analyze the physics of hailstorms and obtain better forecasting methods.

The Atmospheric Physics and Chemistry Laboratory, in Boulder, Colorado, is helping operate a mobile laboratory and specially instrumented aircraft for hailstorm studies in that area.

Hail research has also been carried on by companies and organizations such as the Crop-Hail Insurance Actuarial Association, which is helping finance research to determine geographic distribution of hail. This project involves United Air Lines, the Geophysical Research Division of the U.S. Air Force, and the Illinois State Water Survey. The Hail Insurance Adjustment and Research Association works with experimental stations to determine the susceptibility of crops to hail damage.

Other hail research programs are under way at Colorado State University and the South Dakota School of Mines and Technology.

519. What other nations are active in hail research? At the Swiss Federal Snow and Avalanche Research Institute at Davos, Switzerland, a "hail tunnel" reproduces atmospheric conditions and temperatures in which hailstones are formed. This enables researchers to study the process of how hail develops.

In South Africa, the National Building Research Institute carried out research and experiments on controlling hail by bombarding clouds with different materials.

Scientists at the High Altitude Geo-Physical Institute at Makchik, Russia, have been pursuing several aspects of hail research, with reports of considerable success in suppressing hail by seeding thunderstorms. By locating hail centers with radar and firing silver iodide nuclei into these centers with anti-aircraft guns, Russian scientists claim hail damage has been reduced by 65 to 85 percent in the Caucasian area.

Hail research is being conducted by the Alberta Hail Studies Group of the Research Council of Alberta, Canada, and by the Stormy Weather Group at McGill University, Montreal. Several other Canadian research groups are supporting hail projects.

Projects involving hail research are also under way in France, Bavaria, Italy, Kenya, and Argentina.

V. WINDS

Introduction. Created and powered by the sun and by resulting differences in temperatures and pressures, deflected by the rotation of the earth, molded by mountain ranges and expanses of oceans and plains, winds blow constantly across the face of the earth.

Flowing sometimes like a river of water, the winds rise and fall, form waves and ripples, spill into every nook and cranny of the earth, swirling in gentle eddies or forming catastrophic whirlwinds.

Unseen, sometimes unheard, these moving masses of air, chilled or hot, moist or dry, have brought many kinds of disaster. They whip up enormous ocean waves, capsize ships or blow them against rocks, topple buildings, create fire hazards, and even affect men's minds with fear, irrationality, and anger.

Winds include the large global winds that circle the earth at various altitudes, near the surface or in the upper air—all part of a large world-wide system of air circulation. Winds also include local winds sometimes blowing with such persistent regularity that they have been considered as malevolent spirits or evil manifestations rather than meteorological events.

Many large persistent winds are known the world over, such as the trade winds or the prevailing westerlies. Many of the smaller, local winds, such as the chinook or mistral, are so fascinating in their origins, behavior, and impact that people throughout the world have come to hear about them.

Knowledge of the patterns of these winds is far from complete, even with man's complex array of satellites, airplanes, computers, and weather stations that constantly monitor temperatures, pressures, wind speeds, wind directions, and other factors. Even where they are understood, no simplified description or schematic drawing can be given to explain them without the realization that such explanations fall far short of the complex mechanisms that generate and maintain the winds of the world.

520. What are winds? All winds—gentle zephyrs, breaths of frosted air, or violent storms—are packets of air in motion, usually moving in a horizontal direction over the earth's surface. When air moves vertically, it is commonly called an air current or convection current.

As the rays of the sun fall unevenly on various regions of the spinning earth, masses of air begin to move. They flow on a large global scale as air circulates between the warm tropics and the cool poles or between warm earth surfaces and cooler, higher altitudes. Or they flow on a smaller local scale as air moves across deserts and lakes, over the tops of mountains, or down through valleys.

Whenever packets of moving air circulate, they are called winds.

521. What causes winds? Winds are created by many factors—the uneven heating of the earth's surface by the sun, the cooling of the surface as heat is radiated into the atmosphere or absorbed in energy, resulting in differences in atmospheric pressure.

Somewhat as water flows downhill, air flows from areas of high pressure toward those of low pressure. Cold, dense air flows downward; warm, expanding air rises.

In simple terms, when the sun warms a section of the earth, the air above this area becomes heated. The air molecules become agitated and move faster and away from each other. This causes the air to expand and rise. Since there are fewer molecules, the atmospheric pressure is reduced.

Cooler, heavier air that surrounds the warming packet of air flows in horizontally to replace the rising air and, in its turn, is heated and rises.

522. How does the pressure gradient cause winds? As one of many natural processes that equalize inequalities, the movement of air from one place to another tends to decrease differences of temperature and pressure.

To describe the flow of air from areas of high pressure to those of low pressure, meteorologists use the term pressure gradient, which technically is the difference in barometric pressure between the two locations. The force of the wind is directly proportional to the steepness or difference of the gradient; that is, the more difference there is in pressure between the two locations, the harder and faster the winds blow.

Some of the most extreme pressure gradients in the world are found in hurricanes in the tropics when they are beginning to develop. One pressure drop was reported as 1.14 inches in 30 minutes.

Pressure gradients are measured on weather maps as lines drawn at right angles to the isobars, which are lines connecting areas having similar (or equal) atmospheric pressures at given elevations. Steep pressure gradients are shown by close spacings of the isobars.

523. What effect does the Coriolis force have on wind direction? The force of the rotating earth, the Coriolis force (see Question 79), deflects the movement of air in a way that counterbalances the force of the pressure gradient.

The Coriolis force tends to direct an air mass to flow parallel to the isobars rather than directly from high to low pressure. With isobars generally running parallel in a west-east direction, assume that a parcel of air starts to move northward toward a low pressure. At the same time the Coriolis force tends to push the air to the right of its path, so that winds tend to move at an angle across the isobars. The rotating earth continues to exert its force and turn the wind, so that finally the wind is blowing parallel to the isobars and toward the east.

524. Where does the energy for all weather come from? The ultimate source of energy that powers the winds is the sun. In this huge thermonuclear power plant 93 million miles away, hydrogen, carbon, and other elements are burned and converted to helium, nitrogen, and others as waves, and energy is radiated at the rate of 70,000 horsepower for every square yard of the sun's surface.

Of the vast amount of energy the sun radiates into space, only about half of one billionth is intercepted by earth. Yet even this tiny fraction is equivalent to some 126 trillion horsepower every second—more energy per minute than all men use in one year. About 43 percent of the sun's radiation that reaches earth is changed to heat. The rest stays in the atmosphere or is reflected back into space.

525. How is the sun's energy turned into wind energy? As the sun heats various parts of the earth, its energy creates temperature differences that help form wind energy in somewhat the following manner:

Wind energy is created and sustained throughout the atmosphere mainly by the interaction of masses of air at different temperatures—on a global scale between the equator and the poles, and between the upper and lower atmospheres; or in local regions between smaller packets of heated and cold air.

Wind energy is also created by the effects of the heat exchange of condensation—as warm moist air cools and condenses, heat, which is a form of energy, is released into the atmosphere.

The exact manner in which temperature differences supply energy for the winds is quite complex, scientists have found, and the phenomenon is not yet fully understood.

526. What disasters have winds caused? Every creature on earth has been affected by winds that sweep across the earth, bringing storms that smash with powerful energy. Winds have routed armies with bullets of sand and dust, uprooted shelters and buildings, smashed ships with wind-whipped waves, and twisted aircraft with treacherous turbulences.

Winds have long been destroyers of agricultural crops, flattening fields of grain or corn, scooping up vast amounts of topsoil and eroding the land. Stiff winds have perpetuated destructive fires raging through forests, fields, and cities. Today's winds carry new kinds of disaster in the form of nuclear fallout, from bomb sites across whole nations and continents.

527. What storm prevented an invasion of Greece? Many a battle has been lost because unexpected destructive storms blinded soldiers, panicked horses, or mired vehicles. Storms have sunk countless numbers of ships, pulled planes from the sky, or caused travelers to lose their way.

In the fifth century B.C., Xerxes the Great, King of Persia, prepared to invade Greece from the sea. On the dawn of invasion day, a violent storm came up from the east and destroyed more than 400 Persian ships, drowned countless military men, and turned the remaining invaders away.

528. How did a storm help defeat the Spanish Armada? In August, 1588, a succession of violent gales in the eastern Atlantic Ocean helped shatter the "invincible" Spanish Armada sent to invade England. The Spanish fleet had already been severely damaged by the smaller, more maneuverable British ships, and the remaining Spanish warships were sailing home the long way—up the eastern coast of England, over the top of Scotland, and down the west side of Ireland. The fleeing ships, already damaged and many without anchors, were battered by the storm off the coasts of Scotland and Ireland to such a degree that many more men and ships were destroyed or shipwrecked. Spain's power never recovered from the blow.

529. When did a storm sink a British armada? Little is known about the sinking of the great British armada creeping up on French-held Quebec in Canada in August, 1711, at a time when England and France were not yet at war.

Queen Anne sent Admiral Hovendon Walker to capture Quebec with a fleet of 61 ships, including warships and transports. The British were so confident of taking Quebec that hundreds of women and children accompanied the 9,385 men—families preparing to settle in the captured city. Winter was coming on as the ships sailed northward from Boston, and as the fleet navigated the Gulf of Saint Lawrence, a stiff breeze pushed the ships through the Gaspe Passage and toward the shores of Labrador. Weather conditions became worse, and at 10 P.M. in foggy rough weather on August 22, 1711, ships began to smash into Egg Island. The men-of-war escaped, but 8 huge transports were wrecked with over 1,300 men and an un-counted number of women and children. All these shipwrecked people froze and starved in the land, for although the remaining fleet searched for survivors in the waters, no shore search was ordered. The French learned of the catastrophe 2 months later, when remains of the wrecked ships were accidentally discovered. More than $70,000 worth of materials were salvaged from the wrecks, and more than 2,000 bodies were found strewn over the shores. Other English people died of cold throughout the barren countryside as they tried to reach help.

530. What was the Crimean Storm? On November 14, 1854, a heavy storm, later called the Crimean Storm, moved across the Black Sea and sank many ships lying at Balaklava, destroying vital food supplies destined for allied armies at Sevastopol during the Crimean War. Many soldiers consequently starved during the following winter months.

The French astronomer Urbain Jean Joseph Leverrier was so affected by the catastrophe that he made extensive studies to track the storm's course from west to east across Europe. He submitted his records to Emperor Napoleon III and urged an international system of telegraphic reports for the purpose of issuing warnings of such storms. As a result, the international system of warnings was established in 1855 and has served as a model for setting up other storm-warning and weather-forecasting systems.

531. What was Defoe's storm? One of the most violent storms to strike England was the great storm of 1703, sometimes called Defoe's storm because it is said to have inspired the British author Daniel Defoe to write *Robinson Crusoe*. Reports of this storm state that hundreds of houses and thousands of trees were blown down and that 123 people on land and 8,000 men at sea were killed.

532. What was England's "outrageous wind" of 1095? One or 2 really great storms blow across England every century. One bad storm was given the name the "outrageous wind" in 1095 when it knocked down some 600 houses in London and blew the roof of Bow Church so high that when the timbers fell back to earth, they were said to have rammed some 23 feet into the ground.

533. What was The Wind of the Century? The British called the 125-mile-an-hour wind that blew out of the Irish Sea the second week of January, 1968, The Wind of the Century. Blizzards and winds of hurricane force struck all the way from England to Iran, leaving more than 20 people dead and hundreds injured. Scotland was hardest hit, with at least 16 people dead. Houses were toppled in Glasgow and Edinburgh, and hundreds of people were injured or left homeless.

In England a gust of 134 miles an hour was recorded at Great Dun Fell—the highest wind speed yet recorded in England or Wales. Highly destructive winds swept across Denmark, Germany, and Switzerland and howled over the Middle East and Iran. Heavy snows hit Jerusalem, causing a power failure. For the first time, snow was reported in Beersheba in the Negev Desert.

534. What is England's depressing East Wind? Sometimes in the winter months, a strong circulating wind system called an anticyclone (see Question 574) builds up over Scandinavia, driving a cold persistent wind from the east down across England, bringing depression and moodiness to the British people. This wind has been recognized as evil for centuries. An old English proverb sums up the thoughts of many British:

> When the wind is in the East
> Tis neither good for man nor beast.

535. What did Voltaire write about the East Wind? When the French author Voltaire voyaged to England, he heard reports about England's depressing East Wind that caused many cases of suicide. In one of his letters he described the gloomy situation:

A famous court physician, to whom I confessed my surprise, told me that I was wrong to be astonished, that I should see many things in November and March, that then dozens of people hanged themselves, that nearly everybody was ill in those two seasons and that black melancholy spread over the whole nation, for it was then that the East Wind blew most constantly. [The court physician said] "This wind is the ruin of our island. Even the animals suffer from it and have a dejected air. Men who are strong enough to preserve their health in this accursed wind at least lose their good humor. Everyone at that time wears a grim expression and is inclined to make desperate decisions. It was literally in an East Wind that Charles I was beheaded and James II deposed."*

536. What island was discovered by a storm? In 1609, Sir George Somers, in charge of the British ship *Sea Venture* with colonists bound to settle in Virginia, was shipwrecked by a storm on the tiny mid-Atlantic island of Bermuda. He and his staff built two ships from pieces of the wrecked *Sea Venture* and sailed on to Virginia. Bermuda was included in the charter of the Virginia Company. Afterwards it was settled as a British possession.

Sir George Somers' account of the shipwreck is said to have formed the basis of William Shakespeare's *The Tempest.*

537. What is the Graveyard of the Atlantic? Seafarers used to call the treacherous stretch of coast along North Carolina's outer banks the Graveyard of the Atlantic. Here for centuries the winds and currents drove thousands of schooners, sloops, brigantines, and barks against the many shoals, bights, capes, inlets, shifting sand bars, and beaches of that area. Stretching farthest east into the Atlantic Ocean is Cape Hatteras, against which crash sea currents and shifting winds that trapped many ships and dropped them along with shells, sand,

---

* Slater Brown, *World of the Wind* (Indianapolis: The Bobbs-Merrill Company, Inc., 1961), p. 107.

and sea life onto the dreaded shifting sand bar called Diamond Shoals. To the north lie other shoals, such as Wimble Shoals and Gull Shoal, and other beaches, such as Whales Head, Nags Head, Kill Devil Hills, and Chicamacomico—each with a long list of ships wrecked and sunk, men drowned, and treasures lost. To the south the treachery continues, with names such as Core Bank, Ocracoke, Cape Lookout, and Cape Fear.

Cargo ships and blockade runners used to sail along the outer banks laden with supplies. Larger ships sailing from the Old World to the New by way of the tropical trade winds made use of the Gulf Stream to travel north—as did the Spanish galleons laden with gold, and schooners carrying sugar, coffee, salt, and tobacco.

Many kinds of people—settlers, traders, pirates, adventurers—were shipwrecked along the Graveyard of the Atlantic, and forced to settle along the banks and inlets, where they lived by scavaging shipwrecks of other unfortunate souls.

538. What was it like to be shipwrecked along the Atlantic? Early accounts of the innumerable shipwrecks along the outer banks of North Carolina are fragmentary, for little record was made of the wrecks, or of the few survivors. After the War of 1812, newspapers began listing accounts of the disasters, along with eyewitness descriptions or letters.

On January 15, 1820, the Norfolk *Beacon* and Portsmouth *Advertiser* printed the letter of Captain Hand, sole survivor of the sloop *Henry* sailing from New York to Charleston.

I have a melancholy affair to relate. I am the only one living of the crew and passengers of the sloop Henry. . . .

Saturday morning made Cape Lookout lighthouse, hove about and stood off, wind canting in from the southeast, and the gale and sea increasing so fast that we were obliged to heave to.

Lay to until 5 o'clock P.M., then began to shoal water fast, and blowing, instead of a gale, a perfect hurricane. We set the head of the foresail to try to get offshore, but to no use, it blowing away in an instant; likewise the jib. We then lay to under the balance of the mainsail until we got in 10 and 9 fathoms water, when the sea began to break and board us, which knocked us on our beam ends, carried away our quarter, and swept the deck. She righted, and in about five minutes capsized again, which took off our mainsail. We were then left to the mercy of the wind and waves, which were continually raking us fore and aft. With much exertion we got her before the wind and sea, and in a few minutes after run her ashore on the south beach of Ocracoke Bar, four miles from land.

She struck about 10 o'clock at night bilged in a few minutes, and got on her beam ends, every sea making a fair breach over her. At 12 o'clock her deck blew up and washed away altogether, and broke in two near the hatchway. The bow part turned bottom up, the stern part righted. Mr. Kinley (passenger) and Wm. Bartlett (seaman) washed off. The remainder of us got on the taffil rail, and that all under water. About 2 o'clock A.M., Mr. Campbell (the other passenger) and Wm. Shoemaker (cook) expired and dropped from the wreck. About 4 o'clock, Jesse Hand (seaman) became so chilled that he washed off. At daylight, Mr. Hawley (mate) died, and fell from along side of me into his watery grave, which I expected every moment would be my lot. But thro' the tender mercy of God, I survived. . . .*

539. What storms and inclement sailing weather kept a large continent from being discovered? During the era of the great explorers and navigators of the fifteenth, sixteenth, and seventeenth centuries, the large continent of Australia remained undiscovered, shielded partly by barriers of storms and unknown winds. Adventurous navigators of Portugal, England, Spain, and Holland would sail from east to west or west to east, yet always skirting northward of the southerly massive land. In the sixteenth century voyagers came close, yet it is debatable if any of these early explorers actually saw the continent. Probably the first European who may have reached the coast of Australia was William Jansz who in 1605–1606 landed on the northern coast of Queensland. Reports of desolate coasts and wild savages kept people away. Finally the Dutch navigator Abel Janszoon Tasman, determined to explore the region and seek a southern route along the westerly wind belt, deliberately sailed around the continent in 1642–1643.

540. What are *kamikaze* winds? The Japanese word *kamikaze*, literally meaning the divine or god winds, was given to the violent

* David Stick, *Graveyard of the Atlantic* (Chapel Hill: The University of North Carolina Press, 1952), pp. 9, 10.

storms, possibly typhoons, that wrecked fleets of Mongol invaders approaching Japan in the late thirteenth century.

Near the end of World War II, the Allied powers encountered Japanese suicide bomber pilots called *kamikazes* who deliberately killed themselves by crashing their explosive-laden planes directly onto Allied ships and other targets. They believed that they, like the thirteenth-century *kamikazes*, could stop the attackers.

541. How are wind speeds classified? Wind speeds, ranging from less than a mile an hour to 74 or more miles an hour, have long been classified in a system called the Beaufort wind scale.

Invented in 1805 by Sir Francis Beaufort, an Admiral in the British Royal Navy, the scale was originally designed to help mariners in handling ships and setting sails. It was set up originally as a description of wind effects on sails, of sounds of wind in the rigging, or of noises of the sea. Gradually sail descriptions have been abandoned, and through the years many modifications and additions have been made. As meteorologists become more precise in their analysis of the complexities of wind, the whole Beaufort wind scale will probably become obsolete.

542. What is the Beaufort scale?

BEAUFORT SCALE OF WIND FORCE

Beaufort	Miles per Hour	Wind Effects Observed on Land	Terms Used in U.S. Weather Bureau Forecasts
0	Less than 1	Calm, smoke rises vertically	Light
1	1–3	Direction of wind shown by smoke drift; but not by wind vanes	Light
2	4–7	Wind felt on face; leaves rustle; ordinary vane moved by wind	Light
3	8–12	Leaves and small twigs in constant motion; wind extends light flag	Gentle

4	13–18	Raises dust, loose paper; small branches are moved	Moderate
5	19–24	Small trees in leaf begin to sway; crested wavelets form on inland waters	Fresh
6	25–31	Large branches in motion; whistling heard in telegraph wires; umbrellas used with difficulty	Strong
7	32–38	Whole trees in motion; inconvenience felt in walking against wind	
8	39–46	Breaks twigs off trees; generally impedes progress	Gale
9	47–54	Slight structural damage occurs (chimney pots, slates removed)	
10	55–63	Seldom experienced inland; trees uprooted; considerable structural damage occurs	Storm
11	64–73	Very rarely experienced; accompanied by widespread damage	
12	74 or more	Very rarely experienced; accompanied by widespread damage	Hurricane

543. What is the "law of storms"? The "law of storms" was enunciated by Buys Ballot, professor of physics at the University of Utrecht in Holland in the middle of the nineteenth century. It states that an observer in the Northern Hemisphere, standing with his back to the wind, will find the storm center on his left. Since the winds blow slightly inward, as well as around the storm, the center will be slightly forward of a position directly to the observer's left.

The observer will experience stronger winds on the side of the storm where the winds are blowing in the same direction as the storm is moving. Winds are weaker on the other side where the winds and the storm's forward movement are opposite.

544. What kinds of winds are there? The extremely complex and varied winds can be grouped into 3 basic categories: 1) generally persistent winds, all part of the great global air circulation blowing around the earth at various altitudes (see Questions 545 through 573); 2) irregular, episodic winds such as cyclones and anticyclones (see Questions 574 through 579); and 3) local persistent winds (see Questions 580 through 633).

Powerful individual mavericks, such as hurricanes and tornadoes, have been described in separate chapters.

545. What was an old concept of global air circulation? For many years, meteorologists described the global air circulation in terms of huge vertical air movements driven primarily by the sun's heat—a concept that developed logically from the then known patterns of surface winds, charted for centuries by mariners and explorers.

Primary air movements, it was believed, included the rising of vast masses of warmed air over the equator, spreading out aloft toward the two poles, cooling and sinking over the middle latitudes and the poles where they tended to recirculate across the earth's surface toward the equator to replace the warm rising air. These thermally powered, rising and falling air masses were considered basic factors in forming the large-scale wind belts and areas of low and high pressures such as the doldrums, the trade winds, the prevailing westerlies, and the polar easterlies.

Secondary movements of air were described as more localized wind patterns that included movements of air across mountains, plains, shores of oceans and large lakes, and other geographic regions.

This older concept of world air circulation, long held by many scientists, has now become antiquated and in many respects untenable with new and detailed information collected during and since World War II.

546. What is the updated concept of global air circulation? With today's increasing array of weather stations and instruments, computers and satellites, meteorologists have been revising the basic thermal concept of the world air circulation into one of vastly more complex patterns, driven by thermal and non-thermal forces. With new data indicating various differences in speed, direction, and behavior of winds at various altitudes, meteorologists now agree that the movements of air in the upper atmosphere are quite different from those near the earth's surface.

Meteorologists still agree that the general circulation of air results from the fact that more solar energy is received at low latitudes near the equator, and less in high latitudes near the poles. This produces a basic equator-to-pole motion as warm air rises and flows toward the poles. The earth's rotation deflects this flow so that air currents near the surface spin toward the west or east, depending on the hemisphere in which they originate.

Meteorologists now believe that the general pattern of air circulation is based more on interchanges of polar and tropical air moving in horizontal planes than on thermal-induced vertical convections, as long believed. Non-thermal factors such as large, episodic, horizontal-moving winds further mix the air systems, and large bodies of water, land, and mountain ranges complicate the wind patterns even more. Another significant force in driving the winds are areas of permanent high and low pressure cells, called centers of action. (See Question 550.)

These newer concepts of air patterns are not yet fully defined mainly because precise data is missing for large areas of the world, such as Asia and the Southern Hemisphere. With increasingly efficient instrumentation supplying more data, scientists hope to develop a fuller concept.

547. What winds blow in the upper atmosphere? At increasing altitudes from the earth's surface, winds become freer from the earth's surface friction and change radically as to their speeds, directions, and moisture contents. They are less affected by mountains, valleys, plains, sea coasts, and other topographical features. They begin to blow faster, more directly, and in larger streams than at the surface.

The prevailing motion of upper winds at levels of 40,000 feet is from west to east, as the rotation of the spinning earth directs air rising from the heated equator. At this altitude, westerlies cover most of each hemisphere. Two huge deep whirls of air, the circumpolar vortexes, exist over the polar regions, spinning mostly from east to west in the lower regions, but from west to east in the upper atmosphere.

The speeds of the westerly winds increase at decreasing latitudes from the poles toward the equator, reaching a maximum speed at about 30 degrees latitude. The highest-speed winds of the upper air form a relatively concentrated river of air known as the jet stream. It is interesting to note that below this zone of high-speed winds lies one of the calmest wind belts on the earth's surface, the horse latitudes.

In lower latitudes, toward the equator, speeds of the upper winds decline rapidly to a region where weak easterly winds are generally encountered.

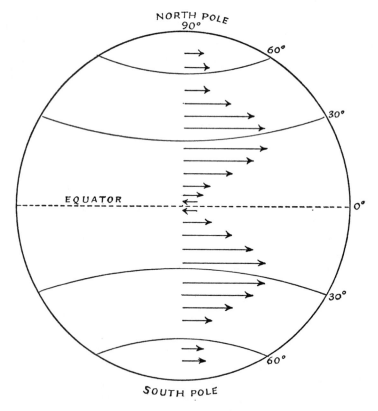

Direction of upper atmosphere winds.
Length of arrows indicate relative speeds of winds.

548. What are upper air waves? In the upper atmosphere, winds within the prevailing westerlies move in undulating, wave-like paths, creating waves of air that are larger and have longer wavelengths than any other atmosphere waves. The smallest wave observed has a length of about 4,000 miles and an amplitude (height from crest to trough) of about 550 miles. Scientists believe these waves form as a result of frictional effects as air blows over the earth's surface, so that waves are oriented in respect to the coastlines and mountain ranges.

In winter, for instance, 3 long standing waves occur at about 40 degrees latitude in the Northern Hemisphere. These waves are about 6,000 miles long—one third of the planet's circumference at that region.

Called standing waves because they generally remain in a relatively fixed position over certain regions, these waves vary in strength and size in different seasons. In summer their average speed is about half that in winter. They are much more pronounced in the Northern Hemisphere than in the Southern.

549. What are troughs and ridges? Meteorologists are finding out more about undulations of upper air waves in the prevailing westerly winds. Just as a flowing river sometimes forms series of dips and rises of water, the westerlies form series of troughs and ridges of wind. Some of these atmospheric waves swell and dip over certain areas to such a degree that they become relatively constant, and the horizontally moving winds become vertical high- or low-pressure cells.

A trough is an elongated area of low pressure, sometimes occurring along a line called a trough line. Many cyclones are generated in these areas.

A ridge is an elongated area of high pressure.

The troughs and ridges of these waves create convergences and divergences that may develop into relatively permanent low-pressure and high-pressure cells or centers of action at the earth's surface.

550. What are permanent centers of action? Within the general wind systems of the world, certain low-pressure and high-pressure cells or centers remain more or less constant over certain regions. Meteorologists believe these are maintained as part of the currents of wind blowing over mountains, continents, and ocean shore lines.

In general, regions of low-pressure cells are moist, for here the air is rising and cooling and dropping its moisture. Regions of high-pressure cells tend to be dry, for the air is descending, warming up, and picking up moisture.

551. Where are permanent pressure centers located? At all times of the year, 5 permanent high-pressure cells or centers remain above all sub-tropical oceans, at about 30 degrees latitude—3 in the Southern Hemisphere and 2 in the Northern.

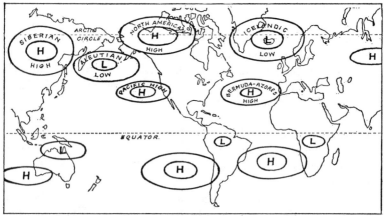

JANUARY (WINTER IN NORTHERN HEMISPHERE)   HIGH PRESSURE **H**...LOW PRESSURE **L**

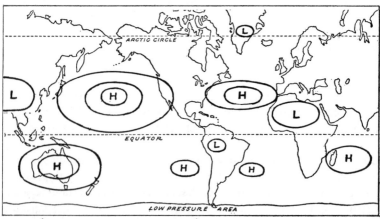

JULY (SUMMER IN NORTHERN HEMISPHERE)

Permanent pressure centers

In winter these marine highs move toward the equator. At the same time of year, high-pressure cells grow taller and wider over certain land masses, especially Asia and North America, and Australia in the Southern Hemisphere. These persistent highs exert powerful effects upon the weather in adjacent areas.

Centers of low pressure exist in the Southern Hemisphere as a belt circling the Antarctic continent. In the Northern Hemisphere, a low-pressure belt exists in 2 main pieces—the Aleutian low and the Icelandic low. These grow stronger in the winter.

552. What are jet streams? Jet streams are ever-present, relatively narrow, streams of high-speed winds undulating around the Northern and Southern Hemispheres. The streams rarely exist as a single air stream encircling the earth. More frequently they are found as segments, some of which range from 1,000 to 3,000 miles in length, about 50 to 400 miles wide, and 3,000 to 7,000 feet thick. There may be two or more such jet streams in existence at the same time.

Within these streams, winds travel at different rates of speed, from some 50 miles per hour at the outer edges of the stream, to some 250 miles per hour at the center. Speeds as high as 300 miles an hour have been reported.

These jet streams were relatively ignored until World War II when B-29 superfortresses flying at speeds of nearly 200 miles per hour were held almost to a standstill when they encountered the jet stream. The winds actually had been measured before by balloons sent aloft at various weather stations, but these observations were generally mistrusted.

553. Where do jet streams occur? Jet streams are usually found embedded in the upper westerly winds at heights of 10,000 to 40,000 feet. They encircle the earth in both hemispheres, usually located in temperate zones, in regions where there are differences in temperatures. The stream usually separates the cold air masses on the poleward side from the warm air masses on the side toward the equator.

Their courses swerve in sinuous paths from north to south and from high to low altitudes. They generally shift farther toward the poles in summer, to about 45 degrees latitude. In winter, they shift toward the equator, to about latitude 25, moving to higher altitudes and increasing the speeds of the winds.

554. What effects do jet streams have on surface weather? Meteorologists believe that the high-altitude jet stream has a close relationship with the low-altitude frontal zone separating tropical air masses from polar air masses near the surface of the earth—a zone intimately associated with the development of storms. Some meteorologists believe the eddies developed by these streams descend to lower altitudes where they may help form violent storms such as tornadoes, direct the path of hurricanes, or create abnormal changes in the weather.

555. What sometimes happens as jet streams move north and south? As the undulating waves of the jet streams move in a north-south direction, they sometimes bring unusual weather to the earth's surface.

Bordered on the north by cold air and on the south by warm air, the jet stream of the Northern Hemisphere sometimes begins bending in waves farther and farther south. In a series of deepening oscillations, called an index cycle, the wind stream forms larger and larger loops that finally break off, leaving cells of cold air in the south and warm air in the north. When this happens, for instance, Alaska may have warmer temperatures than Florida. The cycle may last from 4 to 6 weeks.

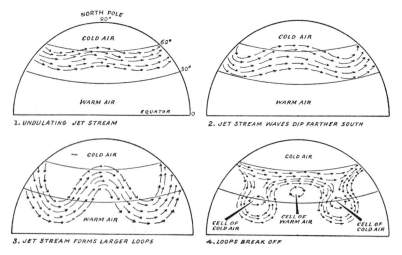

1. UNDULATING JET STREAM

2. JET STREAM WAVES DIP FARTHER SOUTH

3. JET STREAM FORMS LARGER LOOPS

4. LOOPS BREAK OFF

Jet streams undulating north and south

556. What is wind turbulence? Turbulence is an irregular, often invisible, but definite whirl or eddy of air. It is caused mainly by air moving vertically in convective currents, by air moving around or over obstacles such as mountains, or by wind shear. (See Question 560.)

To an aircraft, turbulence ranges from annoying bumps to severe jolts that can injure people and create structural damage to the plane.

557. What are turbulent convective currents? Convective air currents, also called thermal turbulences, are rising and falling air movements, developed as air is heated and rises over such topographical earth features as sandy or rocky beaches and deserts, newly plowed fields, or bodies of warm oceans—or is cooled and sinks over cold lakes and waters, green forests or pastures.

For every rising current there is a compensating downward current. Aircraft flying over the patch-work quilt of terrain often experience convective currents as severe jolts or sickening bumps.

558. What are mountain waves? Another invisible and treacherous disturbance is the mountain wave, a gigantic air ripple in the atmosphere formed in the lee of certain mountains as the wind blows across them. Some meteorologists believe that the wind, pushed vertically up by the mountain, has a strong opposing and restoring buoyancy thrust that shoots the wind down past the initial level where it started to climb up the mountain. This thrust is helped by the downward momentum of gravity. Downward winds over these mountains may blow at 50 miles an hour, with gusts up to 125 miles an hour.

Mountain waves, sometimes called standing mountain waves, are much like waves of water that form downstream from a rock in a rapidly moving river.

Sometimes these mountain waves create visible clouds called stationary lenticular or lens-shaped clouds. Each cloud traces out the crest of a wave. In reality, these graceful clouds hovering motionless downwind from mountain ranges are being formed and reformed rapidly as the wind blows right through them. The wind, cooling by ascent, turns cloudy upon entering the wave, and then, heating by descent, turns clear upon leaving the wave.

559. Why are these mountain waves dangerous? Mountain waves can severely damage aircraft with their strong, often invisible drafts and turbulence. Airplanes can become caught in the downdrafts and crash against the mountain or ground, or suffer extreme damage from the turmoil.

For pilots of glider planes who understand the mechanics of mountain waves, these air currents offer much challenge. All world glider records have been established with these winds, which have shot gliders to heights well beyond 40,000 feet. Sailplane pilots have learned to ride the mountain wave parallel to the mountain, much as a surfboarder learns to ride the ocean wave breaking along shore.

560. What is wind shear? Wind shear is the change in the speed or direction of wind over a relatively short distance, resulting in a tearing or shearing effect.

Wind shear occurs in either a horizontal or vertical direction and produces churning eddies and turbulence. Meteorologists say a wind shear is steep if the differences between wind speeds are great. A wind shear can occur with winds moving in the same direction, but at different speeds. A shear line is the boundary between the winds of different speeds.

561. What is CAT? Several thousand feet high in the atmosphere, invisible and treacherous disturbances occur that scientists do not yet fully understand. This is clear air turbulence, called CAT by aviators and meteorologists. CAT is the acronym from the first letter of each word.

Scientists believe that CAT is caused by temperature instability and wind shear—a turmoil of winds moving at different speeds or in different directions, setting up whirls and eddies of air. These disturbances occur predominantly in two regions—in the jet streams, or in mountain waves. (See Questions 552 and 558.)

562. What disasters does CAT bring? As yet, CAT cannot be detected nor foreseen by pilots flying high altitude planes. At some 40,000 feet in a clear blue sky, a plane will suddenly, without warning, hit an air disturbance that puts enormous strain and pressure on the wings and other structures. Planes have been known to flip over or drop several thousand feet in a few seconds. Plane seats have been ripped from their bases, rivets popped from their sockets, and one plane had an engine twisted off. Pilots have much difficulty controlling their planes, and passengers have been severely hurt by being flung around the cabin.

The sudden crash of a Lockheed Electra in 1960 near Tell City, Indiana, taking 63 lives, has been blamed on CAT. Other unexplained airplane crashes have been attributed to this phenomenon.

563. How are researchers trying to track CAT? Scientists are attempting to track clear air turbulence with instruments recording changes in temperature or pressure, as well as the presence of ozone or static electricity. Experiments at Wallops Island, Virginia, for instance, indicate that the turbulence could be detected by microwave radar equipment in time to warn airline pilots to avoid it. These experiments have not been verified, however.

564. What are the predominant surface winds? Meteorologists describe 3 great basic zonal wind systems moving across the earth's surface, generated as air flows from centers of high pressure toward low pressure and affected by the Coriolis force:
1) the easterly winds near the equator, often called trade winds.
2) the middle latitude westerlies, a region of storms
3) the polar easterlies—regions where scientists have not yet accumulated enough data to define accurately pressures and wind directions.

Prevailing winds are named by the direction from which they come. Hence a wind from the south is called a south wind, winds from the west are called westerlies, and winds from the east are called easterlies.

Many of these surface winds have affected man and history, bringing some changes for the good and also catastrophic disasters.

565. What was one of the most important early charts of winds and ocean currents? One of the first great navigational aids to ancient sailors and navigators was *The Physical Geography of the Sea*, published in 1855 by Matthew Fontaine Maury, a U.S. Navy officer and hydrographer. This book was compiled from thousands of meticulous wind and ocean current entries recorded from countless ships' logs. It included complete charts of ocean winds and proved an invaluable aid to navigators who found they could cut sailing time of long voyages by as much as 25 percent. Similar charts of average winds and currents are used by mariners today.

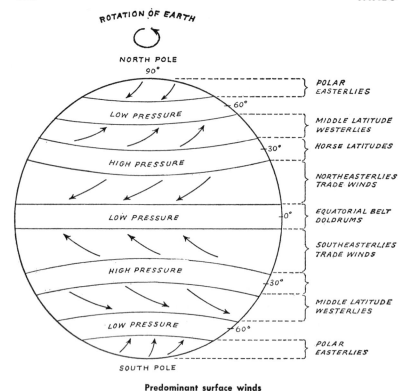

Predominant surface winds

566. Do winds blow the same in the Northern and Southern Hemispheres? The circulation of prevailing winds in the two hemispheres at the surface of the earth is somewhat like a mirror image. The same surface wind patterns exist essentially in both hemispheres but are reversed. Each has its general counterpart in the opposite side of the world.

567. What are the trade winds? Out of the strong, permanent high-pressure marine centers of the world, air flows outward toward regions of low pressure near the equator.

The rotation of the earth gives this flowing air a twist with the result that at low latitudes, roughly from equator regions to 30 degrees latitude, the winds blow more or less steadily from a northeastern direction in the Northern Hemisphere and from the southeast in the Southern Hemisphere.

These are called the tropical easterlies, more familiarly known as the trade winds. Trade is an ancient German and English word for track or path. To blow trade means to blow steadily along the same path.

Considered the steadiest, most persistent and dependable winds at the earth's surface, these winds helped blow explorers and traders from Europe to the New World.

568. What are the doldrums? As the easterly trade winds from the Northern Hemisphere meet and converge near the equator with those from the Southern Hemisphere, an area of variable winds and calm prevails, with low barometric pressure and little horizontal movement of air. Here warmed moist air rises almost vertically into the sky, occasionally causing torrents of rain.

These areas, called doldrums or equatorial calms, are found around the equator. They do not form a continuous belt, as once believed.

In days of sailing ships, sailors dreaded these areas, for ships would sometimes lie becalmed for days on end, food supplies spoiled or ran low, wine fermented, and supplies of drinking water ran low. The frustration of being caught in such a place is described in Samuel Taylor Coleridge's *The Rime of the Ancient Mariner*:

> Day after day, day after day,
> We stuck, nor breath nor motion;
> As idle as a painted ship
> Upon a painted ocean.

569. What are the horse latitudes? Another area dreaded by early sailors was known as the horse latitudes—an intermediate area lying between the trade winds and the westerlies about 30 degrees north and south of the equator. Here in an area of high pressures, winds are relatively calm, weak, and not dependable. Skies are generally cloudless since the air becomes warmed adiabatically by sinking. (See Question 324.)

In the Northern Hemisphere, particularly in the North Atlantic Ocean near Bermuda, this region was given the name horse latitudes, for sailing ships carrying horses from Spain to the New World often ran short of water here. The animals were the first to be rationed supplies, and they often died of thirst or were tossed overboard alive in order to conserve water for the men. Sailors and explorers of the sixteenth, seventeenth, and eighteenth centuries sometimes reported that seas were strewn with bodies of horses.

570. What are prevailing westerlies? The prevailing westerlies on the earth's surface extend in a zone from about 30 to 60 degrees latitude in the Northern and Southern Hemispheres.

These wind belts are the most stormy and boisterous of all such belts and contain the world's most variable surface winds. Strongest winds blow at about 40 to 60 degrees latitude.

Within this belt of westerlies and along the polar front (see Question 573) are formed the turbulent whirling winds of low-pressure and high-pressure that move in endless procession from west toward east, creating many storms.

571. Where are the fastest blowing persistent winds? The westerlies are some of the fastest blowing persistent winds on all oceans, reaching speeds up to 70 miles an hour and more in winter and generating huge waves some 50 to 60 feet high that have battered ships and ripped holes in ocean liners.

These winds are particularly violent in the Southern Hemisphere, where there is less land to generate friction and hence slow down the winds. These winds sometimes blow hard and fast across thousands of miles of ocean.

Whalers, traders, and explorers of the eighteenth and nineteenth centuries named these powerful winds blowing around latitudes of 40, 50, and 60 degrees south the Roaring Forties, the Howling Fifties, and the Shrieking Sixties. These winds are particularly vicious around Cape Horn at the tip of South America, where gale after gale blows at high speeds across long stretches of uninterrupted ocean, bringing raw chilly weather and mountainous waves.

572. What are the polar easterlies? Polar easterlies are generated from regions of intense cold near the poles. Here masses of cold air flow out across the earth's surface and spread toward the equator, but are deflected by the Coriolis force toward the west. Since they tend to blow from the east at the surface, they are called the polar easterlies.

Meteorologists are finding these surface winds are quite indefinite, and shallow, less than 2 miles thick. Higher in the atmosphere, the winds change direction, and form the circumpolar vortex of westerlies of the upper atmosphere.

573. Where do many storms originate? Most of the northern temperate zone's stormy weather originates along the region called the polar front, where two global wind belts, the easterlies and the westerlies, meet and collide. The interaction of these two wind belts constantly creates a condition of atmospheric instability and disturbances.

From this polar front, great eddies of air and whirling winds form sporadically, then move off as isolated masses of rotating air within the general wind circulation. These wind systems may grow and develop, expand and retract, and finally die out. They are episodic storm systems known to meteorologists as anticyclones and cyclones. (See Questions 574 through 579.)

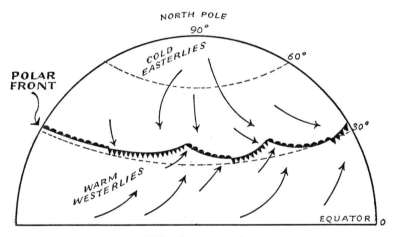

Origin of stormy weather

574. What are anticyclones? Anticyclones are wind systems in which subsiding air settles downward and spins outward from a high-pressure center, clockwise in the Northern Hemisphere and counterclockwise in the Southern Hemisphere. The subsiding air heats adiabatically (see Question 324), tending to dissipate moisture and clouds and, hence, usually bringing clear fair weather.

Anticyclones, the highs of the weather maps, are either migratory or permanent. (See Questions 550 and 575.)

575. What are migratory anticyclones? Migratory anticyclones are those which generally originate in the polar regions and move rapidly south or southeastward in the Northern Hemisphere. They are comparatively shallow, not reaching very high into the atmosphere, and generally do not last for a long time.

576. What are cyclones? Cyclones are rotating wind systems that spin inward toward a center of low pressure, counterclockwise in the Northern Hemisphere and clockwise in the Southern. The converging air currents move upward, expanding and cooling, forming clouds and usually bringing bad weather in the form of rain and snow.

Cyclones, the familiar "lows" of the weather maps, are either migratory or relatively permanent. (See Questions 550, and 577 through 579.)

577. What is the difference between tropical and extratropical cyclones? Cyclones are called tropical cyclones if they originate over tropical areas and move in these regions. Outside the tropics, in the temperate zones, however, they are known as extratropical cyclones.

578. How long do migratory cyclones last? A migratory cyclone or low system may last 3 or 4 days, or it may last for several weeks. There is no regularity about its duration or its movement, although its behavior varies somewhat with the seasons. Sometimes cyclones move rapidly, or sometimes they may move slowly or remain stationary.

579. How fast do migratory cyclones travel? On the average, cyclones generally travel from 200 to 500 miles a day in summer and about 700 miles a day in winter. They have been known to travel 1,100 miles in one day, a speed of about 47 miles an hour.

580. What are local persistent winds? Within the large global wind systems are smaller wind patterns that move over specific earth features such as mountains, valleys, plains, and rivers. These winds are almost invariably small-scale convection winds. Comparatively shallow in depth and limited in range, local persistent winds are relatively simple. They blow from high-pressure areas to low-pressure areas, and are relatively unaffected by other large-scale forces such as the rotation of the earth.

They play an important part in forming the weather of particular localities, often causing severe discomfort or even death. Generations of people living in these areas and having to endure the winds year after year have often personified them, giving them distinctive nicknames and considering them as malicious individuals. There is much confusion and overlapping of these names and often no definite distinction. Some winds such as the sirocco may have 50 different names.

581. How are local persistent winds classified? Researchers generally classify these local winds into many groups. Some of the more common are the sea and land breezes (see Questions 582 through 585) and mountain and valley winds (see Questions 586 through 608). Also there are regional winds that blow in certain seasons (see Questions 609 through 633).

Many winds do not always fit readily into such simple classifications. There are more than 150 of these specific local winds, many of which bring disaster in the form of too much snow or cold or too much heat and dryness. They have been accused of affecting men's minds, of bringing headaches, fever, anger, epileptic attacks, and causing acts of violence including murder and suicide.

582. What are land and sea breezes? Along the coasts and shores of oceans and of great inland lakes, each day winds constantly blow back and forth. These diurnal winds are caused by differences in the temperature of the air over the land and the sea, leading to differences in pressure and hence creating a movement of air—a wind.

In the daytime, as the sun warms the land, the air over this land rises, and the cooler sea air blows across the shore to replace it. This sea-to-land wind usually begins around 10 A.M., reaches its peak around 1 or 2 P.M., and subsides in the mid-afternoon.

At night the movement of air is reversed. The sea becomes relatively warmer than the land, and the air flows from the cooler land to the sea.

583. How fast do land and sea breezes blow? In hot climates, sea breezes blow hardest and reach their highest speed at the hottest time of day—traveling some 20 to 24 miles an hour. In the temperate zones, these breezes travel about 8 to 12 miles an hour, with bursts up to 25 miles an hour.

SEA BREEZE
IN DAY

LAND BREEZE
AT NIGHT

Land and sea breezes

584. Where are strong sea breezes? Sea breezes can become quite powerful in some locations, especially on the west coast of South America. Along the coast of Chile, for instance, a sea breeze called the Virazon blows sometimes at gale force. It blows so hard on some afternoons that it picks up pebbles and hurls them through the air with great impact. People generally prefer to stay indoors when the Virazon starts to blow.

Some sea breezes are quite refreshing and pleasant, bringing a cool change of air. The *datoo* that blows over Gibraltar, the *imbat* of Morocco and other North African coasts, the *ponente* of western Italy, the *kapalilaa* of Hawaii, and the *doctor* of various tropical and sub-tropical regions—all these are mild, refreshing breezes.

585. What are lake winds? Winds along the shores of large lakes blow according to similar principles as those of sea and land breezes. During the day, as the sun warms up the land and sets up convection air currents, breezes blow in from the lake. In the cooling late afternoon and evening, the wind dies down. Summer vacationers sometimes find themselves stranded on mirror-like lakes far from their destination unless they allow for the dying winds.

Some of the more famous lake breezes include those blowing across the Great Lakes of North America, along Lake Geneva in Switzerland, and across the Italian Lakes.

586. What are mountain and valley winds? Small-scale winds basically move on the same principles of thermal and pressure difference. In mountainous and hilly regions, these differences are formed along the slopes, creating mountain and valley winds that are called thermal slope winds by meteorologists.

During the warm daylight hours, the sun beats down on valley slopes and sides of mountains, heating the air above them so that it rises and moves up the slope. Cooler heavier air flows downward, usually in the middle of the valley, to replace the rising warm air. This cool air in turn becomes heated along the slopes and rises. In essence, every valley has its own separate convection system that circulates the air.

At night, the process reverses. Soon after sunset, the mountain sides cool faster than the valley. Downdrafts flow along the sides, and updrafts form over the relatively warm center of the valleys.

587. What was an exceptionally strong canyon wind? On January 7 and 8, 1969, unusually strong canyon winds swept down the eastern slopes of the Colorado Rockies over Boulder and caused $1½ million in damages and 1 death. Wind speeds reached 125 miles per hour—the highest number the anemometer records. Observers estimated that winds were blowing at speeds well above this. More than

100 gusts blew faster than 100 miles per hour. Roofs were blown off, windows damaged, cars overturned, and trees and power lines knocked down. The wind was created by a high-pressure area over the Rockies and a low-pressure area over eastern Colorado.

588. How do these mountain winds permanently bend trees? In large, elongated valleys, such as those of the Alps or the Sierras, wind blows by day up the length of the valley from the wide end to the narrow upland end. Sometimes this wind blows so strongly and steadily that it permanently bends trees and shrubs in its path, forming them into grotesque, unusual shapes, with knotted denuded branches stretching away from the source of the wind. At night, when the process is reversed, the downward blowing winds are so gentle and light that they do not force the trees back into their more natural shape.

589. What are other names for mountain and valley winds? These winds are also called canyon winds, or orographic winds, from the Latin word *oro* meaning mountain.

590. What is a katabatic wind? A katabatic wind is a wind that blows down an incline. Mountain breezes are katabatic winds, from the Greek word meaning to go down. They are called foehns when the downslope wind is warmer than the air in the valley below; and called fall winds or gravity winds if the downslope wind is cooler than the air into which it flows.

Ascending winds blowing up a valley are called arabatic winds.

591. What is a foehn? In certain mountain regions of the world, a hot, dry, violent wind drives down the slopes in a manner different from any other wind in the world.

This is the foehn, or föhn, pronounced firn. This wind has been known to raise the air temperature some 31 degrees Fahrenheit in 3 minutes, to melt several feet of snow in a few hours, to give people headaches and nervous disorders, to ripen grapes and corn, to split wooden walls and floors, to unglue furniture, and to increase the violence of forest fires.

Foehns were mysteries for a long time in the Alps where they were first noticed. Many people thought they were winds driven from the hot Sahara Desert across the Mediterranean Sea.

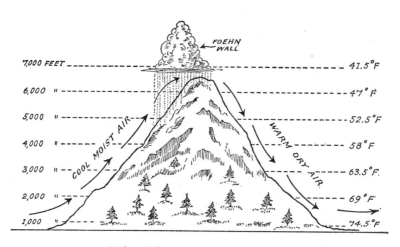

Foehn winds and adiabatic temperatures

592. Why is the foehn warm? Unlike most other warm winds the foehn does not gain its warmth from direct or indirect rays of the sun. It gains its heat by condensation and compression.

When air ascends, it cools and the moist vapor condenses, releasing heat as heat of condensation. As the foehn wind, moving from a high- to a low-pressure area over a mountain, blows up the windward side, it drops moisture but retains a relatively high temperature as a result of the liberation of the heat of condensation. The wind then curves over the mountain crest and drops down the other side, now becoming compressed and heated adiabatically at a rate of 5½ degrees Fahrenheit for every thousand feet of drop. (See Question 324.) By the time it arrives at the lower altitudes in the valley below, it is a very warm, very dry wind.

593. What is a foehn wall? Sometimes a bank of clouds forms at the edge of the mountain top as the foehn blows across it. This bank is called a foehn wall.

594. How fast do foehns travel? Foehns travel sometimes at high rates of speed, as much as 100 miles an hour. Some gusts have been reported at 125 miles an hour.

595. What does a foehn feel like? The unexpected arrival of a foehn in the northern Alps can be quite dramatic, according to Theo Loebsack in his book *Our Atmosphere.*

Let us imagine a winter day. The night before the temperature was well below freezing. All the alpine foothills are coated in crisp, crackling white. Suddenly the power of the frost is broken. High in the mountains a roaring noise begins which approaches on the wind. At the same time the peaks seem to come closer, their violet-black outline clearly showing against the deep blue sky as the föhn sucks up the last remnants of cloud and converts them into vapour.

This wind has an insatiable appetite for moisture. After it has devoured the clouds in the sky, it starts on the snow of the forests and fields. The thaw begins. A task which would take the sun several days is done, almost playfully, by the föhn in a few hours. Beneath its hot, dry breath the winter's snow melts so rapidly that torrents roar into the villages and avalanches thunder from the mountains.*

596. Where do foehns occur? Foehn winds occur in several places in the world, each one with a special name given it by the local inhabitants.

In Austria and Switzerland it is called the *schneefresser*, or snow eater. In Argentina and southern Chile it is called *zonda* as it blows down the slopes of the Andes Mountains. It is known as the *brohorok* of Sumatra, the warm *baru* of the Schouten Islands, and the double foehns on the western coast of Greenland. In the United States, foehn winds are known as the chinook along the eastern slopes of the Rocky Mountains and the dreaded Santa Ana on the western slopes of southern California. (See Questions 597 through 600.)

597. What is a chinook? A chinook is a spectacular foehn wind that blows with warm air from the west over the Rocky Mountains into Montana, Wyoming, and Canada. It was named after the Chinook Indians, for the early pioneers and settlers believed the wind originated in their home territory.

* Theo Loebsack, *Our Atmosphere* (New York: The New American Library of World Literature, Inc., 1961), p. 99.

A chinook may blow at any season, at any hour of the day or night. Usually it blows at speeds of 25 miles an hour.

The ability of a chinook to raise the temperature of the environment in a short time is remarkable. One chinook blew down upon Havre, Montana, after midnight in March, 1900, and raised the town's temperature 31 degrees in 3 minutes.

598. What effects does a chinook have? The chinook is a blessing to many cattlemen along the eastern plains, for as it descends and gains heat by compression it thaws out snows of bitter winters and rescues frozen cattle.

There are several dramatic legends or case histories about chinooks. For instance, a chinook was said to arrive at a range in Alberta, Canada, where a herd of exhausted horses and cattle had been snow-bound for weeks, with temperatures reaching 52 degrees below zero. The owner had just about given up hope for his livestock, when the burst of warm weather arrived, and he was able to rescue all the animals.

A chinook can also destroy. Sometimes its unseasonable warmth causes trees to produce leaves, only to be frozen in a few days as the chinook ends, and cold weather moves in again.

599. Why is the Santa Ana wind so destructive? The Santa Ana wind is a hot dry foehn blowing from the northeast and north in the coastal area of southern California. Speeding with gusts of about 60 to 70 miles an hour, these winds have supported and spread many disastrous fires. They can blow dust and stinging sand at high speeds, break window panes with pebbles, dry up young fruit on trees and vines, rip branches off trees, and sometimes even knock down trees.

During the winter, northeast and east winds often blow from the hot deserts and plateaus of lower eastern California. Already bone-dry, they cross the coast ranges and descend through passes such as the Cajon and Santa Ana, becoming even hotter from compression. They draw up the moisture from bushes, grasses, and leaves, creating dangerous tinder-fire conditions.

600. What was a disastrous Santa Ana wind? On November, 1961, a Santa Ana wind blew over the area around Bel Air and Brentwood in southern California, fostering a fire that destroyed more than 450 houses. With enormous effort on the part of fighting forces of the city, county, and the U.S. Forestry Service, the fire was finally halted—but only after the Santa Ana wind stopped.

601. Why is Death Valley so hot? Death Valley is particularly hot and dry because all the winds that reach that low-lying area in the middle of mountains and hills are downslope winds, becoming hotter as they descend.

602. What are the hot winds of the plains? In the summer, searing hot winds, sometimes classified as foehns, sweep across the plains of Texas and Kansas, bringing excessive dryness and heat, evaporating moisture from the soil and vegetation until the landscape literally looks burned to a crisp. During one spell of recurrent hot winds in Kansas, 10 million bushels of corn were reported seared beyond recovery.

These winds may blow in extraordinarily narrow bands sometimes only a few hundred feet wide. Between these wind bands are calmer areas with lower temperatures, sometimes a few yards, and sometimes a few miles, wide.

These winds bring headaches and feelings of nervousness to many people. In an area where temperatures sometimes rise to 100 or 110 degrees Fahrenheit in the shade, people have difficulty breathing when a channel of hot winds raise temperatures 7 degrees higher in a few minutes.

603. What is a California norther? A California norther is not a true member of the cold winds called northers (see Questions 629 and 630) but is a foehn of hot dry winds descending into lower lands. Dreaded by farmers particularly in the Sacramento Valley, the California norther sometimes arrives in spring and summer with such heat that the grapevines and vegetables of the fertile valley shrivel and dry up.

The great Berkeley fire of 1923 was attributed to a norther.

604. When were California northers considered as excuses for murder? The hot dry winds of the California northers have often affected the temperaments of people, making them irritable and irrational. During the Gold Rush of the nineteenth century, people accused of violent acts, even murder, offered the norther as an extenuating circumstance in court.

605. What is the wind of 120 days? Another member of the foehn family is the wind of 120 days, which is a violent downslope wind blowing from a northwest direction across the Seistan Basin in eastern Iran. It blows from May or June to September so steadily and strongly that trees grow only in the lee of houses, and melon vines and other plants are deformed, spreading their branches toward the southeast so consistently that they could be used as compasses. In this land of winds, houses are built with their backs to the wind, without doors or windows on that side. Human beings are affected by these hot, parchingly dry winds, full of dust, salt, sand, and gravel blowing as fast as 70 miles an hour. The wind drains initiative from people, leaving them enervated and ill at ease.

These winds are also called *bad-i-sad-o-bist-roz* and *seistan.*

606. What is a fall wind? A fall wind is a cool wind that is so dense and heavy that it literally falls downhill under the influence of gravity. It is also called a drainage wind or gravity wind.

Fall winds are usually found falling or draining from high cold plateaus or mountains into a warm plain or coastal region. This wind is so cold at the start that even though it becomes compressed and warmed adiabatically as it blows down the slope, it is still colder than the atmosphere in the valley or plain into which it drains.

607. Where are some strong fall winds? Fall winds are especially strong on the coast of Norway and along the Adriatic and Aegean Seas in the Mediterranean. They are also found in Peru, Greenland, and the Antarctic. Fall winds have been given special names in certain areas. The *bora*, for instance is a well known fall wind in the eastern Mediterranean, as well as the *papagayo* in Costa Rica, the *vardarac* in southeastern Yugoslavia, and the *taku* in southeastern Alaska.

608. What is the *bora*? The *bora* is a fall wind that sweeps particularly hard across the mountainous Yugoslavia coast of the Adriatic Sea. Blowing at speeds of more than 100 miles an hour, this wind brings cold air, snow, and ice to the Dalmatian coast and out along the Adriatic Sea from northern Yugoslavia to Albania, sometimes halting ship traffic with its bad weather.

This wind blows when a low-pressure area forms over the Mediterranean and draws cold polar air from Russia across the Hungarian plains and the high mountain valleys.

The name probably came from Boreas, the Greek god of the north wind.

609. What seasonal winds occur? Masses of moving air, usually traveling from areas of high pressure toward those of low, are often propelled by changes in the seasons. Many persistent local winds are set in motion at certain times of the year, as the sun moves over certain regions, warming up the air over these regions and creating areas of low pressure into which other air masses flow.

Some of the largest seasonal winds are the monsoons that sweep over Asia and Australia, blowing one way for half the year, the other way for the other half year. Other seasonal winds bring hot sands and dust across desert areas or unpleasant cold winds across mountains.

610. What is a monsoon? A monsoon is a large wind system driven by pressure and temperature differences at certain seasons alternating with winter and summer. Each summer, the prevailing winds reverse directions from sea to land and land to sea, involving an enormous area of land and ocean.

The name comes from the Arabic word *mausim*, meaning season. Originally it applied to the winds blowing across the Arabian Sea, generally from the northeast for 6 months and then from the southwest for 6 months.

The Indian monsoons for a long time kept Phoenician and Greek sea explorers from traveling eastward toward India and China. An admiral of Alexander the Great discovered how to use the monsoons by following the coasts from Arabia to India. Further knowledge of monsoon cycles led to more extensive exploration and shipping, until by the fourth century, ships were sailing on monsoon winds from Malobia, east Africa, and Arabia to China.

In the great navigation and exploration era of the sixteenth century, ship captains rounding the Cape of Good Hope learned to use the monsoons on their journey eastward. Since then navigators have constantly used monsoons to speed them on their way.

611. What causes monsoons? The standard explanation of monsoons is that they are a wind system driven by thermal convection and pressures on a seasonal cycle. During the summer, the sun is high over land masses of Asia and heats up the atmosphere over the plateaus. The warm air rises, low-pressure areas are created, and air is drawn in from the Indian Ocean to replace the rising air. As the moisture-laden air is lifted over the land by high plateaus and mountains, it drops rain, sometimes in torrents, over the land. Thus in summer, the sea-to-land air brings rain.

In winter the process is reversed as the sun moves southward over the water, creating areas of low pressure, toward which air is drawn from the cooling land mass.

Meteorologists are now finding, however, that monsoons are far more complex than previously supposed. They develop not only from thermal convections but also involve basic wind patterns and atmospheric disturbances in the form of cyclones, convections, and high and low areas.

612. Where do predominant and frequent monsoons occur? The best developed monsoons in the world are those over Asia—the south Asian monsoon over India and Pakistan, and the east Asian monsoon over China and Japan. Other areas with somewhat weaker monsoons include northern Australia, the Gulf of Guinea in western Africa, parts of eastern Africa, and the Gulf of Mexico.

613. What are etesian winds? For long centuries sailors around the eastern Mediterranean Sea have recognized and used the seasonal winds that blow constantly from June to September across the Aegean Sea, from the north toward the low-pressure area over the hot Sahara. These winds were called the etesian winds, from the Greek word *etos*, meaning season or year. Nowadays, Greek sailors call them *meltemi*.

Prevailing winds also blow across the western Mediterranean, stunting the vegetation along the northern slopes of islands, such as Majorca and Minorca, and permanently bending the trees in a southward direction.

614. What is the loo? The *loo* is the hot dusty wind of Bihar, India, that begins to blow as the sun climbs northward. This wind dries out crops and makes people fretful by its eerie singing through the parched fields. In times of drought, as the old Bihar saying goes, "The *loo* brings death." In April of 1967, under drought conditions, the *loo* destroyed crops and affected 40 million people in the area.

615. What suspension bridge collapsed because of wind? The Tacoma Narrows Bridge over Puget Sound, Washington, nicknamed Galloping Gertie, twisted and heaved on November 7, 1940, when a 42-mile-an-hour wind set up vibrations that finally snapped the suspension bridge. The $6.4 million bridge, nearly a mile long and considered the world's third longest span, had been completed only 4 months before. The engineers had not taken into account the lateral winds which blow down the Tacoma Narrows. These winds turned the bridge into an air-foil that soared and undulated until it snapped.

No one was killed, but several persons narrowly escaped death. A replacement bridge was opened to traffic on October 15, 1950, designed to take the winds into account.

616. What is the sirocco? Each year, during the months of February, March, and April, a hot dry irritating wind called the sirocco blows out of the Sahara and Arabian deserts and flows for hundreds of miles across northern Africa and the Mediterranean Sea into low atmospheric pressure areas found in the Mediterranean area at that time of year.

617. Why is the sirocco disliked? Inhabitants of the Mediterranean countries dread the sirocco because of the heavy clouds of hot dust it brings. A sirocco can become a full-fledged sandstorm, penetrating into small cracks and crevices, abrasing paint off vehicles and buildings, damaging engines and forcing airplanes to stay grounded. It withers field crops and fruit. It inflames the eyes, throats, and lungs of people, cracking their lips and parching their skin. Worst of all, it leaves a person enervated and demoralized, bringing the nervous disorders and uneasiness known as *ghibrilis* in Tripoli.

618. What is it like to be in a sirocco? When the great hot winds of the Sahara blow against the outposts of civilization, men and animals can do nothing about them but endure:

I woke at dawn with the sensation that the hair was being scorched off my head. I sat up and saw that Madani had already left. I scrambled quickly out of the tent myself. The nomads, men and women, were working like feverish ants hammering in tent pegs, tightening ropes. The sheep and the camels and the donkeys huddled together, heads down, their backs to the wind which came in suffocating gusts. I looked up at the sky and found none. A pall of yellow dust obliterated everything. As the daylight increased, so did the heat. My throat felt parched, my eyes smarted. It was like standing in front of a furnace in a glass factory. The wind had ceased to be gusty. It came in a steady roaring rush. . . .

It never lets up, it never lulls; it goes on, and on, and on, until one is as near crazy as it is possible for a sane man to be.

Night and day, day and night . . . wind. There is nothing but wind. One wonders if there was ever a time when there was no wind. The sun is completely shrouded. The dawn is jaundiced, the sunset dark yellow. One can hardly breathe. It is difficult to stand erect. . . .*

619. What hot winds have destroyed an army? There are many ancient stories, some not too accurate, about armies and caravans being scattered and destroyed by windstorms and sandstorms. Herodotus reported a sirocco turning into a violent sandstorm that helped destroy the army of the Persian king Cambyses marching through Libya about 500 B.C. Only a few survivors remained from about 20,000 soldiers.

620. When was the sirocco used to pardon murder? After a sirocco has swept across areas around the Mediterranean, police report more than the usual number of quarrels, murders, and suicides.

During the time of the Ottoman Empire, in the fourteenth century, the law stated that murder was more pardonable if the act was committed during the time the sirocco was blowing.

621. What are other names for the sirocco? People who suffer the sirocco wind as it whistles over the land have given it many different names. When it is particularly hot and insufferable, it is called the poison wind or the *simmoom*. It is called the *khamsin* in Egypt, *ghibli* in Libya, *chili* in Tunisia, *chichili* in Algeria, and *chergui* in Morocco.

People in Spain, Italy, Sicily, and Greece have a particular kind of grudge against the sirocco, for as it crosses the Mediterranean Sea it

* Ronald V. Bodley, *Wind in the Sahara* (New York: Creative Age Press, 1944), pp. 116, 117.

picks up moisture and becomes a humid, sticky, salty mixture that makes breathing difficult.

622. What are showers of blood? Sweeping up clouds of dust from the soils of North Africa, the sirocco sometimes blows across the Mediterranean Sea and dumps its load over Europe.

In the spring of 1901, winds scattered about 2 million tons of dust and sand from the Sahara over Europe. About the same amount was dropped into the Mediterranean Sea.

Sometimes the dust has such a strong reddish tint that, as it precipitates to earth with rain, people call it red rain or showers of blood.

623. What is the *kharif*? The *kharif* (also spelled *karif*) is a wind blowing from the southwest with gale force across the Gulf of Aden. It is often uncomfortably hot and laden with sand from the interior of Africa.

The *kharif* starts about 10 o'clock at night and blows until about noon the following day, reaching its greatest force about 8 A.M., blowing at 25 miles an hour in the months of July and August.

This hot wind strikes particularly hard at the coastal town of Barbera, which was once the capital of Somaliland until government offices were transferred to Hargeisa on the high inland plateau so officials could escape the sweltering wind.

624. What is a *garmsal*? The local name for a desert wind in Turkestan is the heat wind or *garmsal*. After several days of intense heat, the *garmsal* appears from the west or southwest, raising clouds of dust and sand and hurling the particles with stinging force against all objects. Animals unable to protect themselves are subjected to severe flailings all over their bodies especially their muzzles and eyes. Men often acquire headaches and neuralgia. The wind, also called the *tebaad* or fever wind, lasts only a few hours.

625. What is the *haboob*? Among the more spectacular dust storms is the *haboob*, a name derived from the Arabic word *habb*, meaning to blow. In the Egyptian Sudan, this storm moves across the desert like an immense yellow wall, sometimes 10,000 feet high and traveling at speeds up to 60 miles an hour. It moves in any direction, depending on the location of a low-pressure system, toward which it moves, often preceded by dust devils.

626. What are brickfielders? Another kind of dust storm is frequently encountered in southern Australia. Originating in the sandy deserts of the interior, the storm blows into the southern region, sometimes raising temperatures to 120 degrees Fahrenheit. These storms have been often called brickfielders because they raise clouds of dust from the brickfields outside Sydney and cover houses with dusty layers. These storms can last for several days, drying up vegetation and coating everything with dust.

627. What destructive storm did Marco Polo describe? Marco Polo wrote of an army destroyed by the heat and winds in the desert of Kerman. This army had been dispatched by the king of Kerman to subdue the people of Hormuz on the Persian Gulf who were not paying their taxes.

628. Has a sandstorm ever buried a man alive? Contrary to many imaginative stories and movies, no sandstorm has buried a man alive or suffocated whole caravans—although sandstorms have caused many people to lose their way and hence perish from lack of water. Sand does not accumulate on the ground as fast as snow in a blizzard. Even in the sandy deserts, it may take months or even years to cover a dead animal or other inanimate object.

629. What are northers? Northers are a special group of unpleasant cold winds caused when a massive tide of polar air moves toward the equator drawn by an area of low pressure.

630. Where are northers very strong? The norther that hits Texas is particularly unnerving, especially in winter when the polar air sweeps all the way from northern Canada toward the Gulf of Mexico, dropping temperatures as much as 25 degrees or more in one hour. If the climate is wet at the time the norther hits, it turns the rain to sleet, snow, or frozen raindrops which freeze plants and cover objects with ice.

Texans shiver and say the only thing between Texas and the North Pole is a barbed-wire fence. One winter, after unusually warm weather in western Texas, a norther blew cold air down so fast it caught many hunters unprepared. Four froze to death.

Sometimes this blast of cold air reaches across the Gulf of Mexico into Yucatan, bringing a coolness that is invigorating to that warm humid region.

631. What is the pampero? Another norther that brings cool relief to people in a humid climate is called the pampero, sweeping from the South Pole over the Pampas of Uruguay and Argentina and accompanied by a violent roll of clouds.

632. What is the helm? The helm is another cold and disagreeable wind blowing from the northeast across the Lake District in north central England.

633. What is the mistral? Another violent cold wind is the mistral that bursts southward from the cold regions north of Europe, drains across the cold plateau of central France, through the valley of the Rhone, and onto the Rhone delta. It occurs at its strongest in spring and autumn, drawn down by a low-pressure area over the Mediterranean between Spain and Italy.

Suddenly appearing at speeds of 60 miles an hour over the Provence region of France, this cold and dry wind sometimes knocks people down, shatters windows, and sets people's nerves on edge. It is said that the mistral around Arles contributed to the insanity of the Dutch artist Vincent Van Gogh.

The mistral was well known to the Romans in the first century A.D. and was accused of hurling men from their chariots. The word may be derived from the Latin word *magistralis*, meaning masterly.

Another tale of the mistral's strength recounts that the wind once blew a string of freight cars some 25 miles along a railroad track from Arles to Port-St.-Louis.

634. What are some benefits of winds? For all their destructive forces, winds have proved immensely beneficial to man and his evolving civilizations.

As a cheap, inexhaustible source of energy, winds have long filled the sails of ships to help man discover and conquer new lands and develop commerce and civilizations. From the early days of the Egyptians, 3600 B.C., wind energies have been harnessed by windmills, enabling man to grind grain and corn, draw up drinking water, drain inundated regions, or irrigate dry lands. Spring breezes and summer winds have wafted warm weather poleward, helping to ripen fruits and grain and brighten the world with green plants.

Experimenting with the dynamics of air currents, man has developed the arts of ballooning, of gliding, and of flying through the atmosphere, releasing him from the earth's surface.

635. How have storms affected civilization? In the temperate zone, particularly in the Northern Hemisphere, the clash and mixture of cold polar air from higher latitudes with warm tropical air have produced many stormy winds, changing temperatures and invigorating climate. Historians, meteorologists, and other scientists claim this stormy climate has produced people who are more active, determined, and creative than those of the more apathetic tropical zones or the rigorous severe polar regions.

From the time of the Ice Age, primitive people were forced to devise methods for survival against the encroaching cold, snow, and ice storms. Their ingenuity in wrapping skins around themselves for warmth, fostering the light and warmth of fires, seeking shelter of caves and building huts against the hostile environment ensured their survival. Throughout history the restlessness and vigor of men in the great storm belts of the Northern Hemisphere have produced powerful civilizations in Europe, Asia, and North America. Men of these regions have risen to great heights in art, music, literature, religion, science, and technology—as well as plunged parts of the world into warfare and destruction.

636. What are some wind legends? The Greeks believed that Aeolus, god of the sea winds, kept the winds of the world imprisoned in caves of his island kingdom, one of the Lipari Islands of the Mediterranean Sea. Aeolus would release various winds at different times, hence causing changes in the weather.

According to the Greek epic poet Homer in the *Odyssey*, Aeolus presented Ulysses and his crew with an oxhide sack in which were imprisoned all winds except Zephyrus, the gentle west wind which was to blow them home safely. While Ulysses slept, however, curious sailors opened the bag and released all the winds which not only blew them back to the island of Aeolus, but delayed their homeward journey for 10 years.

637. What were some other wind gods? The winds, ranging from gentle to violent, have long been an intimate factor to men through the ages, and they have long been personified and deified.

The ancient Babylonians believed in the gods of the wind Amon and Amannet.

Stribog was the wind god of Slav mythology.

Hindu mythology gives credence to three gods of wind—Rudra, Vogu, and Vada.

One of the Chinese gods of wind, Fei Lien, had a dragon's body and powerful lungs with which he instigated thunderstorms and other fierce winds.

638. What was the Tower of the Winds? The Tower of the Winds was an octagonal marble tower, 40 feet high, built in Athens, Greece, about 100 B.C. Its ruins are still standing today.

On the 8 walls are sculptured figures symbolizing the 8 points of the compass with their names: Boreas, the north wind; Kaikias, the northeast wind; Apeliotes, the east wind; Euros, the southeast wind; Notos, the rainy southwind; Lips, the southwest wind; Zephyros, the west wind, and Skiron, the northwest wind. Inside was a massive water clock that was powered by a nearby spring. The 8 winds were recognized 3 centuries earlier by Aristotle who divided them into 2 surprisingly accurate classes: polar and equatorial.

On top of the tower stood a bronze weathervane depicting the scaly-tailed Triton pointing into the wind with his rod. This may have been the world's first weathervane.

In the ninth century A.D., Pope Nicholas I officially decreed that the weathervane in the shape of a cock was to be set on abbeys and churches in memory of the cock that crowed 3 times when St. Peter denied Christ.

639. What is anemomania? Anemomania is the ancient malady of wind madness. The disease was said to take two forms—one in which the victim became so apprehensive about the shrill wind that he felt there was no escape and voiced his fears in talking. The other form of anemomania was insanity.

640. What protection can be taken against winds? Men have tried many ways to shut out the wind from their lives. Sometimes they build roofs of their houses down to the ground in the direction of the prevailing winds, with windows and doors opening only on the lee side. Hedgerows, rows of trees, walls, and barriers of all sorts have been built to keep the never-ending winds from houses and highways.

In villages and towns, broad streets have been laid out at right angles to the prevailing winds, and other streets built as narrow and twisted as possible to break the wind's flow.

641. How did man start to analyze the wind? The weathervane is perhaps the oldest instrument to indicate the direction of the wind. For centuries curious people have devised simple windvanes of one sort or another to indicate which way the wind blows—a feather, a straw, the lick of a finger.

To measure the speed at which wind blows proved to be more of a challenge. Some men tried to determine the wind speed by seeing how fast a feather cork disk blew along a wire, or by clocking the time it takes cloud shadows to cross an open field or a stretch of water. Some measured the wind speed by observing how limp or straight a flag hangs on a flagpole, or how fast the wind could cool and evaporate a pan of water, or the musical sounds wind makes as it blows across an instrument something like a harp.

In the seventeenth century, British physicist Robert Hooke devised a simple wind measurer called the anemometer (from the Greek word *anemos*, meaning wind) with a spring-attached plate against which the wind would blow. Rotating types of anemometers were developed in the eighteenth century, and in the mid-nineteenth century the cup anemometer was devised. This continues to be used today, with a few changes. The barometer, invented by the Italian mathematician Evangelista Torricelli in 1643, was found to foretell an approaching storm by indicating a drop in pressure.

Man's insatiable curiosity prodded him to investigate the winds of the upper air, and he began to send kites aloft with thermometers attached. Balloons were filled with hot air and gases, and in 1783 the French brothers, Joseph and Jacques Montgolfier, successfully sent a large balloon into the air, inspiring scientists to develop this method more fully. By the end of the nineteenth century, recording instruments had been sent aloft in kites, kite-balloons, and special weather balloons.

642. What modern methods are used to monitor winds and weather? Today meteorologists are keeping track of the winds around the world with highly sensitive instruments and constant monitoring systems working 24 hours a day.

Information is being gathered on temperatures, pressures, wind directions and speeds, cloud formations, and other phenomena from areas all over the world—from land-based national, international, public, and private stations; from private, commercial, and military aircraft. Reports come in from lighthouses and Coast Guard ships, from ships traveling lonely seas, and from remote ocean buoys. Upper air rockets and orbiting weather satellites add data to the flow of material streaming in to give weathermen a fuller concept of the sea of air around us.

Various instruments are used to watch the world's weather:

Standard weather radar scans horizontally and vertically and projects maps of oncoming storms in terms of direction, distance, and intensity of precipitation. Recently developed Doppler radar instruments are being used in attempts to tell which way the storms are rotating. Another new device, the laser radar, scans and identifies weather conditions from the reflections bounced off objects.

Radiosonde is a modern, accurate means of finding out atmospheric conditions such as pressure, temperature, and humidity at various heights in the atmosphere. Developed in the mid-1930's, radiosonde instruments are carried by helium-filled balloons to heights of 20 miles. As they rise, radios transmit data to earth centers.

Dropsonde is an instrument dropped with a parachute from airplanes flying at various heights. In the long, slow fall to earth, these instruments record and transmit weather data from various levels of the atmosphere.

Rawinsonde balloons are used as targets for radar observations of storm and other winds. Carrying instruments to record weather data, these radar-tracked balloons rise continuously to thousands of feet, sending detailed profiles of atmospheric conditions.

The construction of MANIAC (Mathematical Analyzer, Numerical Integrator and Computer), the first electronic computer for weather research, brought new dimensions to meteorologists in the late 1940's by being able to receive enormous amounts of weather data and at lightning speeds turn out analyses. Today meteorologists are working with huge computer complexes.

643. What are weather forecasts? For centuries people believed the weather was caused by the whim of the gods. Violent winds, torrential rains, and strokes of lightning shook the earth because the gods were angry; and fair blue skies and sunlight reflected the satisfied well being of a smiling god. Yet weather was regarded as a natural phenomenon and, as such, observations were made. As sailors and explorers pushed farther out to sea, information on the winds and storms became increasingly important.

As civilization advanced and more instruments were devised—the barometer, the thermometer, the hydroscope, the anemometer—man began accumulating huge records of weather data. Gradually people began to gain an idea of the passage of weather over the surface of the earth—the broad movements of winds, the journeys of storm centers, the flow of hot and cold temperatures. With the invention of the telegraph in the 1840's, man began to understand the pattern of weather on a global scale.

Today weather reports are gathered from many thousands of stations, in an awesome attempt to analyze trends and forecast what might occur. Immense complexities of weather are accounted for— wind directions of upper atmospheric winds and of surface winds; cloud cover; precipitation; pressures; and topographic features. Forecasting has become an immense application of physics principles, supplemented with many varied empirical techniques and with ever-changing statistical data gathered from innumerable locations.

The U.S. Weather Bureau is an immense factory issuing thousands of forecasts, on short-range projections and on long-range forecasts, as well as special warning bulletins on storms and other hazards.

644. What are short-range forecasts? Short-range forecasts are made by mathematical and statistical computations of phenomena such as movements of air masses and fronts, precipitation, humidity, and temperatures. Interpretation of these and other facts relies generally on the experience of the weathermen, who can give a forecast for the next 18 to 48 hours. These forecasts, issued several times a day, are reported in the local newspapers or over the radio and television.

645. What are long-range forecasts? With increasing use of weather reporting equipment such as lasers, radar, automatic weather stations, buoys, balloons, and satellites—all gathering and sending vast amounts of information at incredible speeds—weather forecasting has taken on new scope. At the Weather Bureau's National Meteorological Center in Suitland, Maryland, enormous complexes of computers are able to assimilate vast quantities of data and code out trends and forecasts.

Here is a general idea of computerized forecasting:

Two cycles of machine analysis are run each day on observations taken over the Northern Hemisphere at midnight and midday Greenwich Mean Time. Each machine cycle begins with a preliminary analysis, made after 1½ hours of collection over North America from the surface to heights of 18,000 feet. An operational analysis is made after 3½ hours of collection, when 70 percent of all data has been received. In half an hour the computer describes 10 levels of the atmosphere over the Northern Hemisphere from the surface to 53,000 feet—an operation that would take 5 men 8 hours to complete. The data are reproduced on curve followers, photographic systems, and a facsimile chart transmission.

After these data are collected, prepared, and analyzed, facts are fed into programmed computers that are able to shape composite trends of the weather in the future.

646. What are some weather research projects? The World Weather Watch is an international meteorological program established through the World Meteorological Organization (WMO) for the years 1968 through 1971 and designed to observe and record many facets of the atmosphere on a global scale. The resulting data is communicated, processed, and analyzed in an effort to better understand large-scale weather climates and to work toward long-range predictions and climate modification. Operating with satellites, aircraft, ships, and hundreds of land and sea stations, data are fed into 3 World Weather Centers, located at Washington, D.C., and Moscow, U.S.S.R., in the Northern Hemisphere; and Melbourne, Australia, in the Southern Hemisphere. From these centers, weather data are relayed via satellites to WMO computers where the pooled information is sent out on request.

A weather facsimile experiment, called WEFAX for short, has been set up by the Weather Bureau and the National Aeronautics and Space Administration to develop an economical worldwide system for distributing weather data to all parts of the world. As part of the World Weather Watch, WEFAX uses 3 earth-synchronous satellites, equally spaced around the equator, to relay weather data to and from various stations and centers throughout the world.

The Global Atmospheric Research Program (GARP), coordinated internationally by the International Council of Scientific Unions and the WMO and part of the World Weather Watch, is working toward an intensive research effort on a global scale during the 1970's to understand the interactions of atmospheric motion that influence weather. It is designed primarily to test mathematical models of the atmosphere in order, for instance, to produce reliable forecasts for weather 2 weeks in advance.

The Global Meterological Experiment (GLOMEX), part of World Weather Watch, is set up to acquire the most comprehensive set of global atmospheric data ever compiled.

The Global Horizontal Sounding Technique (GHOST), set up by the National Center for Atmospheric Research, includes a project of many weather balloons stationed in various uninhabited regions throughout the Southern Hemisphere to gather data on many unknown areas.

The Navy Oceanographic and Meteorological Devices (NOMAD) are unmanned floating weather stations set out in 1961 by the U.S. Navy throughout parts of the Gulf of Mexico and the southwestern parts of the North Atlantic Ocean. These buoys automatically send out weather information for 2 minutes every 3 hours.

Establishment of a National Oceanic and Atmospheric Agency (NOAA) has been recommended to coordinate programs of meteorology and oceanography. Federal agencies would be made up primarily of the Environmental Science Services Administration, the Coast Guard, and the Bureau of Commercial Fisheries.

647. What is BOMEX? One of the most intensive scientific investigations of the interrelationships between sea and air is the Barbados Oceanographic and Meteorological Experiment (BOMEX). During May, June, and July, 1969, planes, ships, satellites, and buoys were set up to gather valuable atmospheric and oceanic data from an area covering 90,000 square miles. Located in the Atlantic Ocean east of Barbados in the West Indies, the area stretches vertically from altitudes of 100,000 feet above sea surface to depths at the sea floor.

This is the first of a series of large-scale research projects planned by nations under the Global Atmospheric Research Program and intended to aid the World Weather Watch. Various U.S. government departments and agencies are involved, along with the Government of Barbados and several universities and industrial laboratories in the United States, Canada, and Barbados.

648. What are some international organizations concerned with meteorological research? The International Council of Scientific Unions (ICSU), centered in Rome, Italy, consists of 14 associated but separate unions and commissions and committees. The Council has helped sponsor major international programs to extend cooperation among all nations in attacking worldwide or regional environmental research problems.

The World Meteorological Organization in Geneva, Switzerland, was established as part of the United Nations in 1947 to help coordinate, standardize, and improve meteorological activities throughout the world and to improve communication on these matters among different countries. Any state or territory with a meteorological service of its own can become a member of this organization, which now includes representatives from more than 100 nations.

The International Association of Meteorology and Atmospheric Physics (IAMAP), in Toronto, Ontario, carries on meteorological projects.

649. What organizations in the United States are involved with meteorology? The U.S. Weather Bureau, part of the Environmental Science Services Administration, maintains a growing number of research and recording stations to gather basic data of weather and climate throughout the nation and its possessions. It issues daily weather forecasts and special warnings of severe weather conditions, and supplies special weather services for agricultural, aeronautical, maritime, space, and military operations. The Weather Bureau maintains a national network of surface and upper-air observing stations, aircraft, satellite systems, communications, and computers.

The various Research Laboratories of the Environmental Science Services Administration include the Atmospheric Physics and Chemistry Laboratory, the Air Resources Laboratories, the Geophysical Fluid Dynamics Laboratory, the National Severe Storms Laboratory, the National Hurricane Research Laboratory, the Institute for Telecommunication Sciences, the Wave Propagation Laboratory, the Aeronomy Laboratory, the Space Disturbances Laboratory, the Research Flight Facility, and the Environmental Data Service.

The Office of Aerospace Research, part of the U.S. Air Force, was established in 1961. As part of this, the Air Force Cambridge Research Laboratories, at Bedford, Massachusetts, carry on research in fields such as weather observation, radar and radio astronomy, balloon research, and cloud physics. They develop improved equipment, gather precise data on meteorological phenomena, and advance theoretical understanding of environmental factors. The Office of Aerospace Research also administers the Aerospace Research Laboratories, at Wright-Patterson Air Force Base, Ohio; the Air Force Office of Scientific Research, Washington, D.C.; and the Office of Research Analyses, Holloman Air Force Base, New Mexico.

The U.S. Navy maintains a Navy Aerological Service and the Office of Naval Research and Naval Research Laboratory in Washington, D.C.

The American Meteorological Society (AMS) in Boston, Massachusetts, is a private society that acts as a clearinghouse to exchange information on all phases of meteorology.

The National Center for Atmospheric Research (NCAR) at Boulder, Colorado, sponsored by the National Science Foundation, is a nonprofit corporation to which 21 universities belong. Founded in 1960, it has focused on large fundamental atmospheric problems.

Meteorological research is carried on in various university and college departments, including the Atmospheric Sciences Research Center of the State University of New York, the Massachusetts Institute of Technology, Cornell University, Rutgers University, University of California, University of Chicago, University of Utah, University of Washington, University of Wisconsin, Colorado State University, and the Joint Institute for Laboratory Atmospherics.

VI. FOGS

Introduction. Fog—one of the softest, most silent and beautifully mysterious meteorological phenomena—causes hundreds of deaths and thousands of injuries by its insidious presence.

Created by a cooling temperature that squeezes tiny drops of visible water out of moist warm air, fog creeps, an earth-bound cloud, through forests, up mountain slopes, across beaches and shores, and through cities. By blotting out visibility as it crosses highways and airports, fog helps cause accidents of today's fast speeding airplanes, cars, and ships. It also causes enormous expenditures of money, time, and energy by slowing down transportation and creating long delays for travelers.

Latest offspring of fog is smog, created only recently by modern man, who is already suffering from its destructive and increasingly frequent appearance. Smog is an unhealthy combination of fog and waste products of today's affluent and technical society—smoke, ashes, fumes from factories, buildings, and vehicles. It is causing increased deaths of people and animals, as well as corroding buildings, damaging crops, and possibly changing the climate of the world.

Scientists are attempting to diagnose fog—its formation, its behavior, its locations, and its combination with pollutants. With modern instruments and imaginative ideas, they are trying to disperse it in more places and lessen its disastrous effects.

650. What is fog? Fog is simply a cloud whose base touches ground.

Like a cloud, fog is an air mass of tiny droplets of water or, in rare cases, of ice crystals. These droplets are nearly spherical and vary in diameter from 2 to 50 microns. Some fogs may contain very few droplets per square inch and seem as a thin gray veil, while other fogs are so thick they are called "pea soup," obscuring visibility and causing accidents and collision on land, sea, and in the air.

In recent years with the advent of automobiles, industries, and increasingly populated cities, fog has combined with smoke, soot, and chemical particles blown from factories, incinerators, and exhaust pipes to form an ever more hazardous combination called smog.

651. What causes fog? Fog is caused when warm moist air is cooled to such a point that the air cannot hold moisture in the form of invisible water vapor any longer. The water is squeezed out in the form of water droplets, so fine and buoyant they are suspended in the air and drift softly on slight currents of air.

Conditions for forming fogs can occur in various ways. Two of the most common occur when warm moist air blows over a cold surface of land or water and becomes cooled—or when heat is removed from the atmosphere and ground by radiation, as on a cool still night.

652. Has man caused fogs? During the past several decades, man has inadvertently caused more fogs by building larger cities and factories that pollute the air with quantities of heat and industrial wastes. Tiny particles of dust and soot given off from furnaces, smokestacks, and car exhausts form nuclei on which water condenses to form fogs.

Researchers believe that the incidence, duration, and probably the density of fogs vary directly with the amount of industrial activity and subsequent air pollution.

653. How does thermal pollution create fog? Steel mills, paper factories, sewage-treatment reservoirs, manufacturing, chemical, and many other industrial plants use enormous quantities of water to cool their heated equipment. The water that has carried off the heat from the equipment is thereby heated and is often dumped back into nearby streams and rivers, estuaries, bays, or along the edges of oceans. These waters then become so heated that they begin to steam, throwing enough water vapor in the air to form fogs.

Some industries are trying to dispose of this heated water by evaporating it in massive cooling towers. This method is not only expensive, but may produce more fogs. Recent studies show that plumes of fog from evaporating towers are contributing to increasingly troublesome fogs plaguing nearby areas.

654. What damages are caused by fogs? As soft gentle fog drifts silently across seacoasts, over highways, airports, and cities, it can cause deaths of thousands of people and injuries to thousands more—

merely by impairing the vision of people who drive cars, planes, and ships.

Insurance companies state that in one year (1965–1966), fog caused some 1,000 deaths in the United States and 50,000 to 60,000 injuries.

Some of the worst hazards of fog occur when it combines with smoke and soot spewing from factories, cities, and vehicles. This recent creation in technical civilization brings death and sickness to people and animals, as well as damages crops and corrodes buildings with toxic chemicals. (See Questions 699 through 718.)

655. How does fog affect the airplane industry? Airports throughout the United States are closed down about 115 hours each year because of thick fogs. In 1967, fogs cost the airline business an estimated $75 million because of disrupted schedules.

No one yet has calculated the enormous amount of physical inconvenience and mental anguish that fogs have inflicted on passengers, as business meetings are missed, sick relatives or loved friends are not reached in time, or other essential life and death schedules are not met.

656. What highway damage has fog caused? Fogs spreading across today's huge highway complexes have caused innumerable accidents throughout the world where more and more cars are moving at such high speeds that they cannot stop in time to avoid collisions.

As many as 100 vehicles have crashed at one time in massive chain collisions along the coastal speedways, especially in the Los Angeles area where the combination of sea fog, smog, and fast-speeding cars has proved particularly lethal. Another area where crackups are frequent is the fog-shrouded New Jersey Turnpike, where cars are forced to move at a snail's pace or are stalled for hours as damaged vehicles block the highways.

657. What were some marine fog disasters? Fog has long been the cause of death and destruction at sea, as ships ram into each other in the thick "pea soup," or into icebergs or rocks and shoals.

In old sailing days, many vessels were shipwrecked along lonely fog-shrouded shores. Even today, in spite of modern ship-building techniques and fog-penetrating equipment, marine tragedies continue to occur.

Two of the greatest occurred when the steamships *Titanic* and *Hans Hedtoft* struck icebergs in foggy waters and sank. (See Questions 887 and 888.)

Another fog-provoked tragedy occurred at a few minutes past 11 P.M., July 25, 1956, when the Swedish-American Line motor ship *Stockholm* collided with the Italian luxury liner *Andrea Doria* about 45 miles south of Nantucket Light, on the coast of Massachusetts. In the next 12 hours the *Andrea Doria* foundered and sank, taking 51 lives. So many contradictory versions were given later of the disaster that the full causes still remain unexplained.

Another tragedy occurred on Friday, April 20, 1952, just before five o'clock in the morning, when the tankers *Esso Suez* and *Esso Greensboro* collided in a dense fog about 200 miles south of Morgan City, Louisiana. The fully loaded *Greensboro* burst into flames that completely engulfed her. Only 5 of her crew escaped alive. The *Suez* also burst into flames but was able to pull away with a 20-foot gash in her bow, and the death of only one man.

658. How did fog delay the discovery of San Francisco harbor? The discovery of the harbor now called San Francisco bay was delayed for some 200 years because of thick fogs that hid the entrance from the English navigator Sir Francis Drake and his crew in 1579. The explorers, sailing in the vessel *Gold Hind* for Queen Elizabeth I, passed the harbor entrance and beached in northern areas for nearly a month, with complaints about "those thicke mists and most stinging fogges." The harbor was not discovered until the eighteenth century by the Spaniards.

659. How did fog aid Allied troops escaping from Dunkirk in World War II? During World War II, the surrender of Belgium to Germany in 1940 caused several hundred thousand Allied soldiers to be trapped on the mainland of Europe, surrounded by German forces. British troops and members of the Free French fought their way across northern France to Dunkirk on the English Channel, where from May 26 to June 4 they were picked up by the thousands along the beaches by ships and vessels of all sizes and transported safely to England, despite heavy bombardment by the Germans.

Losses to the Allies during this debacle were relatively light—only about 10 percent. Much of the reason for this was the effective air shield maintained by Royal Air Force fighters, but also because a thick fog rolled over the Normandy beaches and for 2 days screened the evacuation processes from German gunfire and Luftwaffe dive bombing.

660. What kinds of fog exist? Fogs are named and classified in many different ways, some by their density, such as thick or light; some by season, such as summer fog and winter fog; some by temperature, such as warm or cold; and others by their location, such as California fog, Grand Bank fog, and London fog.

Meteorologists generally classify fogs according to the basic processes involved in their formation: radiation fog, advection fog, and evaporation fog. (See following questions.)

661. What is radiation fog? Radiation fog, also called ground fog or land fog, and sometimes summer fog, occurs at night when the heat of the sun-warmed land and air radiates back from earth into space.

This fog usually forms during a long clear night, when the air is calm and winds blow no more than 6 miles an hour. If winds are stronger, fog does not form, because the cooler saturated air near the ground is stirred, lifted, and mixed with warmer layers above.

As the night deepens, the air near the ground becomes gradually cooler, until the dew point is reached and a layer of water vapor is condensed into water droplets of fog. More and more layers form as the night continues.

Radiation fog is nearly always shallow. On windless still nights, it may be only waist deep. If there is a slight wind stirring, enough to mix the cooler ground air upward, the fog may be 30 to 100 feet deep. Sometimes fog is trapped in a valley and becomes quite dense and persistent for several days until a wind blows it away. The fogs of central California and parts of western Europe, for instance, sometimes are extremely deep, dense, and long-lasting.

Most radiation fogs, however, last only one night. They begin to disappear after sunrise when the warming rays of the sun heat up the atmosphere, which absorbs the moisture.

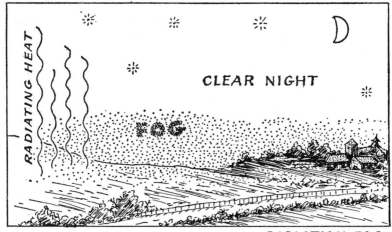

RADIATION FOG

662. Why does radiation fog seem to "rise"? People sometimes notice that a blanket of fog seems to lift off the ground and rise into the air, especially in the early morning hours as the sun rises over the horizon. The warming sun rays first penetrate to the ground beneath the fog, warm it, and then gradually warm the air above it. Convection air currents start to rise, stirring up the layers of fog and wafting some of them higher into the air. At the same time, the added heat near the ground increases the air's capacity to hold water vapor, so the fog droplets evaporate. The fog does not lift, as it seems to, but the zone of warmed air near the ground grows higher, evaporating the fog droplets into invisible water vapor.

663. Where are radiation fogs usually found? Radiation fogs are among the most common fogs. Since cold and heavy masses of air tend to flow downhill by gravity, they usually settle in valleys, marshes, ponds, hollows, or other low places. Here the moisture condenses and forms pools of fog that deepen—become thicker and more dense—during the night. Thick blankets of white fog can be found hovering over surfaces of quiet lakes or calm rivers.

Radiation fogs can cause considerable danger to motorists in the hollows or dips in a road, as cars may suddenly drive into a pool of mist and become submerged in blinding fog.

664. In what seasons do radiation fogs form? Radiation fogs can be seen at many times of the year, whenever conditions of atmosphere and temperature are right. They are found most often in the middle-latitude regions during the autumn and winter, when nights are longer and the ground and air have more time to cool off. In regions of the higher latitudes, they are formed most often in summer. This kind of fog forms during the rainy seasons in the tropics.

Each winter, a lake of radiation fog several hundred miles long forms in California's central valley, between the Sierra Mountains on the east and the Coastal Ranges on the west. This fog may last for many days before a strong wind sweeps it away.

665. What are advection fogs? Advection fogs are dense and persistent forms of condensation formed when warm moist air blows over cold sections of land or water. As the moving warm air cools, it reaches the saturation point and water droplets condense. Sometimes so much moist air is blown over the colder sections that the fog covers large areas and becomes quite deep—some several thousand feet high. The land need not necessarily be frozen nor the water icy to form this kind of fog—there merely has to be enough difference in temperature between the warmer saturated air and the cooler surface over which it blows.

The word advection comes from the Latin word *advecio*, meaning convey or carry.

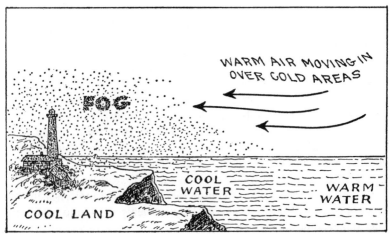

ADVECTION FOG

666. Where do advection fogs occur? Advection fogs are quite commonly found along cool seacoasts, where ocean breezes bring in warm moist air that forms billowing, rolling clouds of fog. Further out at sea, however, the air is clear, for there is little or no temperature difference.

Toward the north, these fogs are most common in the late spring and early summer as warm air blows over cool oceans or over large inland lakes or bays that still retain the chill of winter.

Sometimes these fogs form during a winter thaw, when warm humid air blows from the lower latitudes over a cold frozen land.

667. Where are inland advection fogs strong? The Great Lakes area is particularly shrouded with fogs in spring and early summer, when the lake waters are still frigid from winter and the warm spring air blowing in from the south brings a lot of moisture that condenses to form layers of fog.

668. What is upslope fog? One kind of advection fog is called upslope fog, which occurs when moist wind blows against a gradual slope of a hill or mountain and is forced upward. As it ascends, it cools adiabatically by expansion. (See Question 324.) This condenses the moisture from the air, and a cloud-like fog forms, usually about halfway up a hillside.

This kind of fog is found often along the Appalachians as warm air from the Gulf of Mexico is lifted up and cooled.

669. What are sea fogs? Sea fogs are advection fogs formed when air from a warm ocean current blows over an adjoining strip or current of cold water. The droplets condense and a thick, often sinister, fog forms over the cold waters.

670. What are the largest sea fogs? One of the foggiest stretches of water in the world is the dense marine fog banks of the Grand Banks off southeast Newfoundland, long known and dreaded by sailors. Here the warm moist air traveling northward with the warm Gulf Stream sweeps over the cool Labrador Current flowing from the Arctic. The resulting condensation forms great gray banks of fogs. These fogs are particularly dangerous for ships, for they conceal bulky icebergs drifting down from the north into the shipping lanes. (See Question 886.)

671. Where else do sea fogs form? Thick Nantucket fogs are caused when the colder water from the northern New England coast meets the warmer water flowing from south of Cape Cod.

Another area for sea fogs is along the southern coast from Texas to Florida. Here in winter, the cold water flowing down the Mississippi River into the warm Gulf of Mexico can cause fogs.

672. Where are other fog banks? Along the coasts of southern California, southwest Africa, and Peru, advective fogs rise from the cool ocean currents that flow parallel to these tropical or subtropical coasts. Here moisture-laden air from the warm ocean farther out drifts over the cool current, and the resulting fog blows over the coast onto land, especially in the cooler hours of the day.

673. What is skodde? Skodde is the nickname sailors give to the mantle of low-lying, gray-black fog that covers thousands of miles in Arctic waters, constantly advancing and receding. Sometimes this skodde is so low and thick that sailors on deck cannot see more than a few feet, but lookouts above on the mast can sight the tops of other ships.

674. What is evaporation fog? Evaporation fog occurs when vapor evaporating from warm water rises into cooler air that is already saturated. The resulting condensation forms fog.

Evaporation fog occurs most frequently over inland lakes or rivers in the late fall. At this time of year the lake or river water is still quite warm, while the surrounding air is turning cool. The warm water vapor steams into a very thin fog layer.

675. What is steam fog? Steam fog is an evaporation fog that can be seen rising from such things as engine boilers, city sewers, hot baths, and bubbling pots on the stove. Steam fogs can be seen rising from an asphalt roof or macadam or concrete road after a summer shower, when warm vapor condenses in the cooler air.

676. What is warm rain fog? Warm rain fog is a form of evaporation fog that occurs when raindrops from warm clouds drop through a layer of cooler air near the ground. The evaporating raindrops saturate the cold air layer.

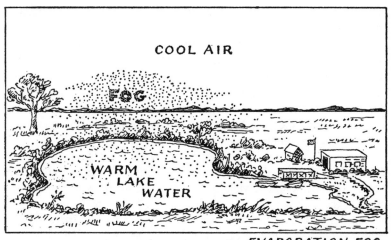

EVAPORATION FOG

This kind of fog is common over airfields in rainy weather. To an observer looking up, the fog often looks like a low cloud.

677. What is Arctic sea smoke? This is an unusual form of evaporation fog, formed in conditions where freezing air of ice fields flows over warm water. This fog is found in the polar regions, especially over the Bering Sea.

As cold air blows over warmer ocean surfaces, the warm water vapor condenses rapidly. A fog results that is churned upward into columns that look like smoke rising. Sometimes these sea-smoke columns rise 50 feet or more.

678. Where is sea smoke seen? Sea smoke is a common sight over patches of open water in ice packs of polar regions. It has often been cited as a warning sign of treacherous melting ice to experienced explorers traveling over the frozen region.

Northwest Russia is a region where sea smoke is often found as warm ocean currents flow over very cold winter lands.

Sea smoke rises in the cold spells of early winter over northern rivers and inland lakes that are still warm from summer.

679. What is ice fog? Fog sometimes consists of ice crystals rather than water vapor. It is then called ice fog. Ice fog usually occurs in the colder regions of the world, where temperatures reach minus 40 degrees Fahrenheit. At these low temperatures, the water vapor on various nuclei sublimates, turning directly into ice without passing through the liquid stage.

Ice fog forms naturally in cold areas as cold moist air blows up a slope. Ice crystals form at certain heights and temperatures. This phenomenon actually is a cirrus cloud on the ground.

Ice fog can be found in cold northern cities such as Fairbanks, Alaska, where laundries or small industries discharge excess water vapor into the cold air. Also clouds of ice fog have been observed around herds of running deer in northern regions.

680. What is pogonip? Pogonip is a particular kind of fog that occurs in the mountain valleys of western United States and other areas in the high latitudes. It is caused when fog in the form of super-cooled water droplets or tiny ice particles condenses. In extremely cold weather, the fog may condense in the form of ice spicules, which can be quite harmful to the respiratory system of animals and people.

This fog is most common during the months from December through February, appearing in the morning and disappearing before noon.

The name comes from an American Indian word meaning "frozen fog of fine sharp ice needles."

681. What is the *garua*? In winter, a thick, damp fog called the *garua* forms along the coasts of Ecuador, Peru, and Chile. This fog creates a raw, cold atmosphere on the coast for several weeks.

682. What is fog drip? Fog drip is the term applied to the water dripping from trees or other objects where moisture has collected from windblown fog.

This fog drip is extremely valuable, for sometimes it is as heavy as light rain, as in the redwood forests of coastal California, for instance.

This kind of drip keeps vegetation thriving in parts of the world where there is little or no rainfall. In southern California, as much as 0.05 inches of water has dripped in a single night, the equivalent of a moderate shower.

Certain regions of England also continuously drip with fog, as well as Table Mountain in South Africa and parts of Ascension Island.

In some regions of the Hawaiian Islands, Tasmania, and South Africa fog drips as much as 10 to 20 inches a month onto the land.

Sometimes more water is obtained from fog in these areas than from rain or snow.

683. What trees best catch fog? Pine trees, fir trees, and other conifers are able to condense more water from damp air than other trees because their needles have greater total surface upon which the water condenses. Also these trees retain their needles all year around.

Groves of these trees, or others depending on the location, are often planted on fog-swept hillsides above valley reservoirs or other places where water is needed.

684. What is the difference between cold fog and warm fog? Researchers also classify fog according to the temperatures of the fog droplets or the air surrounding them.

Cold fog forms at temperatures below freezing—the water droplets remain liquid sometimes at temperatures as low as minus 14 degrees Fahrenheit. Cold fog has been successfully precipitated by seeding (see Question 725), in the northern climates, particularly around airports.

Warm fog is composed of water droplets at temperatures above freezing. About 95 percent of the fog in the conterminous 48 states is warm fog. This is the hardest to disperse.

685. What is the difference between dense fog and light fog? The transparency of fog depends mainly on the concentration of droplets—the more droplets present, the thicker the fog.

There are several means of classifying fogs in this way. The Weather Bureau has revised the nomenclature several times. It now has the following definitions:

Fog is considered dense if visibility is less than 1,000 feet.

Light fog has a visibility of more than 1,000 feet, with no upper limit.

686. What is the difference between fog, mist, and drizzle? A fog can be defined as a cloud on the ground.

A mist is a thin light fog consisting of very small, numerous drops of water that appear suspended and gently floating down air currents. It produces a fairly thin grayish veil over the landscape. Scotch mist is a combination of thick mist or fog and heavy drizzle, occurring frequently in Scotland and parts of England.

A drizzle is a fine rain, heavier than a mist, composed of falling drizzle drops.

Definitions of these states of moisture vary in different countries. *The International Cloud Atlas* defines them according to visibility and the size of the droplets; British meteorologists define mists as drizzles; and the American *Glossary of Meteorology* defines mists as fog.

687. How has a drizzle been described? Some people have described a drizzle in less technical terms: "Take of London fog 30 parts; malaria 10 parts; gas leaks 20 parts; dewdrops gathered in a brickyard at sunrise 25 parts; odor of honeysuckle 15 parts. Mix. The mixture will give you an approximate conception of a Nashville drizzle."*

688. What is the difference between fog and dew? Fog and dew are similar phenomena in that they are both condensations of water from the air, usually formed at night when temperatures are low.

Dew is water condensed on grasses, rocks, and other objects near the ground. Fog condenses not on surfaces but on tiny particles in the air.

689. What is damp haze? Damp haze is a condition similar to a very light fog, when small water droplets give the atmosphere a somewhat grayish color; but visibility is as high as 1¼ miles. Damp haze is commonly observed on seacoasts and in the southern United States, most frequently when strong winds carry saltwater sprays inland along the shore.

Damp haze differs from dry haze, which is a suspension of extremely small dry particles appearing with yellowish or bluish tinges.

690. How large do fogs get? Fogs have a tremendous range of sizes. Some are so small and localized that they hover over a small creek or river bottom, only a few yards wide. They may fill an inland valley, cover a swamp, or drift through trees, extending for 10 miles

* O. Henry (William Sydney Porter), "A Municipal Report," *An Anthology of Famous American Stories* (New York: The Modern Library, Random House, 1953), p. 512. (From *Strictly Business* by O. Henry, Doubleday & Company.)

or more. Sometimes, along the coasts they may stretch for hundreds of miles.

Over water, fogs sometimes grow larger than over land. Some of the largest fogs are the ocean fogs, extending several thousand feet high and covering several hundred square miles.

691. How high do fogs get? Fogs vary in height, depending on factors such as temperature, moisture, stillness of the wind, and length of time they have been forming.

Some fogs over a quiet lake, for instance, are only a few inches thick, or even less. Ducks swimming in the early morning may be hidden in a thick layer of fog close to the water—except for their heads which stick out in the clear air like periscopes.

Sea fogs may extend 40 or 50 feet above the water, concealing hulls of ships but permitting masts to protrude above the fog layer into clear air.

Other sea or valley fogs can be several thousand feet thick. The average depth of sea fogs is about 500 feet.

692. When are radiation fogs thickest? One might suppose that radiation fogs are thickest just before dawn, when the night air is at its coldest. Actually, the fog is often thickest shortly after sunrise. This occurs because the sun's early rays are not yet strong enough to evaporate the fog droplets, and yet they have enough heat to set up convection currents and turbulence that gently swirls the layers of fog higher into the air.

693. How long do fogs last? Some fogs—radiation fogs—last only one night, forming as the cooling night falls and dispersing with the morning sun.

Other fogs—advection fogs—may be more persistent and settle over an area for several days or weeks, before winds or a fresh air mass move over the area and disperse them.

694. How does fog tend to perpetuate itself? Dense fog tends to shut out the rays of the sun. Sometimes it cuts out 60 to 80 percent of the sun's energy. Consequently not enough radiation penetrates the fog layer to warm the ground and air and to raise temperatures enough to burn off the fog. Even more radiation is cut off when pollution particles combine with fog to thicken the sky with smog.

695. What, officially, is a foggy day? The U.S. Weather Bureau defines a foggy day as one in which fog surrounds the weather station for one hour or more.

696. Where are some of the foggiest areas in the United States? The foggiest place in the United States is the Libby Islands off the coast of Maine, with an average of 1,554 hours of fog each year.

Point Reyes, California, averages 1,468 hours of fog each year.

697. What was a record fog year? A record fog year was 1907, when Sequin Light Station, in Maine, reported 2,734 hours of fog.

698. Are there any benefits in fogs? Fogs have several beneficial aspects.

They can be "caught" by leaves and needles and branches of trees to bring down valuable moisture to otherwise dry areas. (See Question 682.)

They also can form a sort of blanket over the ground where fragile crops are in danger of being frozen. This fog blanket actually prevents heat from escaping from the earth's surface at night. Some farmers have reported that fog blankets can extend the frost-free growing season and increase production.

Russian meteorologists report they have successfully used artificial fogs to protect vineyards from frost.

It has been suggested that manufacturers build their new plants with evaporation towers in rural farm areas so that the resulting fogs could benefit farmers.

699. What is smog? Smog is an injurious, sometimes lethal, combination of fog with soot, smoke, dust, and chemical particles given off from furnaces, factories, incinerators, vehicles, and other combustion objects. Smog is becoming an increasingly serious hazard in our crowded cities and congested industrial complexes.

The term smog to describe the combination of smoke and fog was first coined in 1905 by a British meteorologist.

700. What causes smog? The lethal combination of fog and smoke occurs when layers of air are trapped over a city or valley in a condition called inversion. Inversion occurs when a layer of warm air rests over a layer of cold air and forms a lid or ceiling that prevents the smoke and dust from rising and dispersing into the higher atmosphere, thus precluding normal circulation of air. Consequently as fumes and smoke continue pouring into the air with no outlet, the pollutants build up to great densities.

This condition becomes particularly serious over certain industrialized areas lying in valleys or hemmed in by hills and mountains, where the air mass may remain stagnant for several days. The area around Los Angeles is prone to suffer from such conditions, as well as cities such as Jersey City, New Jersey, and New York City.

701. What damage does smog do? Smogs have caused such serious health problems for human beings as chronic bronchitis, emphysema, asthma, and probably lung cancer. People having to endure smog for a few hours or days cough, wheeze, develop shortness of breath, have smarting eyes, sore throats, and nausea. A long exposure to smog can be lethal to older persons or those already afflicted with asthmatic or cardiac ailments.

Livestock are afflicted in the same way as people.

Crops are damaged, as sulfur dioxide, fluorides, and other toxic chemicals dry out plant leaves and bleach them to a light tan, killing leaf tips, as well as grasses and pine and fir-tree needles. In citrus-growing and truck farm states, such as California and New Jersey, crops have been heavily damaged.

The toxic smog can corrode metals, rot wood, and cause paint to discolor and peel. It abrades, tarnishes, soils, erodes, cracks, and discolors materials. Steel corrodes 2 to 4 times as fast in urban areas as in smog-free areas. The erosion of stone statuary and buildings is becoming a disaster to irreplaceable ancient cathedrals and monuments. Smog causes more than $11 billion in property damage each year in the United States.

702. What was the Black Fog of London? One of the worst smog disasters took place in December, 1952, when a heavy winter fog settled over London for 4 days. The fog combined with toxic sulfur dioxide and other chemicals from fuming industries, traffic, and furnaces of the populous city. There was little wind. A strong inversion blocked the removal of the polluted air.

Within 24 hours, tons of smoke, dust, and fumes had turned the fog brown and then black. Visibility was reduced to only a few inches. People wandered around lost, fell into the Thames River and drowned, or died of exhaustion and exposure when they could not find their way home or to friends. Lung and heart disorders increased; eyes were irritated.

An estimated 4,000 people died, particularly older people and newborn babies. An estimated 10,000 more people were seriously affected. Much livestock and many city pets died, and agricultural crops suffered.

703. What was the Big Smog of Donora? For nearly a week, from October 25 through 31, 1948, a damaging smog settled over a section of the Monongahela Valley near Donora in southwestern Pennsylvania. When it finally blew away, it left 20 people dead and more than 2,000 afflicted with respiratory and circulatory disorders. Several thousand more people had suffered nausea or irritations of the eyes and throat. Many pets—dogs, cats, and birds—had died.

The smog was composed of damaging chemicals such as oxides of nitrogen, halogen acids, zinc, lead, and other metals, as well as carbon monoxide and carbon dioxide. Sulfur dioxide was the chief offender. These gases were waste products of the many factories, refineries, smelters, and automobile exhausts of the area.

704. What other smog disasters have occurred? Scientists estimate more than a thousand deaths a year are caused in smog-filled cities.

The smogs of London have been particularly disastrous, as chilled air collects in the Thames Valley and condenses in the pall of oily smoke rising from the city, often locked in beneath the lid of an inversion. (See Question 700.) The yellowish fog may remain for days, even weeks, until a fresh air mass arrives and blows away the dirty air. A smog in 1956 killed an estimated 1,000 people. Another smog in 1962 killed some 300 people.

Europe's mainland also suffers from smog. In early December, 1930, 63 people died from smog conditions near Liege in the Meuse

Valley, Belgium. Several hundred people suffered from respiratory problems.

Pollution problems have also occurred in the heavily industrialized Ruhr Valley, in Cologne, in Venice, in parts of Greece, and in Japan.

705. What other smog disasters have occurred in the United States? In November, 1953, an inversion settled over New York City for 10 days, causing a possible 200 deaths. Another smog in 1963 killed more than 400 persons.

Other areas with growing threats of smog disasters include Los Angeles, Chicago, St. Louis, and the many cities of the megalopolis area along the eastern coast between Boston and Washington.

One of the more seriously polluted cities in the United States was Pittsburgh. With rigorous smoke and fume control, however, that city has cleared its air of dangerous smog.

706. Why is Los Angeles particularly smoggy? The heavily populated and industrial city of sprawling Los Angeles is situated in a hollow with hills on three sides, and the sea on the fourth. Here moist air from the warm Pacific Ocean blows eastward across a cool water expanse along the coast. A fog forms which blows across Los Angeles, combining with factory smoke, soot, and automobile fumes.

This smog is trapped by the hills on three sides, and the lid of a temperature inversion. It cannot escape until a strong wind clears it out or the inversion disappears and allows the murky air to rise.

707. What pollutants are found in smog? About 10 percent of smog is composed of visible particles such as pieces of carbon, ash, oil, grease, and metal, including lead. Some of these particles are large enough to fall out, while most others are so small they remain suspended in the air.

About 90 percent of smog is composed of particles so small they are invisible, such as radioactive isotopes, hydrocarbons, and insecticides; and also of highly lethal gases, such as carbon monoxide, carbon dioxide, oxides of sulfur, and nitrogen oxides.

708. What are natural pollutants? Nature has long thrown pollutants into the atmosphere, in the forms of forest fire smoke, sand, dust storms, and debris from erupting volcanoes. Volatile hydrocarbons can be emitted by trees, and gases rise from decaying animal and vegetable matter in swampy areas. Even pollen grains of flowers can be so numerous as they blow through the air at certain seasons that they contaminate the atmosphere.

709. What is the worst source of toxicity in smogs? Researchers have found that sulfur dioxide issuing from smokestacks is the principal source of toxicity.

Sulfur dioxide is oxidized in the air to become sulfur trioxide, which combines with water droplets in fog to form sulfuric acid—a particularly strong and corrosive acid that irritates throats and lungs and severely corrodes buildings, clothing, paint, and other items.

710. How is lead being added to the atmosphere? One of the most rapidly growing air pollutants is the exhaust from internal combustion engines. In only 70 years the automobile has strongly contributed to the pallor of smog above our heavily populated cities and crowded regions.

Tetraethyl lead was introduced to automotive fuel in 1923, and since then lead has contaminated most of the earth's surface. Increasing amounts of the toxic materials have been found in the oceans, crops, and human blood.

711. What is fallout? Another source of contamination of the earth is radioactivity from test explosions of nuclear weapons in the atmosphere. The fallout, as the radioactive particles are called, now has contaminated every part of the earth's surface and every living thing.

Strontium-90, one of the constituents, is being built into the bones of every living person.

712. Why is strontium-90 so deadly? When a nuclear bomb explodes, it shoots strontium-90 and other radioactive isotopes high into the stratosphere, from where the elements fall back to earth at a rate depending on the size of the particles and on the weather. Many fallout particles are carried to earth by rain or snow, where they soak into the ground.

Since strontium-90 is a chemical relative of calcium, it reacts much like this valuable and harmless element. Like calcium, strontium enters the root cells of plants in the ground or enters leaf cells from raindrops caught there. Strontium thus enters the plant system.

When a cow eats the contaminated plant, strontium-90 again acts like calcium and concentrates in the cow's milk. When people drink the milk, the strontium is absorbed and retained in their bones.

713. How is air pollution making the earth's climate colder? Some scientists believe that the increasing layer of dust, soot, and smog particles being thrown into the atmosphere is acting as a shield to block the sun's rays from reaching the earth's surface. With less warmth received from the sun, the earth may be slowly cooling off. This may result in advancing glaciers, in colder and longer winters, and in a general decrease in temperatures throughout the world.

714. How is carbon dioxide making the earth's climate warmer? The pollutant carbon dioxide has an extremely important role in the weather and climate of the world. Although only relatively minute quantities are present in the atmosphere—about 0.03 percent by volume—they are enough to be a very efficient absorber of infrared or thermal radiation. Like the glass of a greenhouse, a layer of carbon dioxide permits passage of the sun's light waves, but traps the longer infrared or thermal rays within the earth's atmosphere, thus increasing the temperature. This condition is called the greenhouse effect. (See Questions 715 and 716.)

Nearly 6 billion tons of carbon dioxide are added to the world's atmosphere each year. Some of this is absorbed into the oceans, and some is taken up by plants and vegetation around the world. But most of it remains as a shield around the earth. Since the industrial revolution, the percentage of carbon dioxide in the air has increased by 14 percent. One scientist believes the present amount of carbon dioxide keeps the world temperature about 20 degrees warmer than it would be if there were no carbon dioxide.

Meteorologists estimate that as continuing rates of carbon dioxide increase, the world's climate will be about 2 or 3 degrees warmer by the end of the twentieth century. This continual warming would result in significant changes in the wind and air circulation, plant growth, man's adaptation, and other factors in the ecology of the entire world.

715. What is the principle of the greenhouse effect? The temperature inside a conventional greenhouse for plants remains high because the glass permits passage of shorter wave radiation from the sun— the visible light rays—but does not permit passage of longer rays— the infrared and heat waves. In other words, light from the sun passes through the greenhouse glass. Inside the greenhouse, this light is absorbed by soil and plants and converted to longer wave heat energy and infrared radiation. These long waves cannot pass through glass, hence remain trapped within, raising the temperature of the greenhouse. At night, or in cooler days, or when a wind is blowing, this heat is radiated outward from the building through the walls.

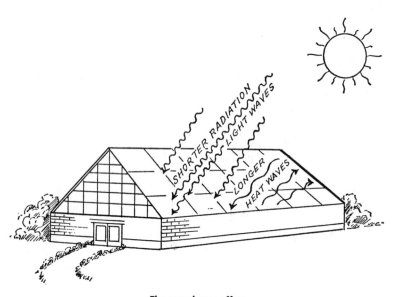

The greenhouse effect

716. How does this greenhouse principle apply to the world? In much the same manner as a glass greenhouse is heated by the sun, scientists believe the world is heated. They think the layer of carbon dioxide surrounding the earth acts as the glass of a greenhouse—it permits the shorter rays, the visible light of the sun, to pass through, but prevents longer rays of infrared heat energy from radiating outward into space. As the amount of carbon dioxide is increased, the temperature of the earth is therefore constantly increased.

717. How long has man been polluting the air? Ever since man first began using fire and sending smoke into the air, he has been contributing to air pollution.

At first, in his early years of evolution, his contribution was negligible. As his species increased on earth and he developed more uses for fire, however, the quantity of air pollution increased. Yet still this caused no problem, partly because there was so little of it and partly because the winds and atmosphere could disperse and clean the pollutants.

The arrival of the industrial revolution about 100 years ago brought about a profound increase in the amount of pollution man put into the air. As man's civilizations expanded and developed into more industrial and technical societies, combustion processes continued to increase.

Today, with vehicles, factories, industries, heating of homes and buildings, burning of trash and solid wastes, generation of electric power, more numerous and more powerful jet airplanes—pollutants are being added to the atmosphere at an alarming rate.

718. What methods have been undertaken to clean up man-made smog? The demand for effectively cleaning the air is steadily increasing, not only from scientists and doctors, but from architects, property owners, and most importantly from increasing numbers of citizens who are becoming aroused as to the growing danger.

Many remedies have been proposed and put into effect, such as better control of combustion itself—for example, more complete burning of materials so less debris escapes into the atmosphere or lowering the content of toxic materials, such as sulfur and carbon monoxide. Many variations of dust and ash filters have been devised to screen out the larger particles. Electrostatic precipitators have been designed to charge the smaller particles, which are then attracted to electrodes and captured.

City and state ordinances have made various legal remedies, such as prohibition of trash and leaf burning and insistence on proper filters for vehicles and smokestacks. The Federal Government has set out air quality standards and methods to achieve them in acts such as the Clean Air Act of 1963 and the Air Quality Act of 1967.

Some cities have had extraordinary success in clearing up the air. London in the early 1920's was inundated with 600 tons of pollution per square mile in a year. Today less than a third of this amount falls from the sky. Pittsburgh, St. Louis, and Cincinnati and other U.S. cities have also carried on successful air clean-up campaigns.

719. How can fogs and smogs be prevented from forming? In addition to the various methods of preventing smog (see Question 718), researchers are experimenting with methods of preventing fogs. For example, chemical films have been spread over swamps and lakes in the vicinity of airports and highways to reduce evaporation and fog formation.

By planting vegetation around areas where shallow radiation fogs form, men have hoped to trap the fog by catching moisture on plant leaves.

720. What methods are being studied to dissipate fogs? Hundreds of schemes and ideas have been searched in the attempt to dissipate fogs. A few have been found to be of practical use.

Scientists have approached the problem in many ways, including evaporating the fog by heating the atmosphere, removing fog droplets by causing them to coalesce and drop, and physically blowing the fog away.

721. What attempts have been made to blow fog away? Some researchers have attempted to disperse fog by setting up wind currents with giant fans. This method is not successful with advection fogs where the wind blows in more fog. Fans have been found somewhat successful for dispersing local fogs. For instance, fans have blown away local fogs in the Arctic where settlements emit water vapor and heat that saturate the air and produce fogs around the area.

722. What experiments have been tried to disperse fog by cold or to evaporate it by heat? Man has tried to inject cooling agents in a fog area in efforts to cool fog particles to a point where they will freeze and eventually precipitate. By seeding fog with particles of a very cold substance such as dry ice or liquid propane, researchers can cause some of the droplets to freeze. Some of these frozen droplets shatter, forming tiny splinters of ice which serve as nuclei for condensation of other droplets. Sometimes water vapor in the air condenses onto these ice crystals, and the resultant drying of the air turns additional fog droplets back into invisible vapor.

Unfortunately, cold fogs are rare in the populated areas of the warmer climates.

Man has tried to burn off fog by raising the air temperature with barrels of burning fuel. Some have considered raising the temperature with large electric heaters or with jet-engine exhausts. These methods have been somewhat effective, but too costly to be practical on a large scale.

723. What was FIDO? A practical method of clearing foggy airfields with heat was called FIDO (Fog Investigation Dispersal Operation), developed and used successfully by the British in World War II. With this method, oil was burned in rows of jet faucets arranged around the airfield. Even with a moderate wind, the heat would clear a large field of fog to heights of 100 feet or more above the runways in only a few minutes. With stronger winds, more heat was required.

During a period of 2½ years, 2,300 fogbound planes were able to land or take off from British airfields because of the success in burning away fogs. Some 15 million gallons of gasoline were used.

Although the method was successful during the emergencies of wartime, the high costs of installing the equipment—$1 million to $2 million—and of operating the process—$100 to $500 a minute—have prevented its more widespread use.

724. What methods have been tried to precipitate fog? In laboratory experiments, scientists have found that fog droplets coalesce into larger drops and precipitate when subjected to intense high-frequency sound waves. As yet, this method has not been successful in any large-scale practical application.

Another method of precipitating fog involves the creation of a "curtain" of salt or other chemical spray, shot through nozzles in a long line to the windward of an airstrip. As the fog blows through the spray curtain, water coalesces on the chemical nuclei and drops to earth. This produces a clear patch of air to the lee of the curtain. The chlorides in salt and other chemical agents, however, have proved too corrosive for aircraft equipment and property.

725. What experiments have been done with seeding fogs? Some of the most successful attempts to break up cold fogs have been made by seeding the fog with various chemicals and particles, in an effort to distribute nuclei upon which fog droplets coalesce and grow heavy enough to drop to earth.

This method has proved successful in particular with cold fogs, for chemical nuclei such as silver iodide, silver nitrate, and others become more activated at lower temperatures, mostly around 0 degrees Fahrenheit.

Unfortunately, cold fogs are rare in the populated areas of the warmer climates.

726. What complicates the dispersal of fog? Winds make a big difference in attempting to disperse fogs. When the air is calm, clearing a fog is relatively easy because the volume of air to be treated is limited. Whenever a brisk breeze or wind is blowing, the problem is compounded by having the cleared space filled up again with fog driven by the wind.

727. What instruments are used to penetrate through fog? Old-fashioned foghorns have disappeared from many busy commercial shipping lanes although they still exist in many places. These resonant sounds were often not successful in penetrating a fog, since sound does not travel far enough or quickly enough for today's fast ships. Also the sound waves have been known to skip over certain areas, leaving pockets of silence and not reaching a ship bearing down upon rocks in the fog.

Lighthouses also are edifices of the past, since the circular areas swept by their flashing lights are too small for modern sea traffic.

The advent of radar in World War II brought considerably more protection to fogbound ships and planes. With radar scanners, pilots can "see," and "blind" landings can be made by aircraft.

On airplane runways, powerful searchlights have successfully penetrated some fogs, enough to allow landings and takeoffs.

728. What modern instrument is more successful in penetrating fogs? A relatively new device, the laser radar, now bounces concentrated rays of light off distant objects in thick fog. Objects can be scanned and identified to a greater degree than with radar.

729. What research institutions are instigating research on fog? Fog has become a matter of such growing concern that increased personnel of the U.S. Weather Bureau and the U.S. Department of Defense, as well as private industries and airlines, have been assigned to study the phenomenon and methods of controlling it.

United Air Lines has had success seeding supercooled fogs in the Pacific Northwest and Alaska. The Air Transport Association of America, composed of 34 airlines interested in solving fog problems, has commissioned a testing service, World Weather, Inc., of Houston, Texas, to study dispersion of warm fogs. Theoretical, laboratory, and experimental work on fogs has been progressing at Cornell Aeronautical Laboratory in a study called "Project Fog Drops."

The Environmental Health and Safety Department of the University of California at Berkeley is studying the properties of smog and other pollutants. The Expert Committee on Atmospheric Pollutants of the World Health Organization has been active in analyzing air pollution and making recommendations on its control. The National Center for Air Pollution Control, under the U.S. Department of Health, Education, and Welfare, has been conducting research.

The Air Resources Laboratories, part of the Environmental Science Services Administration, conduct research on diffusion, transport, and dissipation of atmospheric contaminants in the laboratory and the field.

730. What methods are being pursued by the Russians? Scientists at the Central Aerological Observatory of Russia have been trying various methods of fog modification, including heating by burning various fuels and by increasing absorption of solar radiation. The methods of water vapor absorption being tested include the use of cooling reagents, ice-forming aerosols, soluble substances, chemisorbents, and fine-pored absorbents. The Russians are also investigating forced filtration of fog, effects of spraying water, use of charged particles, and electrical or acoustical fields.

VII. RAINSTORMS, SNOWSTORMS, AND ICE STORMS

Introduction. Falling, spinning, cascading, floating out of cloudy skies come countless numbers of water drops in all forms, sometimes as raindrops or drizzle drops, sometimes as snowflakes or ice pellets. These drops, either in solid or liquid form, are vital elements of the never-ending circulation of water throughout the earth and atmosphere.

This perpetual precipitation of water, always occurring sometime at some place on earth, is one of the world's most important meteorological events. It is the only system by which lakes, streams, ground water, oceans, and other natural water sources are replenished with fresh water.

Yet excesses of this precipitation can bring destruction to mankind: inundating plains and fields with floods, setting off mudslides and landslides, paralyzing cities and highways with snow and ice, and crushing houses and toppling trees with tons of snow avalanches that cascade down mountain slopes.

In today's burgeoning growth of cities, highways, airports, dams, factories, and other complexes, these storms of summer and winter, spring and fall, are constantly harassing mankind.

731. What are the storms of earth? The great wet storms of the world occur when wind, temperature, and condensation forces work together to create precipitation of water in forms of rains, drizzles, snows, sleets, hails, ice pellets, frosts, and freezes.

These storms are all part of the great hydrologic cycle in which water is constantly circulated from the atmosphere onto the earth, through the soil, into the oceans, and back into the sky again—bringing life-giving water to living creatures of this water planet. (See Question 732.)

Yet the excesses—or dearth—of any of these forms of precipitation often bring grief and disaster to man and his complex civilization.

732. What is the hydrologic cycle? The hydrologic cycle is the vast circulation of water from the earth into the atmosphere, down to earth, through the soil and seas, and up into the sky again.

For some 3 billion years, since the earth's atmosphere was formed, about 326 million cubic miles of water have been constantly in motion in this enormous cycle.

In general, the cycle functions in the following manner:

Water is evaporated into the atmosphere from surface waters such as oceans, lakes, streams, and wet soaking ground—even from people and animals. An enormous amount is also transpired from every type of vegetation—from leaves, stalks, petals, and other parts of plants. This water vapor rises into the atmosphere, propelled mainly by warm rising convection currents—until it begins to condense into water droplets to form clouds and fogs. Coalescing into larger drops and pulled by gravity, the droplets precipitate as rain, snow, hail, and ice. These particles fall to earth over land and sea, and sooner or later the water sinks into the ground as groundwater, or runs off into lakes and rivers, and eventually back to the oceans.

733. How does precipitation begin? Precipitation of water from the air may begin with the cooling and condensation of water vapor, which is invisible moisture in the air, usually upon tiny particles of ice or chemicals, called condensation nuclei. (See Question 734.)

A simple relationship exists between temperature and molecular activity: an increase in temperature excites the molecules. This means the higher the temperature, the more active the molecules.

In warm air, water-vapor molecules are in a highly excited state, too active to stick to one another. As the air cools, the molecules slow down and tend to adhere to each other or to condensation nuclei in the air. As they stick and condense, they grow larger into visible particles—cloud droplets.

734. What are condensation nuclei? Condensation nuclei are tiny, often microscopic, particles floating throughout the atmosphere upon which water vapor can condense to form water droplets. They may be molecules, pieces of molecules such as ions, or larger particles of material such as dust or ice. They are composed of chemicals such as sodium, potassium, ammonia, sulfur, nitrogen, chlorine, ozone, and oxides of many elements.

Many of these particles come from the ocean—blown into the air when winds churn the seas into breaking waves, fountains of spray, or bursting bubbles, all spewing bits of chemicals and particles into the air. Other particles are whirled into the air when land is plowed on farms, when wind stirs up dust storms over deserts and arid lands, or when volcanoes erupt.

Other nuclei include coal smoke, soot, fly ash, and other particles rising from industrial and domestic combustion and from the exhaust gases of aircraft and automobile engines. Grass fires and forest fires also contribute particles of coal dust, carbon black, oils, and tars.

Depending upon their physical and chemical nature, particles are suspended throughout the atmosphere for varying lengths of time—sometimes only for minutes, sometimes for years.

The concentration, size, distribution, and nature of the particles determine the amount and rate of condensation that occurs.

735. How does air become cool enough to start precipitation?
Warm moist air may become cooled in several ways. For instance, moist air may rise by convection currents over heated ground, as a cumulous cloud rises and develops to form a towering thundercloud. (See Question 291.) Or it may be lifted as it blows up and over hills and mountains. Cooling in these two cases comes about by expansion of air and resulting lowered temperatures as the air pressure decreases with height.

For instance, air also becomes cooled when winds converge in a cyclone or low-pressure area and slowly circle upward. When a large block of cold air wedges beneath a mass of warm air, the warm air is forced to rise and cool. Warm air also becomes cooled when it is overridden by a mass of sinking cold air and loses its heat as the two masses mix.

736. How do cloud droplets grow into raindrops? Minute cloud droplets grow into larger raindrops by a process called coalescence, which literally means growing together into one body.

Meteorologists believe coalescence may come about in a number of different ways.

One way, for instance, is simply by the collision of one droplet with another as they drift around in the air currents. The larger droplet absorbs the smaller and keeps growing larger with each collision until enough water exists in one drop for it to have mass and weight enough to fall.

Coalescence may also come about through a process of ionization. Many droplets in a cloud are ionized—either with a negative or positive charge of electricity. As these ionized particles drift around, they come in contact with one another. Just like magnets, particles with opposite charges attract each other. They merge together and form larger and larger drops.

737. What is the shape of a raindrop? The traditional concept that a raindrop is shaped like a pear as it falls through the air has been discarded as erroneous. Actually, photographs taken with high-speed equipment show that a large raindrop is shaped somewhat like a doughnut with a hole not quite through it.

738. What disasters can rain cause? The greatest curse from rain is from too much water. The torrents that flood down from clouds can fall so heavy and fast that the soil cannot absorb the moisture. Rain falling at the rate of an inch or more an hour does not soak into the soil. It runs off, causing serious destruction by eroding the land, undermining buildings and highways, destroying crops, and causing the deaths of thousands of people and millions of animals.

Throughout the history of civilization, floods have inflicted terrible disasters in loss of life and property. The constant patter of rain, saturating the ground and spilling over from streams and rivers, is causing increasing damage to property as people continue to build larger cities and more industrial complexes in flood plains.

Floods can result from factors other than rain—high winds can bring excessive ocean surges onto coastal areas; spring thaws can melt mountain slopes or high plateaus of snow; and rivers can become blocked by large landslides or ice jams and back up over hundreds of square miles.

739. What is considered the most disastrous flood in history? During the months of September and October, 1887, the Hwang Ho or Yellow River of China rose so high it flowed over 70-foot-high levies and inundated some 50,000 square miles of farmland. Reports of deaths from this great flood vary, but some claim that 6 million people were killed that autumn, 300 villages were inundated, and 2 million people were left homeless.

This great river, known for centuries as China's Sorrow, winds across China's plains for 3,000 miles, often flooding over its banks and destroying villages.

740. What were some other catastrophic floods? Floods in over-populated regions of Asia and the Middle East have often taken the lives of thousands of people in one inundation.

One of the worst floods of recent years occurred in southern Viet Nam on November 12, 1964, causing the deaths of some 5,000 people.

Floods in Teheran in August, 1954, killed about 2,000 persons, and a flood occurring the same year in Tibet killed about the same number.

In August, 1951, a flood in Manchuria drowned 1,800 people.

741. Has America had heavy floods? Many huge floods have over-flowed the banks of America's large rivers, but fortunately the loss of life has been relatively low. For many years, population was scarce along the water fronts. Today crowded cities have developed and spread out on the flood plains, often in the paths of annual floods. Yet river stations keep such constant watch over the rise of the waters and warning systems are so advanced, that people are able to evacuate endangered regions before the flood waters arrive. Property damage, however, continues to increase and will continue to do so until people become more aware of the nature of rivers and their flood plains, and until strict laws of building zones are enforced.

742. What was the worst flood tragedy in U.S. history? On May 31, 1889, heavy rains swelled the Little Conemaugh River in Pennsylvania so high it broke through the South Fork Dam and sent a wall of water 30 to 40 feet high surging down the valley at speeds of 22 feet a second onto the industrial town of Johnstown. No one knows for certain how many people were killed that day, but one official record reported 2,100 people dead as the mass of water swept down the valley, razing every tree, house, and building in its path—including train locomotives and coaches and 50 miles of track.

743. What was another disastrous U.S. flood? A sudden warm spell in March, 1913, arrived with heavy rains over the Ohio River basin, melting deep winter snows and causing a flood that killed 467 people —the second highest death toll from floods in the United States.

744. What are flash floods? A flash flood is a sudden burst of water that pours down narrow mountain valleys or desert gullies, often washing out bridges, roads, and nearby houses in its path. Caused by a sudden rainstorm, these floods do not cover a large area and do not last long. But they arrive so fast that people have little or no warning and many are drowned.

One disastrous flash flood occurred in 1903, when a cloudburst caused a 20-foot-high wall of water to rush down Willow Creek, Oregon, hitting the town of Heppner. In less than an hour, more than 200 people were drowned, and nearly one third of the town was washed away.

Another disastrous flood occurred in 1955 when a cloud suddenly dropped 14 inches of rain over New England. The Connecticut River rose nearly 20 feet, killing 186 people and damaging property worth more than $2 billion.

745. What are hurricane floods? Some of the heaviest and most destructive floods result from thousands of tons of water dropped by hurricanes, or typhoons as they are called in Asia. Deaths of thousands of people have been caused by these storms, particularly in Asia. (See Questions 6 through 9.)

746. What kinds of property damage do floods cause? Floods can be enormously destructive to any kind of property lying in their paths. Bridges and buildings collapse, towns and fields of valuable crops are inundated, and machinery and materials are soaked and corroded beyond repair. Much damage is caused by debris such as mud, silt, branches, rocks, and sand that is carried along by the waters and deposited in thick packed layers over the land.

One of the most costly floods in the United States was caused by a stationary rainstorm over the Kansas River basin in July, 1951. For 4 days, rains fell on an already soaked area, creating a flood that swept over heavily populated areas such as Topeka, Lawrence, Kansas City, and Jefferson City. Fortunately, only 28 deaths occurred, but damages ran as high as $935 million.

747. How do rains cause landslides? Heavy rains are often the cause of some of man's persistent property destroyers—landslides and mudslides. As more slopes are undercut and leveled to develop real estate property and highways, they become unstable and subject to slides. Rainwater saturates the soils, loosening them and lubricating each grain so that friction is reduced and thousands of tons of soil may suddenly slip to the bottom of the slope. Houses, walls, trees, everything on the surface is carried along with the sliding earth.

Landslides are not always caused by rainfalls. Other forces such as earthquakes, ocean waves, eroding rivers, excavation mines, even changes in temperature can start tons of earth thundering downward.

748. Where have rain and man caused landslides? As men continue expanding their communities, they often sprawl over hills and mountains where careless excavation and construction can create hazardous conditions. Some of the worst property damages have occurred in California, where increasing populations have crowded the hills.

In January, 1969, the heaviest rainfalls in 31 years fell over southern California, setting off landslides that killed 88 people, stranded tens of thousands, and caused about 9,000 persons to evacuate their mud-filled, toppled homes. Property damage was estimated to be some $35 million.

Tons of gravel and mud slid down on slum hovels in Salvador, Brazil, on June 17, 1968, killing 20 persons.

More than 100 women and children were killed when two sight-seeing buses were caught in a landslide and slid into flooded river waters near Honshu, Japan, on August 18, 1968.

749. What was the Aberfan landslide disaster? Heavy autumn rains of October, 1966, weakened a huge mountain of black coal mine waste in south Wales and sent 500,000 tons sliding into the mining town of Aberfan, killing more than 144 people, 116 of whom were children trapped within the doomed school house. The waste pile had been heaped up during the past 50 years right behind the town, in spite of warnings that it was getting too high and dangerous.

750. What kinds of clouds produce rain or snow? In order to produce rain, clouds must contain combinations of water droplets, condensation nuclei, and/or ice crystals and be thick enough—usually several thousand feet—to permit the condensation-to-precipitation process.

Fibrous, striated, uniform-textured clouds called altostratus (from the Latin word *stratum*, meaning layer or sheet) usually form at middle altitude heights of about 6,500 to 20,000 feet. When these become thick and drop lower to form a nimbostratus cloud, steady rains or snows are produced. This precipitation usually persists for several hours, sometimes for a day or more.

Clouds of vertical development, having a piled-up or cauliflower

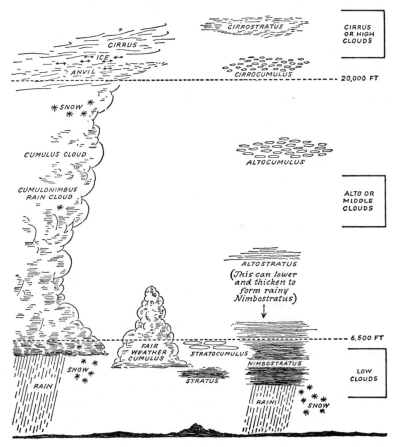

Cloud formations

appearance and called cumulus (from the Latin word meaning mound or heap), may grow tall and thick enough to produce intermittent showers of rain or snow. These cumulus clouds sometimes build up to form the towering thunderclouds producing heavy thundershowers.

751. Why doesn't rain fall from all clouds? In general, clouds that are made up only of ice crystals at heights above 20,000 feet—the cirriform clouds—will not produce rain or snow.

The thin lower clouds at altitudes below 6,500 feet—the stratus, strato-cumulus and small cumulus—made up only of water droplets, will not produce precipitation of any significance. Sometimes they produce fine drizzles.

Sometimes cumulus rain clouds pass overhead, loaded with moisture that seems ready to drop and soak the earth, yet not a single drop of rain reaches the ground. This is because the raindrops are so small, they are pulled only minutely by gravity toward earth. The uprising convection air currents are so strong that the drops are buoyed up and kept aloft.

Some raindrops are so small by the time they leave the bottom of the cloud that if they fall through warm air they may evaporate before they reach the ground. It is not uncommon to see streamers of rain falling from the bottom of a cloud, and yet feel no rain. This condition does not usually last long, however, for as the drops evaporate, they remove heat from the air, and so reduce the updraft current of air. When this occurs, the raindrops fall all the way to the ground.

752. How much rain falls in a storm? Storms bring enormous quantities of water to earth. Scientists calculate that, with a storm dropping 7 inches of rain over a region, more than 189,000 gallons of water falls on 1 acre alone. This weighs more than 100 tons.

753. What regions of the world have heaviest rainfalls? Meteorologists believe some of the greatest amounts of constant rainfall may fall upon the equatorial oceans, where the northeast trade winds converge with those of the southwest to produce rising cyclones of moist warm air. As these air currents rise, the moisture falls in heavy torrents. As yet scientists have not been able to take adequately accurate measurements of rainfall along these oceanic regions.

Heavy rainfalls have also been recorded along slopes of mountains and hills standing in paths of moist warm winds, usually in the tropical climates. As winds are pushed up along the windward side of the hills, they drop their moisture in torrential rains.

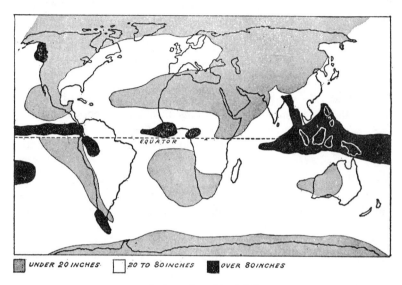

UNDER 20 INCHES   20 TO 80 INCHES   OVER 80 INCHES

Regions of world rainfall

754. What are some record rainfalls? Some of the greatest rain records have been taken on the slopes of Mount Waialeale, Kauai, Hawaii—a mountain that stands in the path of the moist northeast trade winds and receives some 476 inches of rain each year.

Another extremely rainy spot of the world is Cherrapungi at the end of a funnel-shaped valley in northeastern India, where the southwest monsoons start rising to cross the Himalaya Mountains. Some 450 inches of rain fall in this area each year. In one year—from August, 1860, through July, 1861—1,041.78 inches of rainfall were measured.

The Olympic Peninsula at Wyoochee, Washington, catches about 155 inches of rain each year.

755. What can be done to prevent excessive rainfall and flooding? As yet, little can be done to prevent excessive rainfall. Scientists continue experimenting with cloud seeding (see Question 150) in an effort to precipitate rain in smaller quantities before the clouds build up to such size they might release destructive downpours.

Much is being done to prevent rain-swollen rivers from flooding and destroying valuable property and land. The primary idea is to establish flood zone areas on flood plains so that man will be discouraged from building expensive houses and factories in sites where floods are expected. More research and experiments are being conducted on flood-proof buildings, materials, and equipment.

756. What rain warnings are issued? In cooperation with weather report and river surveillance stations throughout the country, the U.S. Weather Bureau and other agencies keep close watch on the passage of rainstorms across the country, the amount of water they drop, and the resultant rise of streams and rivers to possible flood stage. When the rainfall or river waters become excessive, warnings are sent by radio and television to endangered areas, where people have time to prepare for the storm or evacuate the area if necessary.

757. What is a snowstorm? Fragile snow crystals fall to earth in many forms, sometimes creating scenes of transient shimmering beauty but often causing misery and disaster.

Snowfalls can be continuous or intermittent, light or heavy, lasting or brief. The U.S. Weather Bureau has defined degrees of snowfall with definite terms that aid in forecasts and warnings.

758. How is snow formed? Snow crystals form in clouds much the same way as do raindrops—by coalescence of tiny water droplets, usually upon a minute particle of dust, chemical, or upon an ice crystal.

Water droplets of clouds may remain in supercooled liquid form until temperatures drop to 20 degrees Fahrenheit or lower. In laboratory experiments, supercooled drops may remain liquid as low as minus 40 degrees Fahrenheit, at which point they all suddenly freeze. At these low temperatures, complex heat exchanges take place, and the supercooled droplets sublime (evaporate, then freeze) directly on nearby ice crystals or other nuclei. As more droplets coalesce, the crystal grows in size until large enough to drop by gravity. Scientists estimate that a million supercooled cloud droplets come together in about 10 minutes to form one snow crystal—which is the same as a snowflake.

759. What causes a snowstorm? Snowstorms can be formed when series of gigantic cyclones, or low-pressure systems, and anticyclones, or high-pressure systems, move across the country.

In the United States, a cyclone with a center of low pressure brings in large masses of warm moist air from the south and southeast. As a cold front coming from the north and northwest moves across the land, it wedges in beneath the warm moist air, pushing the moist air upward to altitudes and temperatures where the moisture condenses and falls to earth as snow. Cold-front snowstorms are usually rough and windy, followed by clearing skies and frigid temperatures.

A warm-front snowstorm occurs when a warm front is displacing a cold front and overrides the cold air, being forced to rise. The precipitation freezes into snow as it falls through the cold air. This snowstorm is more widespread and persistent than the cold-front snowstorm.

760. What kinds of damage does snow cause? A snowstorm can take lives of individuals unfortunate enough to be trapped in its numbing ice and cold. It can kill thousands of unprotected cattle, sheep, or other livestock.

Snowstorms can cause millions of dollars of damage to highways, runways, buildings, and other structures.

They can also cause incalculable, intangible damage by delaying or preventing busy officials, scientists, business men, and other top authority people from getting to their offices or conferences. Cities in the mid-coastal regions of the United States, for instance, can become immobilized for hours by merely a light snow—mainly because people are not used to it and feel they cannot cope with it.

761. How do snowstorms kill people? Many people die during snowstorms from automobile and other vehicle accidents. Many others have heart attacks resulting from exhaustion and overexertion from shoveling snow, pushing vehicles, or even walking in deep snow. Many more die from exposure and freezing if they happen to lose their way. Miscellaneous accidents such as home fires, carbon monoxide poisoning in stalled cars, falls on slippery walks, electrocution from fallen electric wires, and bruises from collapsing buildings can occur during snowstorms.

In the high mountains, snowstorms build up masses of snow that may suddenly thunder down slopes as avalanches, engulfing everything in their paths.

Of the 3,000 people killed by winter storms in the United States in the past 30 years, more than one third died from automobile and other accidents, about 800 died from heart attacks, and 350 persons froze to death or died from exposure.

762. What was the blizzard of 1888? On Sunday evening, March 11, 1888, 2 huge storm systems clashed over the eastern coast and began dropping their load of snow on an area extending from Washington, D.C., to Maine, from New York to Pittsburgh, piling snow drifts 40 to 50 feet high, isolating towns and cities, burying houses, stranding buggies and trains, striking ships at sea, and causing the deaths of more than 400 persons.

One of the 2 storms was a cold winter cyclone stretching from the Great Lakes to the Gulf of Mexico, sweeping eastward at speeds of some 600 miles a day. The other storm was moving up from the south, from Georgia northward along the Atlantic coast, bringing warm rain and winds. The 2 collided and stalled for a day and a half. A high-pressure area over Newfoundland kept them from moving out to sea.

From 40 to 50 inches of snow fell in Connecticut and Massachusetts, but the greatest concentration of the storm centered on New York City. Winds blew as fast as 70 miles an hour, and 21 inches of snow fell with drifts more than 20 feet high. Thousands of people were marooned as horse cars, cable cars, and elevated trains stalled. There were remarkably few accidents and deaths, as New Yorkers faced the disastrous event with courage and spirit.

On the ocean, from Chesapeake Bay to Nantucket, 200 ships were reported blown ashore, sunk, or severely damaged.

● **763. What was the blizzard of 1966?** One of the worst storms in the United States lasted 4 days in February, 1966, disrupting activities of millions of people in the eastern part of the country and causing at least 208 deaths. The Northern Plains were submerged by a blanket of snow that stranded travelers and killed much livestock. In South Dakota alone, 96,000 cattle, sheep, and hogs were reported dead.

The main part of the blizzard hit central New York State, depositing 53 inches of snow on Syracuse and more than 28 inches on Rochester.

Yet this destructive storm also brought a blessing: it helped break the serious 5-year drought in New England.

764. What was the tragedy at Donner Pass? In October, 1846, early snow began to fall over the Sierra Nevada mountain range in northern California, trapping a wagon train of 87 pioneers traveling from Springfield, Illinois, to California under the leadership of George Donner. The group had already been delayed on the salt flats of Utah and Nevada and had difficulties in starting to cross the Rockies.

As the party attempted to drive their wagons through the Truckee mountain pass a few miles north of Lake Tahoe, they encountered early autumn snows lying some 5 feet deep. They made camp and for weeks attempted to survive the severe winter storms that piled up snow drifts 30 to 40 feet high. As their food supply dwindled and they could get no outside help, in desperation they began to eat their own dead. By February, 5 months later, when rescuers finally discovered them, only 47 people were alive.

765. What is the dreaded purga blizzard? One of the fiercest blizzards of the world is the purga—or poorga—that sweeps over northern Siberia. This severe storm usually arrives every winter, sweeping down from the north or northeast tundra regions and blowing with extraordinary violence over the northern open plains onto the Kamchatka Peninsula. Sometimes it blows into southern Russia.

The powerful purga is so filled with wind-driven snow that people report they cannot open their eyes and have difficulty breathing and standing upright. During the blizzard's onslaught, people have been frozen to death only a few yards from their homes, having completely lost their direction in the blinding snow.

766. What is the buran? The buran is a violent snowstorm that drives from the northeast over south Russia and central Siberia. Similar to the purga of northern Russia and the blizzard of North America, the buran fills the air with snow. Winds blow so violently that the cold seems unendurable, although the actual temperature is not especially low. Men and animals become lost in the blinding storm and often perish.

767. What are avalanches? High in the snow-filled mountain ranges of the world, masses of snow may suddenly break loose and slide down slopes, crushing houses and whole villages in their paths and suffocating or crushing men and animals.

Depending on the quality, quantity, and condition of snow, the angle of the slope, the amount of wind and stability of the existing snow cover, the angle of the sun's rays, and other factors, these white juggernauts may weigh hundreds of thousands of tons and spill down the mountain at speeds as great as 225 miles an hour.

Some kinds of avalanches are composed of wet, heavy snow and travel close to the ground. Others are made of dry, loose, powdery snow and lift off the ground, sometimes swirling through the air at heights up to several thousand feet.

768. Do avalanches cause much destruction? Compared with other natural hazards, avalanches do not kill many people. On an average, about 150 persons a year are reported killed by snow or ice avalanches throughout the world, although there have been terrible exceptions. (See Questions 770 through 772.)

Annual property damage is also relatively small.

The relatively low number of deaths and slight destruction stem from the fact that these slides generally occur in remote mountain regions, where not many people have settled.

769. Where do avalanches occur? Avalanches occur wherever snow falls on mountains or steep slopes. Most of them occur in the large mountain ranges—the Himalayas, Alps, Rockies, Caucasus, Andes, and Pyrenees. They also occur on snow mountains in Japan, Persia, New Zealand, and the polar regions. In North America they occur in the Sierras, the Cascades, and in the Alaska ranges.

770. What was a disastrous avalanche? One of the worst single avalanche disasters occurred on January 10, 1962, when a huge mass of ice weighing 3 million tons broke from a glacier on the Nevada Huascaran mountain in Peru. The avalanche rolled and ricocheted over 6 villages and damaged 3 more. More than 4,000 people were crushed by this catastrophe, and an estimated 10,000 animals died.

771. What other avalanches have caused much loss of life? The worst avalanche disaster, in terms of numbers of people perishing, was the series of avalanches that killed some 10,000 Austrian and Italian troups in the Alps during World War I. This heavy loss of life occurred during a 24-hour period on December 13, 1916.

Reports estimate that during that war, more than 60,000 soldiers were killed by avalanches in the Alpine region.

772. What was a disastrous avalanche in the United States? The most disastrous avalanche in the United States occurred at the Wellington train station, Washington, in the Cascade Range. On March 1, 1910, an avalanche swept 3 large locomotives with several train carriages and the train station itself over a ledge into a canyon 150 feet below. More than 100 persons were killed, and damage was estimated at $1 million.

773. How is a blizzard defined? A blizzard is a cyclonic storm in which strong horizontal winds blowing at speeds of 35 miles per hour or more drive large amounts of fine drifting snow, mostly picked up from the ground. Temperatures are 20 degrees Fahrenheit or lower. This storm fills the air so completely with blinding powdery snow that people and animals easily lose their sense of direction and become lost. Blizzards are the most dramatic and perilous of all winter storms, especially in high mountains and polar regions. Most severe blizzards occur in Antarctica where some of the strongest winds on earth blow across the ever-present snow.

A blizzard is called severe when winds blow at speeds of 45 miles an hour or more, when large amounts of snow are falling or blowing, and when temperatures drop to 10 degrees Fahrenheit or lower.

In the Northern Plains of the United States, the combination of blowing and drifting snow after a large snowfall has ended is often called a ground blizzard.

774. Where does the word blizzard come from? The word blizzard may have come from the German word *blitz*, meaning lightning. In America, before 1870, a blizzard meant a violent blow, possibly a rifle shot. Davy Crockett in 1834 wrote about taking a blizzard at some deer. By 1846, a blizzard meant a cannon shot, and during the Civil War, it meant a volley of musketry.

The first time the word was used in a meteorological sense may have been when a newspaper editor in Iowa described a violent snowstorm as a blizzard. Soon the word was used in this sense in the

Middle West. By 1880 the word had spread to the rest of the country and to England. Today the word means a severe snowstorm in nearly every English-speaking country.

775. How is a heavy snowfall defined? A snowfall is considered heavy when snow reaches depths of 4 inches or more during a 12-hour period, or depths of 6 inches or more during a 24-hour period. The term heavy is relative, depending upon the region where the snow falls. For instance, in a southern area, where snow is rare, a fall of 2 or 3 inches may cause unexpected difficulties and hence could be termed heavy. In another area toward the north, where snow occurs frequently, heavy snow may mean a fall of 6 or more inches.

776. How are snowfalls measured? Snowfalls in cities are usually measured in inches, while those in remote wilderness areas are measured in feet. Records of very heavy snowfalls have been taken only where measuring devices have been set up. Some of the world's snowiest regions have not yet been studied.

The amount of damage snowfalls bring to mankind may bear little or no relation to the actual measured snow depth. A snowfall of several inches in a city can cause far more damage than a snowfall several feet deep in a forest. Death and damage depend on how well man and his civilization are prepared to cope with the falling and fallen snow.

777. What were some heavy one-day snowfall records in the United States? Various records have been set in the United States for different amounts of snow falling during one day, one week, one year, or one snowfall. Many of these records are being changed as more accurate measuring equipment becomes available.

The deepest snow in a single 24 hours was recorded at Silver Lake, Colorado, on April 14 and 15, 1921, when 75.8 inches of snow accumulated.

On January 19, 1933, 60 inches of snow fell on Giant Forest, California.

On January 21, 1935, 52 inches of snow covered Winthrop, Washington.

778. What were the greatest seasonal snowfalls in the United States? The greatest seasonal snowfall in the United States was officially recorded as 1,000.3 inches—more than 83 feet—during the winter of 1955–1956 at Paradise Ranger Station in Mount Rainier National Park, Washington. At this same spot the record for the greatest average annual snowfall record has been retained for more than 23 years—582.1 inches.

Runner-up for deepest seasonal snowfall is Thompson Pass, Alaska, where 974.4 inches fell in 1952–1953.

Third place honors go to Tamarack, Alpine County, California, where 884 inches were received in 1906–1907.

Other locations that have accumulated more than 600 inches of snow in a single season are Crater Lake, Oregon, which had 879 inches in 1932–1933; Alta, Utah, with 663 inches in 1951–1952; and Ruby, Colorado, with 644 inches in 1894–1895.

779. What was a destructive Boston snowstorm? The 4-day snowstorm that struck parts of New England in late February, 1969, deposited 26 inches over Boston—the most snow fallen from a single storm in 99 years, when records began. There were 44 deaths in the city, and damage was estimated to cost $150 million. Snow removal cost $1,200,000.

780. What was a heavy New York snowfall? A heavy snowfall that started about 5:30 in the morning of December 26, 1947, dumped 25.8 inches of snow over New York City. This added up to about 100 million tons of snow—and it took a working force of 30,000 people to remove it from the city, at a cost of more than $8 million.

Another severe snowstorm striking the northeastern United States on February 9 and 10, 1969, covered New York City with 15 inches of snow, killed 25 people, and cost $5 million for snow removal.

781. What were some serious snows in 1949? On January 2, 1949, a series of snowstorms came down from the north, striking Wyoming, Colorado, the Dakotas, and Montana with blizzards of lashing, suffocating snow. The storms fanned over the Rocky Mountains into Idaho and Utah and Nevada, then drove eastward into Kansas and Oklahoma, Arizona and New Mexico.

For a span of several weeks, the storms brought 6 major blizzards and several smaller snowstorms, marooned 50 trains, stranded several families, and killed livestock. The National Guard and the Fifth and

Sixth Armies were called out to rescue people with snowplows, bulldozers, and other equipment.

Farmers were hard hit by the snows of 1949. Cattle, numbed by cold and dazed by pelting snow, wandered aimlessly and died in heaps along fences and gulches, or were smothered in snow-sealed barns. No total figures were given for the nation's livestock losses, but estimates from Wyoming alone reported that 125,000 sheep and 25,000 cattle perished.

782. What other severe snowfalls have occurred in the United States? On January 27, 1922, 28 inches of heavy wet snow fell upon Washington, D.C., causing the collapse of the roof of the Knickerbocker movie theater, killing 96 people, and injuring 125.

In late March, 1957, a snow blizzard piled drifts 30 feet high over the central Great Plains regions, marooning 2 trains and thousands of vehicles on the highways for several days. More than 40 deaths were attributed to this blizzard.

Heavy snows in January, 1952, marooned more than 200 people for 3 days in a passenger train in Emigrant Gap, California.

783. What are snow flurries? Snow flurries are defined by the Weather Bureau as bursts of snow falling for short durations at intermittent periods. During these flurries, snow may be falling so thick and fast that visibility is reduced to an eighth of a mile or less.

784. What are snow squalls? Snow squalls are brief but intense falls of snow, accompanied by gusty surface winds. They are comparable to the rain showers of summer.

785. How much snow falls on various states? During the winter, regular snowfalls occur over nearly four-fifths of North America, with Canada receiving most of them.

Even though the amount of snow falling in different winters varies considerably, there is a generally consistent pattern of snow cover that increases in depth the farther north one is located.

Meteorologists have recorded average number of days each state retains a snow cover—about one inch of snow. (Snow cover is different from a snow day, which is a day when snow falls.) Florida and other states along the Gulf of Mexico have less than one snow cover day. The middle states, including Virginia, Arkansas, Oklahoma, and northern New Mexico, have 40 snow cover days. States near the Canadian border have as many as 160 days with snow cover.

786. Where in the world does snow fall? Only about a third of the earth's surface is touched by the white coldness of snow.

The greatest amount of snow falls across the temperate zone in the Northern Hemisphere. Here the heaviest snowfalls occur in mountainous regions—the Rocky Mountains and Sierra Nevadas, the Alps and the Himalayas.

In winter the great snow cover spreads down over the northern top of the world, encompassing Greenland, the ice floes of the Polar Sea, Canada, northern United States, Japan, Korea, northern China, Siberia, and northern parts of Russia, the northern half of the European mainland, and parts of Scotland.

Less snow falls south of the equator than in the north because the land masses are much smaller, and most of them are below 40 degrees latitude, in the warmer climates. Aside from the high mountain ranges, very little snow falls in South America, Africa, and Australia.

787. Where is snow always present? Snow is always present in all seasons and over all regions. In the tropics or in midsummer, it is confined to the high cold clouds.

Snow is eternal wherever massive mountain peaks or ranges of the world tower several thousand feet high. The summits of South America's Andes, of Mexico's Mount Ixtacihuatl, and of Africa's Kilimanjaro, Mount Kenya, and Mountains of the Moon gleam with banks of snow, even though their feet may stand in tropical forests.

788. What percentage of the earth's water is snow and ice? Of the 326 million cubic miles of water in all forms on earth, only 9 million cubic miles consist of fresh water. About 80 percent of that is in the form of snow or ice—7 million cubic miles.

Permanent snow and ice cover about 12 percent of the earth's land surface in this present era—a total of about 8 million square miles.

789. When do snowfalls usually occur? In the temperate zone of the Northern Hemisphere, snows begin falling in October and November. In the United States, February is usually the snowiest month of all. In many parts of the country, January is the month when most snow is expected.

790. Is it ever too cold to snow? Snow rarely falls at temperatures below minus 40 degrees Fahrenheit. A basic unchanging law of nature decrees that as atmospheric temperatures decrease, the density of the water vapor decreases also. For instance, there is 50 percent less water vapor in the air at minus 40 degrees Fahrenheit than at 50 degrees Fahrenheit. With less moisture present, naturally there is less chance of snow.

Many of the cold regions of the world are actually arid regions, where only a thin scanty precipitation occurs. Parts of Canada and Siberia, the Arctic, and the Antarctic are arid lands.

In the United States, more snow seems to fall when temperatures are between 24 and 30 degrees Fahrenheit.

791. What shape does snow take? A snowflake is actually a snow crystal. The individual crystal may be quite simple in appearance, or intricately complex and varied.

Most crystals are symmetrical in shape, with 6 sides or points. The 6-sided shape results from the basic construction of a water molecule —2 atoms of hydrogen and 1 of oxygen—in triangular form with roughly equal sides. As 6 water molecules join together, held by bonds of attraction between the oxygen atoms, a hexagon forms. The crystal grows as more supercooled droplets sublime and develop crystal buds at 60 degree angles from the hub of the group of 6 molecules.

Snow crystals are found in many other shapes: hexagonal columns, pyramids, and plates; triangular plates; and 12-sided plates. They occur in shapes of needles or splinters, bullets, cones, rods, and even cuff links with caps on both ends.

As the snow crystal falls through the air and encounters various degrees of temperatures and moisture, the design and size varies. If the atmosphere remains cold—below freezing—they fall as single flakes. If the temperatures are above freezing, many crystals collide and stick together, forming larger snowflakes sometimes several inches in diameter.

792. Why is snow white? When sunlight falls on snowbound fields and snow-covered rooftops, the snow seems the most dazzling white of any white on earth.

Snow appears white because billions of tiny microscopic prisms in each snow crystal reflect and refract the light to produce rainbow colors too fine for the human eye to distinguish. The general sensation to the viewer is that of intense whiteness.

Yet individual snowflakes can often appear an icy gray, when the snow absorbs rather than reflects the light.

793. What is a whiteout? Sometimes so much snow falls and clings to objects that every line of the landscape is blotted out with whiteness. The condition is accentuated when the sky is overcast with white or light-gray clouds. A traveler cannot perceive distance, depth, or the nature of the terrain in this condition and may tumble over cliffs or into crevices. This condition, called a whiteout, can occur quite often in snowy regions of the Arctic and Antarctic.

794. What is snow blindness? Snow blindness, also called niphablepsia in medical language, is a temporary blindness that occurs from intense and prolonged glare of the sun on snow. Snow blindness can occur also on an overcast day, where there is no apparent sun but the glare can be just as intense.

The person's eyes become red and start to burn and water, and the person feels dizzy and faint. This blindness may last for a few hours or even for several days. Rest and darkness are essential to cure this condition. It can be avoided by wearing dark glasses or goggles.

The same condition can occur when the sun reflects from sand or rock on a desert. Here it is called sun blindness.

795. Who has the greatest vocabulary relating to snow? The Eskimos may well be the most snow conscious of all people. They have evolved the most complex vocabulary for subtle variations of snow. Here are a few words of their snow vocabulary:

Ah-put	snow in general
Ah-ki-lu-kak	soft snow
I-mu-gak	snow water (water made from snow)
I-ya-go-vak-juak	snow crystals
Ka-nik	falling snow
Sa-ki-ut-vuj	snow falling straight down
Ku-ah-li-vuk	snow freezing as it falls
Pu-ka	snow like salt, not cleaving
Ma-sak	wet snow
Mang-uk-tuk	snow that is getting soft

Msu-yak	soft, deep snow
I-glu-vi-gak	snow house
I-gluk-sak	snow for snow house
Kag-mak-sak	snow for banking house
Ti-vi-gut	snow drifts
Piek-tuk	drifting snow
Ne-ta-go-vi-ak	ground drift
Ne-ta-ku-nak	hail
Ki-mik-vik	uneven snow after drifts
U-nik-ti-vi-yut	holes or hollows made in snow by wind

796. How much energy is in a snowstorm? Scientists have estimated that a single snowstorm can drop about 40 million tons of snow. If this snow could be melted to water by applying heat, it would take heat equivalent to that of 120 atom bombs.

797. What are some benefits of snow? The deep blankets of snow laid down each winter across fields and plains, over mountains and forests not only give frosty beauty and splendor to the earth, but are extremely important and useful in many ways.

More and more recreation industries and resorts depend on this cold precipitation for the sports of skiing, sledding, tobogganing.

Snow cover is a valuable source of stored water, locked up in winter and slowly released in spring and summer as melted water to replenish reservoirs for cities and rural communities, to generate hydroelectric power, to supply irrigation water for farms and forests, to raise the water level of lakes and rivers so ships can navigate and more people can enjoy boating, swimming, and other warm-weather sports.

Snow cover also protects plants in winter, by acting as an insulator and preventing alternate freezing and thawing.

798. What countries in particular derive their water supplies from melted snow? Most of the rivers of Russia and central Asia are supplied principally by water melted from snow.

In the United States, the water systems of many towns and several large cities in the Midwest depend entirely on snow-fed mountain streams. In 17 western states, more than 30 million acres of land are irrigated with melted snow water.

In these regions, snow is called white gold.

799. Can snow be used for construction? Snow has been used as material for houses and other buildings by Eskimos, Laplanders, and explorers in polar regions where other materials such as wood and stone are scarce.

The colder the snow is, the stronger it becomes and the more weight it can support. At temperatures of about minus 20 degrees Fahrenheit, snow can support about 80 pounds per square inch. At temperatures of about 25 degrees Fahrenheit, snow supports only about 18 to 25 pounds per square inch.

800. Why is snow good insulation? New fallen snow is an excellent insulating material. Snow may contain 10 parts of air—or even as much as 35 parts—to one part of ice. This air is held motionless within the snow. There is little or no circulation, hence the cold air of the atmosphere is kept out and the warmth from the ground is retained. This keeps roots of plants and trees from freezing, and keeps rabbits, squirrels, and field mice alive and snug in their burrows. This explains why sheep buried deep in snowdrifts have been found alive even after several days.

This insulating quality is well known by the Eskimos, whose snow igloos may be as warm as 60 degrees Fahrenheit with a small fire, even though temperatures outside may be 50 degrees below freezing. With an extra foot of fallen snow, the igloo can become too warm, even when heated only by body heat.

801. What are temperature variations in snow? Temperatures vary surprisingly in layers of snow cover. The top layer tends to lose its heat to the atmosphere through radiation and evaporation. The lower layer is protected by the outer layer and keeps a fairly stable temperature.

The following temperature differences were recorded during a cold snap:

3 feet above snow surface	$-19°$ F.
at snow surface	$-27°$ F.
7 inches below surface	$+24°$ F.

This represents a difference of 51 degrees.

802. Will man be building cities upon or beneath the snow? With an increasing population needing more land upon which to build, scientists are looking to the vast ice and snow area of the world as increasingly valuable real estate property.

Already, permanent scientific stations have been set up in Antarctica, and tunnels and large storage chambers have been excavated in Greenland.

In the Greenland ice sheet near Thule, a town called Camp Century was built in 1960, then covered with a 5-foot layer of snow. Each year additional snowfalls bury it deeper. This town includes streets, laboratories, dormitories, a gymnasium, a recreation center, and a nuclear power plant.

803. What snowstorm warnings are in operation? For years the U.S. Weather Bureau has been issuing forecasts and bulletins on snowstorms as they sweep across the country. In 1966 the NAD-WARN System (see Question 277) was set in motion to warn people of severe blizzards and storms.

In 1968 for the first time, Weather Bureau stations began using the terms "watch" and "warning" for snowstorms and blizzards in much the same way they are used for tornadoes and hurricanes.

Snow watch means that a storm has been formed and is approaching the area. People in the alerted area should keep listening to latest bulletins and advice on radio and television and begin to take precautionary measures.

Snow warning means a storm is imminent and immediate action should be taken to protect life and property.

804. What should a person do for protection against a snowstorm? When a storm watch or warning is issued by the Weather Bureau, people in the alerted area should take the following precautions:

Check your battery-powered equipment, especially flashlights and radios, and emergency cooking facilities. Be sure you have enough heating fuel to last through the emergency. Stock up on extra food, including some that requires no cooking or refrigeration. Take precautions against the danger of fire from overheated stoves, fireplaces, heaters, or furnaces.

During the storm, stay inside unless you are in excellent physical condition. If you do go outside, remember that several layers of loose-fitting, lightweight but warm clothing are best protection against cold. Mittens, tight at the wrists, are warmer than gloves with fingers.

805. What should a person do if his car stalls in a blizzard? Many unnecessary tragedies have occurred when people in blizzard-stalled vehicles tried to leave the car or did not know how to conserve heat and energy. Several rules of good sense can help a person survive:

First, do not panic. Beware of overexertion and overexposure. If you try to push your car through heavy snow drifts or to shovel, work slowly.

If you cannot get your car out of the snow, stay in it, where rescuers can more easily spot you. Do not attempt to walk for help, for you can quickly lose direction and get lost.

Run the motor and heater sparingly, and make sure you keep a window open enough for fresh air. Carbon monoxide from the motor is a lethal gas that is difficult and sometimes impossible to detect.

Do not stay in any one position for too long. Clap your hands and move your arms and legs vigorously from time to time to stimulate blood circulation and keep muscles from getting cramped.

Wrap any scraps of cloth or rags around your feet and legs to insulate them from the cold. Layers of newspapers or wrapping paper are excellent insulation.

If there are other people in the car, keep shifts for sleeping and watching guard. Someone must be awake at all times.

806. How can animals be protected in blizzards? Wild animals and livestock die in blizzards mainly from lack of water, rather than from cold or suffocation. Body heat from an animal trapped in a snow drift can keep it warm and alive, but the animal—cattle in particular —cannot lick enough snow to satisfy its water requirements.

When a blizzard sets in, stockmen should provide their livestock with water in any way possible, from heaters, water tanks, stoves, or any other source.

In areas where blizzards appear frequently, sheltered zones should be planned in advance, especially for young animals, and emergency supplies of food and water be kept ready.

807. What methods are used to control drifting snow? Snow researchers experiment with many methods to prevent snow from drifting and clogging vital areas, particularly roads and highways. The most common and effective method is to set up snow fences to the windward of highways. These can be made of many materials, such as brushwood, straw, rope, wood, or metal.

More substantial snow barriers are used in the highlands and on mountain slopes to deflect high piles of snow. Snow sheds are sometimes built over railway tracks in certain mountain areas to keep tracks clear.

Engineers have sometimes flattened off the tops of hills so the wind will blow the snow over or past a road. In flat country, the roadway is sometimes elevated above the surrounding land.

Rows of trees or hedges along roads are very effective in stopping snow drift. Planting these living fences costs only one third to one half as much as annually setting up and taking down regular snow fences. In Russia, rows of trees have been planted along railroads for miles. Forest belts have been planted along the windblown steppes, not only to stop drifting snow from covering highways but to hold it as a source of water for spring melting. Germany and Hungary also practice the relatively cheap and attractive method of planting trees as snow fences.

808. What was some early snow equipment? Until only a few decades ago, human beings were resigned to the snowfalls and ice of winter and did not try to change or remove them. They simply trudged through the snow or packed it down and rode over it as best they could. Only relatively recently did the idea of removing snow in mass occur in a practical manner.

In the United States, from about 1880 until as late as 1925, the snow roller was a familiar sight in northern parts of the country. Huge weighted drums, 6 feet in diameter and 8 to 10 feet in width, were drawn by 2 or 3 teams of horses through the streets of towns and cities after a snow. The roller left a street hard packed with snow, over which horses pulled sleighs on snow runners. When the car was invented, for a long time the man fortunate enough to be an owner had to jack his vehicle up in the barn until the spring thaw made roads usable for tires again.

Blade snow plows were introduced shortly before 1915, when cities began to experiment with broad flat blades pulled by teams of horses or secured to the front parts of some early trucks.

809. What methods are effective to remove snow and ice? As civilization expands and more cities, highways, and other structures are built, more complex and varied equipment is devised for fast removal of snow.

Snow plows have long proved effective for work in open country roads and along railroad tracks. Push plows, rotary choppers, and heavy V plows are some of the standard plows.

Snow blowers are used along high roads where snow can be sucked in through wide pipes, then blown high and clear of the lanes.

Sanding trucks spread sand over the road area, thus providing friction between the ground and the moving object. Salt, ashes, or dry chemicals are also used. These materials are not very adequate for roads where speeding vehicles may kick the material to one side. Also, it is sometimes difficult and costly to remove the bulky, unsightly material at the end of winter.

Snow and ice can be effectively melted from sidewalks by means of electric coils or conducting pipes installed within the pavement. Heat can be conducted through the coils, and steam, hot water, or anti-freeze chemicals can be pumped through the pipes. This is, however, an expensive method.

810. How does salt remove snow and ice? Salt has a physical reaction with ice, lowering the freezing point and hence liquefying it. Only enough salt to raise the liquid water content 50 percent is necessary, for the salt dissolves, creates heat, and lowers the point at which the mixture freezes. Researchers figure that one eightieth pound of salt per square yard of road for each inch of snow and for each degree Fahrenheit of temperature below 32 degrees Fahrenheit is necessary.

Rock salt (calcium chloride) and magnesium chloride have been used, but these chemicals tend to cause damage by corroding the metal underparts of cars, damaging carpets and floors as shoes and boots drag it into homes and offices, and injuring plants and trees along the roadsides.

The U.S. Highway Department is attempting various anticorrosive combinations to be used with concrete—either by impregnating roads with a petroleum oil distillate mixture or by mixing fresh portland cement with a waterproof agent such as asphalt.

811. What was a great snow removal emergency in the United States? The enormous job of removing the heavy snow that tied up

Washington, D.C., on the eve of John F. Kennedy's presidential inauguration in January, 1961, may have been the greatest snow removal project to date. With a late afternoon snow beginning to fall the evening before the ceremony, traffic was snarled for hours. Some 3,000 men and 500 city vehicles, aided by the Army's 87th Engineer Battalion, were mobilized to clear the parade route and center of the city. More than 3,500 cars were towed out of the area.

812. What are snow surveys? Snow surveys are made for various reasons—to keep track of the depth and water content of the snow cover, to locate potential avalanches, and to determine the supply of frozen water for the following spring and summer water supply.

The Department of Agriculture first began making snow surveys in 1935. Today snow cover in North America is measured and monitored in different parts of the country by the Weather Bureau and the Army, as well as the states and private industry. Snow surveyors travel in power-driven sleds called snowcats, on skis or snowshoes, by plane or helicopter. Automatic radio-telemetering snow gauges can transmit information on the water content of a snow pack. Scientists are considering the use of remote-sensing devices mounted on orbiting satellites.

Data on snow are sent to snow survey supervisors where they are processed, and calculations are made to forecast water supply for each state involved when the snow starts to melt. Reports are sent to federal, state, and private agencies and released to the public through television, radio, and newspapers.

813. What is an ice storm? An ice storm occurs when rain from overlying layers of warm moist air falls through a shallow layer of freezing air that has chilled everything near the ground. Thus each raindrop falling in liquid form freezes quickly when it hits a tree, a rock, or some other object, creating a coating of ice over the surface.

An exquisite fairyland of silver shining branches appears the morning after an ice storm, when light from the rising sun reflects from millions of glazed twigs. When the ice is thick, severe damage can be caused by branches and wires breaking from the weight. Damage is even greater when the freezing rain is accompanied by high winds.

Ice storms are incorrectly called sleet storms.

814. How do ice storms cause death? During an ice storm, more than 85 percent of the deaths are caused by automobile and other accidents related to traffic, as cars skid across slippery, glazed roads and highways. Other deaths are caused when power lines collapse under the heavy weight of ice, and people freeze to death from lack of heat. Emergency cases in hospitals may also suffer when the electric power fails and machines cannot operate, although most hospitals have auxiliary generators to supply emergency electric power.

815. Where do ice storms strike? Ice storms occur in all but a few southern states of the United States—and in only a few provinces of Canada.

In the United States, only Florida, New Mexico, Arizona, Utah, Nevada, and the southern part of California are free from these storms.

Strangely enough, parts of Canada are too cold for the formation of the fronts that create these ice storms. The Yukon, Northwest Territories, and the area north of the Arctic Circle timber line do not have ice storms.

The worst ice storms in the United States are encountered in an L-shaped belt extending from central Texas northward to Kansas, then eastward across the Ohio Valley and the lower lakes to New England and the Middle Atlantic states. In this belt there is a 3-to-1 chance of an ice storm occurring each winter. Severe ice storms are also found in the northwestern regions.

816. What was a particularly destructive ice storm? Probably the most damaging ice storm in the United States struck various southern states on January 28 to February 4, 1951. When the final tally was made, the storm had killed 22 people and cost Mississippi $50 million in damages, Louisiana $15 million, and Arkansas nearly $2 million.

This storm started when a mass of polar air swept down the Mississippi Valley almost to the Gulf of Mexico. More than 5 inches of water, mostly ice and freezing rain, fell on western and central Tennessee, Kentucky, and parts of West Virginia. Communications were disrupted, and transportation came to a halt. Thousands of homes dependent upon electricity were without heat for more than a week.

817. What was another disastrous ice storm in the south? Damages of more than $7 million were caused by an ice storm over Georgia and South Carolina on December 31 to January 1, 1963. Freezing rain coated roads, wires, poles, and trees over a large area. After the rain, sleet and snow fell and stalled traffic for 1 to 3 days in different areas.

● **818. What was a great New England ice storm?** The severe New England ice storm of November 26 through 29, 1921, is considered the worst ice storm in the memory of local people. More than $5 million in damage was inflicted on the telephone, telegraph, and electric companies, and from $5 million to $10 million worth of trees were ruined. Several people were seriously injured by falling branches and by slipping on the treacherous ice.

819. What ice storm hit New Jersey? One of the worst ice storms on record in New Jersey and Pennsylvania occurred on January 8 to 11, 1953. The storm raged also over Maryland, New York, and parts of New England. It caused the deaths of 38 people, while many others were injured in traffic accidents and by slipping and falling. The storm caused several million dollars worth of damage.

820. Why are ice storms so hard on wildlife? Ice storms can be particularly hard on wild birds and animals that find themselves trapped in fast-forming ice with no way to extricate themselves. Birds have been found with their feet frozen to branches of trees or telephone wires. They die of starvation and cold before they can thaw out or wrench themselves free. Pheasants and ducks have been found with their wings so frozen they cannot fly, helpless captives for a predator. Even nimble animals such as foxes or wolves can have their paws frozen fast to the ground, and have to rip off pieces of flesh to release themselves.

821. What happens when ice coats telephone wires and trees? Freezing rain pellets can coat twigs, branches, telephone lines, and other objects with layers of ice that may be very thin, or several inches thick.

Deposits as thick as 8 inches in diameter were reported on wires in northern Idaho during a storm on January 1 through 3, 1961.

During one heavy ice storm in England at the end of January, 1940, telephone wires supported 11 tons of ice before they finally broke. Individual wires were found carrying as much as 1,000 pounds of ice.

Trees suffer seriously during an ice storm—even though they can look beautiful when the sun transfigures them into shimmering torches of silvery ice.

It has been estimated that a 50-foot evergreen tree may become coated with 5 tons of ice during a severe storm. Older trees generally suffer more damage than younger trees, and tall trees break more often than shorter ones.

Certain tree species break easier and suffer more damage than others—for instance, poplars, aspen, fruit trees, linden, and elms. Some trees are better able to withstand these storms—for instance, yellow birch, hickory, hawthorn, chestnut, beech, spruce, and oak.

822. What is freezing rain? Freezing rain or freezing drizzle is rain falling as liquid but turning to ice when it hits a freezing object and forming a smooth, tenacious coating of ice known as glaze. Freezing rains usually last only a short time, as a transitory condition between rain or drizzle and snow.

823. What is glaze? Glaze is a smooth coating of transparent or translucent ice that forms when supercooled water droplets crash and rupture against walls, trees, roads, and other objects in below-freezing weather. Glaze contains no air bubbles and looks smooth and clear as glass. It is relatively heavy and sticks tenaciously to the object it coats.

If the freezing rain is very heavy and the object it strikes very cold, glaze may build up several inches thick, causing dangerous driving conditions on highways and broken trees, lines, and poles.

In England, glaze is known as glazed frost.

824. What is sleet? Sleet is frozen or partially frozen rain in the form of hard, clear pellets of ice. These frozen raindrops are formed after they have fallen from a warm layer of air and passed through a freezing layer near the earth's surface. The tiny pieces of transparent or translucent ice are so hard and hit the ground or an object so fast that they bounce off with a sharp click. The particles are only about

1/25 to 4/25 of an inch in diameter. Some people consider sleet as a mixture of rain and snow.

When sleet pellets fall in quantities and build up on roads and highways, they cause hazardous driving conditions.

825. What is the difference between sleet and freezing rain? Sleet is often confused with freezing rain. Sleet does not stick to trees, wires, and objects—freezing rain does.

Sleet is frozen rain pellets—freezing rain is liquid raindrops that turn to ice when they strike any object on the ground.

826. What is rime ice? Rime ice is a white, opaque ice that is porous and feels somewhat granular to the touch. It is formed when water droplets freeze without rupturing.

Rime ice is often seen as a white layer of ice crystals deposited on the windward edges of fences, trees, and telephone poles when a supercooled fog or mist is present. Sometimes it appears as a feathery or stippled coating on cold earth objects. Rime is the kind of ice crystal that forms inside a freezer or refrigerator. It also is a common form of icing on aircraft.

827. What is a freeze? A freeze is a meteorological condition that occurs over a large area when the surface temperature of a whole air mass remains below freezing for a certain length of time. A freeze can bring frost, rime ice, and other icy conditions.

828. What is frost? Frost has two connotations: one meaning a crystal deposit, and the second meaning a general condition when the earth's surface and atmosphere have temperatures below freezing.

Crystalline frost is a light feathery deposit of small thin ice crystals formed on objects whose temperatures are at or below freezing. Frost is formed by the condensation of atmospheric water vapor directly into crystalline form—a process called sublimation. The process is the same as that of dew formation, except the surface on which the water vapor condenses is below freezing—even though the air temperature may be above freezing.

829. When does frost appear? Frost usually occurs on clear calm nights, especially in early autumn when the air above the earth's surface is quite moist.

830. How destructive are frosts? Frost can be quite destructive to plant life since it freezes the aqueous solutions of plant cells and kills them. Plants with a lot of water in their leaves, stems, and fruits are most susceptible to frost. Frosts are particularly destructive when a wind blows across the crops or through the orchards, or when freezing temperatures last for a long while.

831. What are different kinds of frosts? Farmers and meteorologists use various terms to describe several kinds of frost:

Light frost has no destructive effect except on tender plants and vines.

Heavy frost is a heavy deposit of crystallized water, but does not necessarily kill the sturdy vegetation.

Killing frost is destructive to vegetation.

Black frost or hard frost occurs in late autumn when both air and earth objects are cooled below freezing. Under such conditions, leaf edges and plant tips turn black, looking almost as if they had been burned.

832. What is permafrost? Permafrost is ground that is frozen permanently in the chilly areas of the world such as the polar regions and high mountains. Here the ground never thaws out completely, although the surface may melt and become soft in spring and summer. The frozen ground can include bedrock, soil, ice, sand, gravel, and any other material usually found in the earth's crust. Areas of permafrost vary, depending on the air temperature, exposure of ground to wind and radiation, the vegetation, nature of the terrain, and other complex factors.

Nearly 10 percent of the earth's land surface is underlain with permafrost. Half of Canada has a permafrost base, as well as much of Alaska and Russia. Permafrost can also be found on the tops of the Alps, the Rockies, the Himalayas, and other mountains.

833. How deep does permafrost extend? Permafrost has a deep zone below which the heat of the earth's interior thaws out the frosted ground. Depth of permafrost varies in different parts of the world. At Point Barrow, the northernmost point of Alaska, permafrost has been measured to depths of 1,030 feet. On the Taimyr Peninsula, between the Lena and Yenisei Rivers in Russia, permafrost is more than 2,000 feet deep. Some coal mines in Svalbard, Norway, extend through 1,000 feet of permafrost to reach the unfrozen rock mines beneath.

Some scientists believe that the permafrost line, extending through Canada, Alaska, Russia, and the Scandinavian countries, slowly retreats at times when the climate becomes warmer, just as glaciers retreat and then advance in colder climates. In Russia, remarkable recessions of the permafrost line have been reported.

Permafrost is a very interesting material in terms of constructing buildings and even whole cities in the polar regions. Experiments are under way to determine the different consistencies and strengths of these various materials.

834. What is hoar? Hoar, or hoarfrost, as it is usually called in England and Europe, is the same as frost.

The word hoar comes from the old Anglo-Saxon word *har* which means white or venerable, as does the German word *herr*, meaning mister. The word frost comes from the Anglo-Saxon word *forst*, meaning what is frozen.

835. What research projects are under way to control or moderate precipitation? Men have long been trying to control the clouds to make them rain or snow less in certain inundated regions, or to bring rain and moisture to other areas suffering from drought.

To channel these misty rainclouds, or tap their cargoes of moisture, scientists have been working on several ideas.

The most promising studies are under way in the field of seeding. (See Question 150.) By causing rain to fall at the right time and place, researchers can relieve drought areas. By causing it to drop over lakes or lonely areas, researchers can prevent disaster from striking a city or community.

As yet experiments have not been decisive, and more data and experimenting are needed. Seeding can enhance or inhibit rainfall, depending on certain conditions.

836. What is the Great Lakes Project? Studies are under way over the Great Lakes area to see if severe snowstorms can be seeded and moderated as they flow across the relatively warm lakes and dump loads of snow on the lee shores. The Great Lakes Project is operated under the Environmental Science Services Administration and several universities.

837. What North American research institutions are involved in studies of cold, ice, and snow? There are many diverse institutions throughout the world carrying on research on the nature and effect of winter storms and cold weather. Because information is sporadic and difficult to obtain, the following is only a brief and incomplete list of some of the work in the United States and other countries.

In the United States, for several years the research organization Snow, Ice and Permafrost Research Establishment (SIPRE) of the U.S. Army was headquartered at Wilmette, Illinois. In 1961 it was replaced by the Cold Regions Research and Engineering Laboratory (CRREL) at Hanover, New Hampshire. The Atmospheric Physics and Chemistry Laboratory, part of the Environmental Science Services Administration, conducts research on processes of cloud physics and precipitation.

Federal and state governments maintain cooperative snow survey systems in Nevada, California, Oregon, Washington, Montana, and in British Columbia. Snow research is carried on also in the Central Sierra Snow Laboratory, in Soda Springs, California, and at the Mount Washington Observatory, a private research institution.

In Canada, the Arctic Institute of North America with headquarters at Montreal, Canada, has many active research programs and includes personnel from the United States, Canada, and Greenland.

The National Research Council has conducted many snow surveys such as the Canadian Snow Survey under the Associate Committee on Soil and Snow Mechanics, with subcommittees to deal with snow and ice and cold weather materials; the Division of Mechanical Engineering; and the Associate Committee on Soil and Snow Mechanics.

Other agencies also do research on snow and ice: the Dominion Water and Power Bureau, the British Columbia Department of Lands and Forests, and the Ontario Hydro-Electric Power Commission.

838. What overseas organizations are researching ice and snow? The British Glaciological Society in London and the Scott Polar Research Institute in Cambridge carry on research work.

The Japanese Society of Snow and Ice Research in Tokyo and the Institute of Low Temperature Science of Hokkaido University in Sapporo have conducted extraordinarily fine studies on snow and ice.

The German Society of Polar Research at the University of Münster is a center of snow study.

Snow and ice research is under way at the Polar Institute, in Oslo, Norway.

The International Commission of Snow and Ice, founded in 1894 and part of the International Association of Scientific Hydrology, coordinates and stimulates research and data collection on glaciers, seasonal snow cover, avalanches, river and lake ice, and ground ice.

As part of the contribution of glaciologists to the International Hydrological Decade, a project is under way to determine more accurately the quality of ice and snow on the earth's surface.

The Swiss Federal Institute for Snow and Avalanche Research, on Weiss Fluhjoch, near Davos, Switzerland, is one of the world's leading institutes in snow and avalanche research.

VIII. EXTREMES OF COLD, HEAT, AND PRESSURE

Introduction. Waves of cold and heat are constantly flowing across the earth's surface from the cold high pressures of lands and seas of the polar regions or from the hot low pressures of deserts and sun-basking plateaus of the tropical regions—and from other places in between. As these masses of air sweep across the cities and communities of man, they sometimes have enormous impact upon his physical and mental activities.

Cold-blooded creatures, the reptiles, cannot withstand extremes of temperatures, for they cannot regulate their body temperatures but must accept and acquire the temperature of their physical environment. Mammals, however, have developed self-heating and self-cooling regulators in their bodies that give them greater advantages to endure extremes of cold and heat and to survive in a wide geographical range. Yet even animals and men cannot exist if body temperatures fall below 80 degrees Fahrenheit or rise above 110 degrees Fahrenheit for any length of time.

In order to survive the rigors of temperature extremes, man has devised innumerable methods to keep alive and even comfortable. Nearly half of man's civilized waking hours are spent in adjusting in some way to variable patterns of cold and heat—or in designing, producing, selling, delivering products for the never-ending activities related to those patterns such as putting on more clothes to keep warm, taking them off again to get cool; finding warm shelter against icy winter winds or a breezy place in summer; building fires, stoking furnaces, setting up the thermostat against winter chill; turning on fans and air conditioners if the days are warm.

Yet even with man's ingenuity to combat the forces of extreme temperatures, the bitter cold of winter and the oppressive heat of summer can still cause much hardship, grief, and even death.

Other weather phenomena, more subtle and less understood than heat or cold, may affect living organisms in serious ways. The insidious forces of pressure, ionization, and sound and shock waves can bring adverse psychosomatic effects such as moodiness, irritability, sudden sweating. Other forces can make a person feel spirited and confident. The fascinating relationship between weather and living creatures involves the relatively new field of biometeorology.

839. What is cold? Cold is a condition of low temperature, a lack of heat. As heat or energy is taken away from an object, it is said to become cold. Hence, the penetration of cold through a medium—metal or water, for instance—is actually the withdrawal of heat. Yet even though cold is a negative condition in terms of physics, it is positive in terms of its impact on man.

Man has long been subjected to cold winter weather with its varying processions of cold air, snows, and freezes. Without the natural accessories that other animals and plants have to withstand cold—such as thick fur, blubber, toughened skin, or the ability to hibernate or adjust in some other way—man has learned to combat cold in various ways and with increasingly sophisticated and complicated methods as his knowledge and technology have advanced. A history of man's battle against cold is a history of man's evolution—from the first primitive fires and skins for warmth, to the development of fireplaces, furnaces, heating units, synthetic fabrics, and nuclear power for heat.

840. What is winter weather? During the winter months, essentially December through February in the Northern Hemisphere and June through August in the Southern Hemisphere, waves and bulges of cold heavy air pour out from high-pressure, cold areas of the world and flow toward the equator.

This is the season when the natural world seems to die, when trees stand bare, and animals lie dormant in burrows. Even man tends to remain inside buildings seeking warmth, light, and comfort that has fled the outside world.

In the Northern Hemisphere, where more land and people exist than in the Southern Hemisphere, winter can be particularly severe. Vast tracts of land and water are locked under snow and ice for long periods of time, immobilizing streams, rivers, and oceans, disrupting communications and traffic, and causing accidents that injure and kill many living things.

841. What causes air to get cold? As the planet earth spins through space in orbit, its axis maintains a nearly constant orientation in space, inclined about 66½ degrees to the plane of its orbit.

In its long journey around the sun, part of the year one of the poles becomes gradually pointed away from the warmth- and light-generating sun, causing the rays to fall obliquely over the land and cast long cooling shadows. Heat radiates from that section of earth faster than it is coming in. The atmosphere and land cool down.

As winter deepens, masses of chilled air, devoid of warming sun rays, become colder, denser, and larger. The temperature of the snow surface continues falling as heat radiates outward, especially at night, and the feeble rays of the sun do not raise its temperature during the day. High-pressure domes of chilled air build up to great heights over certain land areas—3 in the Northern Hemisphere and 1 in the Southern Hemisphere. From them, masses of cold dry air break out and flood across the earth's surface toward the equator, cooling the regions they pass over.

842. Where are the world's cold spots? The coldest spots on earth are—not the geographic North and South Poles, as once commonly believed—but about 4 special regions in the polar areas.

In the Northern Hemisphere, the coldest regions are located in northeastern Siberia, northwest Canada, and Greenland. From these high-pressure, domed, cold areas, masses of cold air flow over Eurasia, the Pacific Ocean, North America, and the Atlantic.

Wind-swept icy Antarctica may have the coldest regions in the world. From here, bulges of cold air flow over the oceans toward the southern tips of South America, South Africa, and Australia.

843. What are the cold poles? A cold pole is the assumed location which has the lowest mean annual temperature yet recorded in its hemisphere. These poles are only theoretical, for not enough data have been collected to determine such locations in reality.

In the Northern Hemisphere, the cold pole has been located in the region around Verkhoyansk, Siberia (about 67 degrees North, 133 degrees East). Here the annual mean temperature has been calculated to be around 3 degrees Fahrenheit.

In the Southern Hemisphere, the cold pole is a region around 80 or 85 degrees South and 75 to 90 degrees East. The lowest temperatures in the world have been reported from this region.

● **844. Where are the coldest temperatures on record?** The coldest place on earth yet recorded is Vostok in Antarctica, where temperatures of minus 127 degrees Fahrenheit were recorded at the Russian station on August 24, 1960. This was at an elevation of 11,220 feet.

Other low temperatures have been reported in Siberia. At Oimekon on February 6, 1933, temperatures dropped to minus 90 degrees Fahrenheit. This region, which is near Verkhoyansk, is considered the cold pole of the Northern Hemisphere. (See Question 843.)

In Greenland, a record of minus 87 degrees Fahrenheit was taken on the icecap at Northice on January 9, 1954.

In North America, a record of minus 81 degrees Fahrenheit was taken at Snag in the Yukon, Canada, on February 3, 1947.

These temperature extremes are determined by a number of factors, including altitude, latitude, and the physical characteristics of the area. There is little doubt that lower temperatures have occurred somewhere on earth, but have not yet been recorded. Very few observing stations around the world have existed more than 100 years. Also, areas with such frigid temperatures are often remote and uninhabited.

845. What is the polar front? The polar front is a turbulent area where cold polar air encounters warmer air from the lower latitudes. This front is more or less stationary and is a major factor in forming weather, particularly in the Northern Hemisphere. As the northern land areas cool in winter, and masses of cold air break out and converge with the warm air of the prevailing westerlies, unstable weather conditions and storms are created.

This turbulent polar front occurs in bands between 50 and 60 degrees north and south latitude. It is particularly strong in the Northern Hemisphere. During the winter, the cold air regions become stronger and push the polar front as far south as 40 and 45 degrees.

846. What is a cold air mass? A cold air mass is a large body of air that is generally uniformly cold and moving across the country.

Air masses can be moist or dry, cold or warm. They are essentially vast bodies of air, in which the temperature and moisture content are fairly constant throughout, and are among the most important moving parts of the atmosphere.

Air masses may sometimes cover hundreds of thousands of square miles, taking on the temperature and moisture characteristics of the land and water over which they move.

847. How are air masses classified? Meteorologists classify air masses basically by the regions where they originate and by the moisture they contain. Air masses come from either of two sources: tropical regions or polar regions. On weather maps, tropical air masses are marked T and polar ones P. Those formed over land are marked C for continental, and those formed over the oceans are marked M for maritime. Air formed over continents is usually dry; over oceans, usually moist.

Thus there are 4 basic types of air masses:

> warm, moist air or tropical maritime (symbol MT)
> warm, dry air or tropical continental (symbol CT)
> cold, moist air or polar maritime (symbol MP)
> cold, dry air or polar continental (symbol CP)

Air masses over North America

848. What is a cold wave? A cold wave is a form of winter storm, bringing rapid and often disastrous drops in temperatures within a 24-hour period. It occurs when a mass of heavy cold air from the polar regions pushes rapidly southward with strong, bitter northerly winds, dropping temperatures at least 30 degrees in only a few hours. The colder the incoming air and the faster it moves, making the pressure gradient steeper (see Question 522), the more violent is the cold wave. It is also called a cold snap.

The definition of a cold wave differs for various parts of the country and also depends on the season. What is considered a cold wave hitting Georgia may appear merely a cold ripple in Massachusetts.

Most true cold waves strike New England, where many cyclone tracks converge and exit from the continent across the Atlantic Ocean.

Over land, cold waves can be forecast a full day in advance, as weather men track the waves of dropping temperatures and rising winds moving eastward.

● **849. What were some record cold waves?** One of the fastest dropping temperature records in the United States arrived at Rapid City, South Dakota, in early morning on January 12, 1911. At 6 A.M. the temperature was 49 degrees Fahrenheit. At 8 A.M. it was minus 13 degrees Fahrenheit, a drop of 62 degrees in 2 hours.

Another outstanding cold wave hit Kansas City, Missouri, on November 11, 1911, dropping temperatures from 76 degrees Fahrenheit at 10:30 A.M. to 10 degrees by midnight, and to 6 degrees by 7 A.M. the next morning.

One cold wave from Canada arrived at New York for New Year's Eve, 1933, and dropped temperatures as much as 55 degrees.

850. What is a cold front? The leading or advancing edge of a moving cold air mass is called a cold front. As a bulge of cold air advances over the earth's surface, its leading edge slides under the warmer air, pushing it up and creating storms with heavy but brief precipitation. The angle of the wedge is steep and the front moves relatively fast, so the storm band is relatively narrow. After the front passes, the air rapidly clears, temperatures drop, humidity drops, and pressure rises.

Thus cold fronts bring dry and cold air in their wakes.

851. Where do cold air masses travel across North America? The North American continent is unique among continents of the Northern Hemisphere in that the great mountain ranges run in a general north-south direction—a condition that tends to block the usual west-east flow of air and channel much of the flow to north-south directions.

Most of the cold air masses pouring out of the high-pressure cold region in Canada flow southeasterly, over the Dakotas and Minnesota, then curve over part of the Great Lakes and drive toward the east. Sometimes they push farther south over the Great Plains toward Texas and Mexico and Florida, partly influenced by powerful jet streams.

Sometimes cold air pushes southwest from the Canada high, through the low-lying passes of the Canadian Rockies in the Peace River area, then southward along the middle trough between the Rockies and the Cascades and Sierra Nevadas. Another cold channel may flow down California's central valley, with the Sierra Nevadas to the east and the coastal ranges and Pacific Ocean to the west. This wind sometimes brings squalls, snows, and freezes to southern California.

Maritime polar air masses ride in from the Pacific Ocean and the Gulf of Alaska, bringing rain showers or snowfalls to the coastal ranges and sometimes farther inland.

852. Where do cold air masses travel across Europe and Asia? From the cold, high-pressure area in Siberia, winter winds sweep southward, southeastward, and southwestward across the large lands of Europe, Russia, and Asia, becoming blocked eventually by a long wall of mountains stretching in an east-west line from the Atlantic Ocean to the Caspian Sea and from west Pakistan into China. The Pyrenees, the Alps, the Carpathian and Caucasus Mountains all protect lands to their south from the chilling Siberian winds. Farther to the east, the Pamirs, the Hindu Kush, and the high Himalayas block the passage of the cold winds.

European winters are mostly affected by the polar maritime air masses that blow eastward with the predominant westerly winds from the Atlantic Ocean, crossing the warm Gulf Stream and bringing generally mild, humid, and unstable air to the coastal areas. When the prevailing westerlies over Europe are weak or absent, and pressures are unusually high to the northeast, Europe is blasted with frigid Siberian winds and snows.

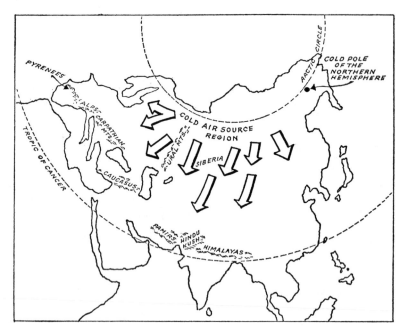

Paths of cold air masses over Europe and Asia

853. What are the Great Ice Ages? For several billion years, ice caps have alternately advanced toward the equator and then receded, forming periods of glacial and interglacial stages. Throughout the earth's history the best known Ice Ages are: 1) the Huronian in Canada, occurring early in the Proterozoic era, perhaps a billion years ago; 2) the pre-Cambrian ice ages occurring about 530 million years ago; 3) the Permian-Carboniferous ages occurring from 275 million to 225 million years ago and extensive in the Southern Hemisphere; and 4) the Quaternary or Pleistocene Ice Age that began about one million years ago and may not yet be ended.

854. What was the last Great Ice Age? The last Great Ice Age, which began in the Pleistocene epoch about a million years ago, was a period of extreme cold, when glaciers and icecaps of the poles and the high mountain areas spread downward and equatorward to cover almost one third of the land areas of the world. During this time, the oceans receded, as increasing amounts of the earth's water was stored as ice in the ice sheets. At their greatest period of advance, ice sheets were some thousands of feet thick and held more than double the amount of water frozen today. In the Northern Hemisphere, they reached as far south as central Missouri, northern France, central Germany, and Poland.

This last Great Ice Age included several smaller ice ages, the last of which receded about 10,000 years ago.

855. Is the world still in an Ice Age? Scientists disagree as to whether we are existing in the last stages of an Ice Age, or in a temporary interglacial cycle of a great glacial epoch.

In historical times the world's climate has fluctuated in interesting ways. From about A.D. 1000 to 1200, the world's climate seemed to be warming up. In Norway and southwest Greenland, men grew crops and raised livestock in regions formerly and now too cold. Then came a Little Ice Age, perhaps lasting from A.D. 1400 to 1700 or 1800, some scientists believe. For about 100 years, from 1850 to 1950, the mean annual temperature of the world rose—generally about 2½ degrees Fahrenheit in the United States. Since 1950, some meteorologists say the earth as a whole may be cooling off again. Yet there is no uniform tendency, for some regions show rising temperatures.

Any discussion of world climate trends is only theoretical, for scientists have only short periods of statistics from localized regions. Very little is known about trends over the oceans, which cover nearly 70 percent of the earth's surface.

856. How is man making the world colder? Scientists believe that a cooling trend in North America and Europe bringing harsher winters and milder summers has developed because of the vast amounts of dust and pollution particles that man is putting into the earth's atmosphere. A murking veil is being spread around the world from the smokestacks and exhausts of man's growing industries, cities, and transportation vehicles. This layer of pollution particles absorbs and holds some of the sun's radiation and reflects some back into space. The net result is that it prevents sun radiation from reaching the earth's surface, hence causing the environment gradually to cool.

Paradoxically, scientists also believe that another kind of pollution

—carbon dioxide produced by burning coal and oil—is also being thrown into the atmosphere, circling the earth in a layer and causing the world to heat up. This gas, like the glass of a greenhouse, permits the sun's radiation to penetrate through, but prevents the longer waves of heat from radiating back into space—a phenomenon called the greenhouse effect. (See Question 715.) Hence, the world may be slowly heating up. A recent study indicates that carbon dioxide keeps the world nearly 20 degrees warmer than if the gas were not present.

857. What damages does cold bring? The cold blasts of winter and the chills of polar regions have caused death and disaster for centuries. Local battles and even national wars have been lost. Industrial, economic, and social activities have been abruptly terminated or disrupted. Exposure of the human body to cold has resulted in pain, maiming, and death.

Spectacular disasters have occurred when cold winter air freezes rivers and oceans. Enormous problems are created when river ice presses against the bridges, dams, and buildings of mankind; when sea ice blocks harbors; and when icebergs threaten busy shipping lanes.

858. What winter defeated Napoleon's army? The bitter winter weather of Russia was a major factor in the defeat of Napoleon Bonaparte's army in 1812.

At the height of his career, Napoleon decided to capture Moscow, at that time a large arsenal and supply center for the Russian Army.

After a successful battle at Borodino, the French Army reached Moscow on September 14. The following night the Russians burned most of the city, destroying the supplies and shelters Napoleon was counting on for his own troops. He ordered a retreat on October 19. Then began a series of hardships from the drastic cold and the guerilla warfare of the Russians. Actually, frost arrived later than usual in that region, and the weather was considered dry and "bracing." Yet to the French, the march was too long and hard, and the army suffered the first knell of downfall.

859. When did a cold flow of air lead to the capture of a fleet? In the winter of 1794–1795, a rare surge of extremely cold air pushed down from the interior of Siberia and froze the Dutch fleet near Texel

Island, off the Netherlands' coast. A handful of French cavalry galloped over the ice to the locked-in fleet and browbeat the crews of the well armed battleships into surrender.

860. What was Europe's coldest winter in a century? During the winter of 1955–1956, large blasts of frigid air flowed down from Siberia and blew westward across Europe instead of taking the usual path eastward. Weakening and erratic westerlies and the southward bending of the jet streams were causes of this cold winter wave that froze taxicabs in Brussels, halted milk deliveries in Holland, necessitated the dropping of food to isolated communities in Scotland, brought rare snow to Rome and to Jerusalem, froze orange crops in Spain, and generally snarled traffic and halted communications throughout the European region. A total of 140 people were reported killed by this cold blast.

861. How much cold can a human body endure? The normal temperature of a person's body is 98.6 degrees Fahrenheit. Any variation as small as 1 or 2 degrees from that throws the body mechanism out of order, causing an increase in discomfort, inefficiency, suffering, and eventually death.

Human beings have survived when their body temperatures were lowered to about 75 degrees Fahrenheit, more than 20 degrees below normal temperatures—but only for a few minutes.

There are no set figures for how long a person can survive at low temperatures. Individual tolerance varies greatly, as do conditions under which a person may be subjected to cold. For instance, a person's ability to endure cold depends on whether his clothing is wet or dry, whether a wind is blowing, or whether he is immersed in cold water. When a person is exposed to cold in wet clothing, he loses heat more rapidly than in dry conditions, especially when a wind is blowing. A person immersed in cold water loses body heat faster than under the most severe conditions of cold air.

862. How long can people survive in cold water? The ability of a person to survive in cold water depends on many factors, such as the amount of clothing he is wearing, his health and stamina, his desire to live, and the amount of shock.

Some people have survived for 14 hours in water that is just above freezing. Others have died in only a few minutes.

In a British research project, one 29-year-old man stayed afloat in 30 degree Fahrenheit waters of the Bering Sea for 14 hours. Other people could not survive these waters longer than 30 minutes.

Data on the length of time people can survive in waters of varying temperatures have been compiled into an informative map, of the U.S. Hydrographic Office (now the Oceanographic Office). In waters extending north of Labrador and the Bering Straits, most people will not survive more than three quarters of an hour. In waters extending as far south as Florida, Western Africa, and parts of the South Pacific, a man can expect to live for 1½ hours, or as long as 12 hours. In warmer tropical waters, a person can survive for an undetermined length of time, depending upon factors such as stamina, food, fresh water, and hazards such as sharks.

863. What was a record survival in cold wilderness? On February 2, 1967, Canadian bush pilot Robert Gauchie ran out of fuel near the Arctic Circle and was forced to land. He was rescued 58 days later, having lost 54 pounds and 5 frostbitten toes in temperatures reaching 48 degrees below zero. Surviving on emergency rations and raw fish, Gauchie stayed by his plane and was finally spotted by a persistent rescue pilot, long after other extensive searches had been called off. Gauchie had spent most of his time under 6 sleeping bags inside the plane. He survived during 2 of the coldest months of the year, February and March, and one of the worst blizzards in that area on record.

864. How have men fared in the bitter cold of the polar regions? Tragic stories of man against the cold were not uncommon during the early days of Arctic and Antarctic exploration. The diaries and journals of Captain Robert Falcon Scott, British explorer who made a courageous attempt to be the first to reach the South Pole, record hardships of that brutal journey in 1912 against gales and headwinds, difficult ice formations, blinding snow, frostbite, and temperatures of 50 degrees below zero. The papers were recovered 8 months later from the camp in which Captain Scott and his remaining crew died, only 11 miles from the One Ton Camp basic depot. This was his last entry:

78° S. Thursday, March 29. Since the 21st we have had a continuous gale from W.S.W. and S.W. We had fuel to make two cups of tea apiece and bare food for two days on the 20th. Every day we have been ready to start for our depot 11 miles away, but outside the door of the tent it remains a scene of whirling drift. I do not think we can hope for any better things now. We shall stick it out to the end, but we are getting weaker, of course, and the end cannot be far.

It seems a pity, but I do not think I can write more.

R. Scott*

Among the many farewell letters Captain Scott wrote in that last camp was a Message to the Public, which ended:

Had we lived, I should have had a tale to tell of the hardihood, endurance and courage of my companion which would have stirred the heart of every Englishman. These rough notes and our dead bodies must tell the tale.**

865. What happens as a human body becomes colder? As a person's body becomes colder, a series of reactions sets in. The body endeavors to decrease its loss of heat by contraction of blood vessels at the skin's surface and at extremities such as fingers and toes. This tends to reduce the blood flow at areas where heat would be lost to the environment. As temperatures continue to drop, the body attempts to generate more heat by shivering—a condition that increases muscle tone and heat. During these reactions, automatic messages are sent from cold-sensory nerve endings to the thalamus in the forebrain where they are relayed to the muscles of the skin and body.

Yet under extreme conditions of cold, these reactions cannot protect the body for long. The first parts of the body to suffer are the extremities of fingers, toes, ears, and nose. Actual freezing of skin tissues takes place at skin temperatures of 26 to 30 degrees Fahrenheit.

* Capt. Robert F. Scott, *Scott's Last Expedition*, arranged by Leonard Huxley (New York: Dodd, Mead & Company, 1913), Vol. I, p. 410.
** *Ibid.*, p. 417.

866. What is frostbite? Frostbite is an actual freezing of skin tissue, usually in the extreme parts of the body such as toes, fingers, nose, and ears. If exposure continues too long, tissues are permanently damaged, and gangrene may set in.

For a cure, the exposed parts should be warmed, gently and immediately. The frozen parts may be thawed in warm (not hot) water or wrapped in blankets or warm clothing.

The frozen part should not be rubbed or exercised and should not, as people used to think, be rubbed with snow, ice, or cold water.

867. What is immersion foot? Immersion foot, also called trench foot, is an injury usually on feet, but sometimes on hands, exposed for a long time to wet cold at temperatures just above freezing. Soldiers of World Wars I and II, as well as of Korea, suffered from this disorder. With immersion foot, the affected area becomes red, swollen, and painful, but not so severely as from frostbite. The limb should be warmed, and the patient placed in a horizontal position to increase blood circulation.

Immersion foot and also frostbite can be prevented by keeping feet and hands warm and dry and by exercise. Several kinds of wax, grease, and plastic appliances have been tried to prevent it, but unsuccessfully. In the early stages, the feet and toes appear pale or bloodless and feel cold, stiff, and numb. Walking is difficult. If cold conditions continue, the feet will swell and become very painful. Some feet have become so injured that amputation was necessary.

868. What is hypothermia? Hypothermia is a medical term for the physiological condition caused when the entire body is exposed to cold, and the body temperature is lowered. This can occur when a person is immersed in cold water for a short period of time or when he is exposed to low air temperatures for a long period of time. To recover, the person should drink hot liquids, and his entire body should be rewarmed within a sleeping bag or with a warm bath.

869. What is the wind chill index? Arctic explorers, military doctors, and Weather Bureau researchers have developed a term called the wind chill index, which indicates the cooling effect of various combinations of winds and temperatures on the exposed flesh of people. In general, the stronger the wind, the lower the temperature seems, and the more harmful are the effects of cold.

The index is a figure that reflects how low the equivalent temperature would be under specific wind-temperature conditions.

For instance, if the actual temperature is 20 degrees Fahrenheit, a 35-mile-an-hour wind will produce a wind chill equivalent to 20 degrees Fahrenheit below zero.

WIND CHILL TABLE

ACTUAL TEMPERATURE, DEGREE FAHRENHEIT

			35°	30°	25°	20°	15°	10°	5°	0°	—5°	—10°
			EQUIVALENT TEMPERATURE									
							(Equivalent in cooling power on exposed flesh)					
W	M	Calm	35°	30°	25°	20°	15°	10°	5°	0°	—5°	—10°
I	I		VERY COOL									
N	L	5	33°	27°	21°	16°	12°	7°	1°	—6°	—11°	—15°
D	E		COLD									
	S	10	21°	16°	9°	2°	—2°	—9°	—15°	—22°	—27°	—31°
S			VERY COLD									
P	P	15	16°	11°	1°	—6°	—11°	—18°	—25°	—33°	—40°	—45°
E	E				BITTERLY							
E	R				COLD							
D		20	12°	3°	—4°	—9°	—17°	—24°	—32°	—40°	—46°	—52°
S	H	25	7°	0°	—7°	—15°	—22°	—29°	—37°	—45°	—52°	—58°
	O					EXTREME COLD						
	U	30	5°	—2°	—11°	—18°	—26°	—33°	—41°	—49°	—56°	—63°
	R	35	3°	—4°	—13°	—20°	—27°	—35°	—43°	—52°	—60°	—67°
		40	1°	—4°	—15°	—22°	—29°	—36°	—45°	—54°	—62°	—69°
		45	1°	—6°	—17°	—24°	—31°	—38°	—46°	—54°	—63°	—70°
		50	0°	—7°	—17°	—24°	—31°	—38°	—47°	—56°	—63°	—70°

870. Can man adjust to cold climates? Yes, man can adjust to cold climates, but the acclimatization is difficult. When a person moves into a cold climate, his blood vessels constrict and his blood volume decreases. The sweating processes change—sweating starts to take place at higher temperatures. It is possible that thyroid glands enlarge and thyroid activities increase in the cold regions of the world.

Man can help make his adjustment to cold easier with the use of proper clothing, food, and shelter. For instance, since the human body loses its heat by radiation and conduction to the cold environment, clothing should be insulated against the wind and made to hold in as much body heat as possible. The extremities—hands, feet, ears, and nose—should be especially protected.

People in cold regions eat larger amounts of food to make up for loss of heat. Fats and carbohydrates especially are helpful in replenishing energy. Since vegetable crops are hard to raise in frigid climates, the staple diet of Eskimos, Lapps, and other cold-dwellers has long been fish and animal meat—until the advent of imported packaged and fresh vegetables and other foods.

871. What is a degree-day? A degree-day is a concept that indicates deviations of atmospheric temperature from normal. It is helpful to specific professions or industries concerned with critical weather conditions—heating industries, for instance, or farming.

The outside air temperature of 65 degrees Fahrenheit is considered a critical temperature for heated buildings. Above this temperature, inside heating is not essential. If a particular day, for instance, has a mean temperature of 45 degrees Fahrenheit, heating engineers subtract this figure from the standard number 65 and obtain the balance of 20. That particular day is said to have 20 degree-days.

A continuous record of degree-days is kept from September through April with each day's total added to the preceding grand total. This gives heating engineers an idea of the deviation from normal or average conditions.

If the mean temperature of the day is equal to or higher than the standard of 65, then the day is called a zero degree day and adds nothing to the total.

872. Why is river ice so destructive? In the spring thaws, when broken pieces of ice move down a river, they exert enormously high pressures against bridges, dams, spillways, and other structures. Pressures may range from 4,000 to 20,000 pounds per square foot. Chunks of ice often cause much damage by grinding, ripping, shattering, and crushing objects in their paths.

Problems of river ice become especially difficult if the ice cover melts somewhat during a warm day, forming fissures which fill with melt water. At night the mass freezes again and expands, thus putting added pressures against the banks and any intervening structures.

Ice jams seldom cause loss of life or inflict serious damage to mankind, for they can easily be broken apart by dynamite before they become too large and dangerous.

873. Where have severe river ice jams occurred? River ice and subsequent ice jams are severe problems particularly in regions of the far north, as in Russia and Canada. Here rivers flow from the south northward toward the North Pole. In spring, the upper part of the river thaws first and begins to flow onto the still-frozen lower part of the river. Chunks of partially thawed ice pile up in enormous masses on the lower reaches of the rivers, causing ice jams and floods.

In Canada, rivers such as the St. Lawrence, Mackenzie, and Nelson have severe ice jam problems during spring thaws. In Russia, the Dvina, Yenisei, Ob, and Lena rivers break up with enormous ice piles.

874. What was the Susquehanna ice bridge of 1852? The winter of 1851–1852 was one of the most severe winters of that century in the eastern United States, bringing temperatures that were almost continuously below freezing.

At that stage of early American development, railroad engineers had not yet solved the problems of building bridges across wide expanses of rivers, including the Susquehanna sweeping down through Pennsylvania. The Philadelphia, Wilmington, and Baltimore Railroad crossed this great river by means of a ferry.

The river froze during the cold winter, and the ferry was blocked. The railroad began to suffer losses, since mail, freight, and passengers could not move across the river. Chief engineer Isaac R. Trimble decided to build an ice bridge strong enough to support a train. Ice hummocks on the river were chopped level and depressions were filled with ice. Larger pileups of ice cakes, some 12 feet high, were circumvented. Railroad ties and rails were laid across this packed ice, with ends inclined against the shores, which were some 10 to 15 feet higher than the level of the river ice. The bridge was completed on January 15 and was used every day until February 24, when a general thaw set in. As many as 40 train cars were moved across the river in one day.

875. What is frazil ice? Icy fresh water streams sometimes flow so swiftly through narrow steep valleys that solid ice does not have a chance to form in sheets. Instead, fine, needle-like spicules of ice called frazil ice are formed in the churning water. The same kind of ice, formed in turbulent salt water, is called lolly ice.

Frazil ice is formed when the water is supercooled to temperatures below freezing. Sometimes there are so many ice spicules that the stream seems a sluggish mass of mushy snow that is difficult to remove, channel, or do anything else with. This can be very troublesome and clog the stream, stop water wheels, endanger bridges, locks, and power plants, and cause floods.

876. What ice is formed in oceans? Dangerous masses of ice are sometimes encountered in the oceans. One kind of ice is true sea ice, formed by the freezing of top layers of salty water. (See Questions 879 through 884.) Another kind is fresh water ice, or land ice, which includes the icebergs broken off from glaciers or frozen ice shelves. (See Questions 885 through 900.)

Although true sea ice constitutes about 95 percent of the ice in the oceans, icebergs bring the greater hazard to ships because of their immense size, most of which is hidden beneath the water surface.

877. How do oceans freeze? Salt water usually does not freeze over, for it takes much lower temperatures to freeze water with a solution dissolved in it than pure water. Yet in the far north and south, temperatures drop so low that large expanses of the ocean freeze over.

In the first stages of ocean freezing, the surface of the water starts to have an oily, opaque appearance, caused by the formation of tiny needles and thin plates of ice known as lolly crystals or lolly ice. These crystals increase in number until the top of the sea is slushy, like thick soup. As this slush thickens, it is broken up by wind and waves into round pieces of ice, with turned-up edges, looking somewhat like lily pads or pancakes. Ice pieces usually take this shape when they are blown around by the wind and collide against one another. As temperatures continue to lower, the pieces gradually grow larger, freeze together, and form large sheets of ice called floes.

878. Why doesn't the world freeze? Frozen water is unique among solids—a fact that keeps the world from freezing. Instead of contracting when it solidifies, as all other solids in liquid state do, water expands.

This one basic property has kept the world from slowly freezing to death.

As water becomes colder, the density or weight increases, down to 39.2 degrees Fahrenheit—some 7 degrees above freezing point— when it decreases. Then the chilling water molecules become lighter than those of the surrounding fluid water. Near freezing, the cooling molecules expand, then form chunks of ice that float, with about nine tenths of the bulk submerged beneath the surrounding water.

Because of this unusual property, the world continues to have fluid water. For if the ice continued to become more dense, it would sink to the bottom of deep ponds, rivers, wells, and oceans, and settle there in dark frigid silence where the warm rays of the returning spring sun could not reach it. Year after year, the layers of ice would thicken from the bottom of the waters, until all water sources were frozen solid except for the surface melted by the sun.

879. What is sea ice? True sea ice consists of frozen sea water and is therefore somewhat salty. This ice gradually loses its salt in time, as the salt and other impurities seep slowly downward in periods of partial thawing, and fresh water ice is left at the top.

There are two main kinds of sea ice—fast ice that is generally immobile and remains fast to the land; and pack ice that is constantly moving or drifting with the tides, winds, and currents.

880. What is fast ice? Fast ice is any type of ice attached to the shore. If directly attached it is called ice foot or ice shelf; if beached, shore ice; if stranded in shallow water or frozen to the bottom, anchor ice.

881. What is pack ice? Pack ice may consist of several broken pieces of ice or floes, or of a large slab of sea ice, held packed together. In summer in the Arctic, cracks appear through which open stretches of water flow. In the winter these cracks are again frozen shut.

Pack ice—also called ice pack—has different names. A piece of ice larger than 5 miles is called a field. A giant floe is a piece from 3,000 feet across to 15,000 feet—the size of a small city. A medium floe is 600 to 3,000 feet across, the size of a city block. A block is 6 to 30 feet, the size of a volley ball court; and a brash is 6 feet long, the size of a pool table top.

882. How thick does sea pack ice get? Sea ice spread over the polar oceans is rarely thicker than 5 or 7 feet in summer or 10 to 12 feet in winter.

When it first freezes, it may become 4 or 5 inches thick in the first 48 hours. After that, the rate of freezing becomes slower.

Sea ice becomes thicker by "rafting" when one piece of ice is pushed up over another. Wind, tides, currents, and other pressures may push up chunks of ice into ridges or hummocks. Tides or waves may rise, flow over the ice sheet, freeze, and add to the thickness.

On the Great Lakes, ice sheets become about 36 inches thick.

883. What are ice islands? Ice islands are blocks of ice some 18 miles long, ¼ mile wide, and 100 or 200 feet thick. These islands drift slowly around and around the North Pole in the Arctic currents. About 85 to 100 have been counted. They are thought to be pieces broken off from the ice shelves around the northern edge of North America and Asia.

Each island has a flat top upon which airplanes can land. Broad parallel low ridges with shallow troughs cover their surfaces, in a conspicuous striped pattern that is distinctive from the air. Unlike an ice pack, these islands are thicker and carry boulders, sometimes 10 feet in diameter, as well as clay and other earth material. Plants such as grasses and lichens grow on them. Some of the islands are inhabited by polar bears, foxes, and other Arctic animals.

They float through the ice pack of the Arctic for years with practically no change in their size or shape.

884. What use is made of these ice islands? These stable ice islands are considered valuable sites for permanent and semipermanent stations for scientific observations. They are useful as long as they remain in circulation. When they run aground or drift out of the Arctic, the station simply is shut down.

The United States and Russia maintain research stations on ice islands where studies are conducted in glaciology, oceanography, meteorology, marine life, and other aspects of the Arctic.

The first U.S. station was called Fletcher's Ice Island. Other islands include T-1, T-2, ARLIS II (Arctic Research Lab Ice Station). The Russians maintain a station on ice island Nordpol 6.

885. What is a glacier? A glacier is a river of ice, a large moving mass of land ice that creeps slowly downhill by gravity into valleys, plains, or oceans. It is formed from accumulations of snow and ice that for centuries have become packed layer by layer. Scientists believe these huge ice masses move downhill because the tremendous weights and pressures compress the ice at the bottom of the mass, altering the molecular structure so the crystal layers flow.

Glaciers may either advance or retreat, depending upon the climate. As the environment becomes colder, more ice and snow remain frozen and the size of glaciers increases. As the climate warms, the glaciers begin to melt, particularly the leading edge which is generally farther south and at lower latitudes. Then the glacier appears to retreat.

The largest glaciers, called ice sheets or icecaps, cover thousands of square miles. They are so thick they sometimes bury entire mountains except for the peaks.

Icecaps and glaciers constitute some 7,000,000 cubic miles of ice—about 2.15 percent of the total amount of water in the world.

As the glaciers move to the edge of the sea, they break off in huge chunks—icebergs that crash into the sea and begin a long journey before melting in warm waters. (See Questions 886 through 895.)

886. What are icebergs? Icebergs are huge pieces of fresh water ice broken off the forward edge of a moving glacier slowly flowing down to the sea. Most icebergs originate in Greenland or on the ice shelf of the Antarctic. With only about one tenth of their bulk showing above the surface, these ice masses may drift in the icy sea currents for years, sometimes becoming stranded in shallow waters or frozen fast to land ice, or moving into warmer waters where they melt and disappear.

These large chunks of ice often transport gravel, rocks, or large boulders that have been picked up as the rivers of ice grind their way to the sea. These rocks may be carried thousands of miles from their place of origin before they are dumped as the berg melts.

The U.S. Navy has several descriptive names for bergs. They are called bergy bits if they are about the size of a small cottage and growlers if they are as large as a piano.

887. What was the tragedy of the *Titanic*? Just before midnight on April 14, 1912, one of the worst peacetime maritime disasters in history took place as the White Star 46,000-ton luxury liner *Titanic* crashed into an iceberg 95 miles south of the Grand Banks of Newfoundland with a loss of 1,517 persons.

The 882½-foot-long ship, the largest liner afloat at that time, was on its maiden voyage from Liverpool to New York City. With a double-bottomed hull divided into 16 watertight compartments, the *Titanic* had been declared unsinkable—yet the huge ship sank at 2:20 A.M. on April 15, only 4 hours after a 300-foot gash had been ripped by the iceberg in its left side.

The liner *California,* only 20 minutes away, might have helped to save a lot of lives, but the radio operator was off duty and did not hear the distress signal. The Cunard liner *Carpathia* arrived 20 minutes after the *Titanic* plunged beneath the waves and picked up the only survivors from the icy waters.

888. What was the *Hans Hedtoft* disaster? On January 30, 1959, despite radar, up-to-date patrolling, and other modern aids to navigation, the ultra-modern Danish passenger-cargo ship *Hans Hedtoft* collided in fog and heavy seas with an iceberg off the southern tip of Greenland. The crew radioed the ship's position and condition. Although ships in the area immediately went to the rescue and aircraft patrolled the region, no trace of the ship or of her 95 passengers and crew was found, even after several day's search.

This ship, like the *Titanic,* was considered unsinkable. The hull had a double steel bottom, with an armored bow and stern, and it contained 7 watertight compartments. She was equipped with the latest inventions—radar and modern electronic equipment. Like the *Titanic,* the *Hans Hedtoft* was completing her maiden voyage—on the last lap from Greenland to Copenhagen.

889. How big are icebergs? An iceberg may tower several hundred feet above the water surface, but because of the nature of ice, this represents only a small fraction of its size. Nine tenths of it lies beneath the surface.

Icebergs have been reported 30 to 40 miles long. Such ice mountains may have about 1,000 square miles of surface area.

The largest icebergs have been found in the Antarctic region, where they break off with huge splashes from the towering ice shelf. One iceberg was reported about 60 miles wide, more than 200 miles long, and about 800 yards thick.

890. How many icebergs are formed each year? No one has taken actual count of the thousands of icebergs formed each year, but U.S. Coast Guard researchers estimate that about 7,500 large icebergs break off the thick glaciers along the west coast of Greenland annually.

About 400 icebergs are counted each year as they drift south of the 48th parallel into the shipping lanes.

891. Where are most icebergs formed? Most of the world's icebergs break off from more than 100 glaciers flowing slowly into the Northeast and Disko Bays along the western coast of Greenland. An estimated 5,400 icebergs each year are formed from 12 large glaciers in this area alone.

The process of an iceberg's being born by breaking off from a glacier in huge chunks and splashing into the ocean is called calving.

In the Southern Hemisphere, immense icebergs are calved from the Antarctic ice shelf.

892. What routes do icebergs travel? Icebergs breaking off from the glaciers of Ellesmere Island and northern Canada sometimes circle

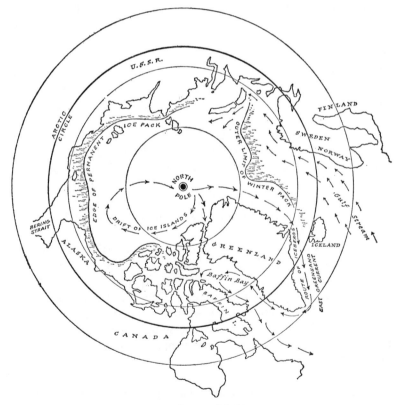

General route of icebergs

in currents around the North Pole in the Arctic Ocean. Some drift slowly southward along the eastern side of Greenland, where many glaciers are calved. The East Greenland current carries them around Cape Farewell, the southern tip of the island, where they may meet the northward drifting West Greenland current which takes them into Davis Strait. Here many become grounded upon the coast. Others move northward for several hundred miles into Baffin Bay, where they swing westward toward the shore of Baffin Island. Caught by the Labrador current flowing southward along the east coast of Labrador, some eventually drift south of Newfoundland and into the Atlantic Ocean, across the active North Atlantic shipping lanes. Here they rapidly melt in the sunshine and warm ocean water. They break up into small packs called ice floes and dwindle in size until they melt completely about 400 miles south of Newfoundland, approximately two weeks later.

893. How long do icebergs last? Some of the icebergs in the north may travel for 3 years, from the time of their calving to their final melting and disappearance in warm waters. In these journeys, these icebergs may cover distances as great as 2,500 to 3,000 miles.

894. Where are icebergs most dangerous to man? The most dangerous iceberg area for ships is south of Newfoundland where the North Atlantic shipping lanes are most crowded. Many icebergs, after their long journey in the northern bays, are large enough to cause considerable damage to ships. The menace of icebergs is increased by heavy sea fogs that frequently hover south of Newfoundland.

895. When is the most active season of icebergs? In spring, the warming sun starts thawing the great icebergs and releasing them from their winter ice moorings. Some of them, propelled by wind and current, manage to drift southward along the coasts into the North Atlantic Ocean and into the trans-Atlantic shipping lanes. The greatest numbers of these icebergs reach the shipping lanes in April, May, and June. At this time of year, many captains pilot their ships on a more southerly route across the ocean.

896. Can icebergs be controlled? Men have tried to break up or destroy threatening icebergs by means of firebombs, gunfire, chemicals, and other methods. To date, no destructive weapon has been able to melt or break up these chunks of ice, or even steer them into harmless areas. It is estimated that nearly two thousand tons of TNT would be needed to break up an iceberg of average size. More than 2 million gallons of gasoline would be needed to melt it down to a size where it could cause no damage.

People have also tried to tow or push icebergs—either to clear a shipping route of the menace or to bring a huge potential source of fresh water close to a city. Because they are so bulky, with 90 percent beneath the water, attempts to move icebergs have generally been unsuccessful.

So far, the best method of defense is to keep close watch on the movements of these ice mountains and warn ships of their precise position so they can be avoided until they are destroyed by the only force powerful enough to combat them—the warm waters of the more southern seas.

897. What instruments are used to detect icebergs? Because the bulk of icebergs are submerged under water, detection of these huge masses of ice is sometimes difficult. In the old days, the fog horn sometimes could bounce sound waves against the sides of an iceberg, warning a ship of its fog- or night-shrouded presence.

The development of radar during World War II helped iceberg detection, but identification was often confused and indistinct. A major advance in ice surveillance was the use of laser beams that permit faster and more precise identification.

Working on the theory that all matter emits electromagnetic impulses of varying wavelengths, scientists have constructed a radiometric iceberg detector that picks up and defines the particular natural impulses sent out from ice.

898. Is it possible to monitor icebergs from satellites? Scientists are studying the use of earth satellites for tracking the movements of the great drifting ice mountains. The NIMBUS satellite launched in 1968 carried improved sensors, including a television camera and transmitting system and an infrared spectrometer giving atmospheric and surface temperatures. Within several years, scientists hope that such satellite monitoring will eliminate the need for aerial reconnaissance.

899. What is the International Ice Patrol? The year after the *Titanic* disaster in 1912, the First International Conference for Safety at Sea was held in London. Here an International Ice Observation and Ice Patrol Service was recommended and, in 1914, was delegated to the U.S. Coast Guard.

With ships, aircraft, and stations on land, sea, and icebergs themselves, this patrol keeps close watch on more than 33,000 square miles of the North Atlantic Ocean.

The iceberg surveillance starts each year on the first of March and continues through the season of most iceberg drift, usually through June or July.

To keep closer track of individual icebergs, researchers splatter them with large spots of blue, red, or other colors.

900. What nations are members of the International Ice Patrol? The International Ice Patrol includes 17 nations—Belgium, Canada, Denmark, France, Germany, Great Britain, Greece, Italy, Japan, Liberia, Netherlands, Norway, Panama, Spain, Sweden, the United States, and Yugoslavia.

901. What are some myths about winter? The ancient Greeks and Romans believed that winter covered the earth for 6 months of the year because Persephone, daughter of Demeter, goddess of harvest, ate 6 pomegranate seeds in the underworld realm of Pluto, king of the dead. The legend goes that Pluto carried off Persephone, causing her mother to mourn so that plants and flowers died and cold winds blew across the land. Pluto agreed to let Persephone return to earth and her mother for 6 months, but the rest of the year she was to remain with him as his queen.

The Norsemen had many great myths about the ice, snows, and winter, contained in 2 collections called the Eddas, one of poetry, the other of prose.

902. Can animals or plants predict severe winters? For generations, people have believed that the behavior patterns of certain animals and plants predict the severity of a coming winter. Nearly all are false forecasts, as animal and plant behavior depends largely upon events of the past and present, rather than upon those of the future.

A few of the superstitious signs believed to forecast a cold winter include:

more berries on trees and shrubs

abnormally thick coats on mule deer and bears

exceptionally thick and furry hairs on wooly bear caterpillars

larger supplies of acorns stored and buried by squirrels

deeper tunnels dug in the ground by ants and other burrowing creatures

903. What is heat? Heat is a form of energy, sometimes called thermal energy. It is associated with the activity of atoms. Heat is measured in terms of temperature, which expresses the amount or degree of atomic activity.

Energy radiated from the sun and changed to heat upon the earth is a major force that motivates all weather phenomena.

904. Where does the world's heat come from? Each second, the earth receives the energy equivalent of some 126 trillion horsepower's worth of solar energy from the sun.

This solar energy is transmitted as radiation of various wavelengths. Some are visible light waves, others are invisible infrared waves or ultraviolet waves. About 47 percent of the radiation reaching the earth's surface is changed to heat, most of which is found in tropical areas.

905. Where are the world's great heat spots? Some of the hottest places of earth are the great subtropical deserts—the Sahara Desert of Africa, the Arabian Desert, the Thar of India, and the Great Australian Deserts. These all occur in low latitudes, near the equator, beneath the direct rays of the sun most of the year long. Even in winter they continue to have long sunlit days and masses of hot air.

906. How hot are these deserts? As the sun beats down from clear skies onto the hard clay, rock, and sand of the great deserts, temperatures build up to more than 100 degrees Fahrenheit near the desert floor. Several feet above the floor, temperatures are less, and they decrease with altitude. Temperatures on the Sahara Desert, possibly the hottest of all deserts, reach 135 degrees Fahrenheit.

Nights on the deserts are extraordinarily cold, for the warmth is radiated directly into the atmosphere. Temperatures in the Sahara and other deserts drop to just above freezing at night.

907. Where are other large hot spots? As the sun moves farther to the north, bringing warmth to the Northern Hemisphere, certain barren deserts in the higher latitudes become even hotter than their surroundings—for instance, the North American deserts in southwestern United States and Mexico, and the Kara Kum, Takla Makan, and Mongolian Desert of Asia.

Some of the semiarid high plains and plateaus of continents also become sources of hot air masses in summer—such as the Great Plains of the United States, the Deccan Plateau in India, and the Great Plain of China.

In the Southern Hemisphere heat sources include the Atacama and Patagonia Deserts of South America, and the Namib and Kalahari Deserts of South Africa.

908. What are some of the hottest temperatures on earth? The world's highest official temperature was 136 degrees Fahrenheit, registered on September 13, 1922, at Azizia, in Libya, Africa.

Another high temperature was recorded as 134 degrees Fahrenheit at Furnace Creek Ranch in Death Valley, California, on July 10, 1913. This was recorded at a depth of 178 feet below sea level.

A temperature of 129 degrees Fahrenheit was recorded at Tirat Tsvi, Israel, on June 21, 1942—at 722 feet below sea level.

Another high temperature was 128 degrees Fahrenheit on January 16, 1889, at Cloncurry, Queensland.

909. How do hot air masses move? Hot dry air does not move in as well-defined masses as cold air, which is denser and flows close to the ground.

Hot air streams outward in broad or narrow currents. Because heat tends to rise, it flows up over colder, denser masses of air and often may travel for long distances without being felt on the ground at all.

Hot air expands and rises from a heat source in separate cells and segments, forming a region of low pressure, into which blows cooler air from the surrounding environment. This inward flowing air in turn is heated, and rises. At times, from sources of great heat, bursts of dry hot air move horizontally across the ground, drawn toward a low-pressure area.

910. What are some routes of hot air? In summer, as the sun moves northward, it creates areas of low pressure by heating certain regions and drawing up air. In some cases, hot air masses move in to fill the vacuum. The dry hot sirocco, for instance, moves northward in spring across the Mediterranean toward a low-pressure area. Other hot winds blow across certain regions of Africa and Asia, causing discomfort, pain, and unnatural behavior. (See Questions 623 and 624.)

In winter, the hot air of the Sahara moves southward with the retreating sun. Warm air pushes from Liberia to southernmost Sudan.

911. What is a heat wave? A heat wave is a period of abnormally and uncomfortably hot and usually humid weather, lasting more than one day. During this time there is little wind, and the sun beats mercilessly from sunrise to sunset. Relatively little of the sun's heat is radiated back into space during the brief summer nights. Most of the day's radiation is absorbed, heating the ground and lower air levels, and raising the general temperature.

Heat waves in lesser degrees of intensity are called hot waves or warm waves.

912. What causes heat waves? Heat waves can occur when a high-pressure area stalls over a region, slowly spiraling down and outward for thousands of miles from a radiant cloudless sky. Along the western and southwestern edge, hot air is brought in from hot dry or hot and humid climates.

Sometimes heat waves occur when a warm front edges in a humid cyclone from the southern seas.

913. What causes severe heat waves in the United States? Some of the worst heat waves in the United States have occurred when a high-pressure area stalls over the Atlantic coast, and warm moist air flows from the Gulf of Mexico over the Midwest, causing extreme discomfort and even death with its heat and humidity.

Long-lasting relentless heat waves also occur in the eastern United States when the Bermuda-Azores high-pressure area is strong and somewhat west of its usual position toward the middle of the Atlantic

Ocean. This clockwise-rotating high can keep warm, dry air spinning in from the arid southern regions, flowing north and northeast over the central and eastern part of the nation for many long suffocating days.

● **914. What was the severest heat wave in the United States?** The most severe heat wave in recent U.S. history took place in the summer of 1934, part of a great widespread drought. From the middle of June to the middle of August, temperatures soared to 100 degrees Fahrenheit and higher in the Midwest during the day, and remained above 80 degrees Fahrenheit at night. Ponds and lakes dried up, trees died, fires consumed forests and scorched communities, the water level in the Great Lakes dropped by a foot or more, glaciers receded, and many other drying processes spread across the land during the era that produced the Dust Bowl.

According to records, during the month of August alone, 1,500 people died from this heat and affiliated causes.

915. What was the 1936 heat wave? One of the worst heat waves of the central United States started in early July, 1936, when abnormally high temperatures settled for 1 or 2 weeks across the Great Plains from the Rocky Mountains to the Appalachians. In many areas temperatures rose to 100 degrees Fahrenheit and stayed there. Some temperatures rose to 120 degrees. In areas where people were not used to such heat, heat exhaustion, fatigue, and death occurred. The worst disasters occurred in Michigan, where 679 people died—nearly 300 in Detroit alone.

This heat wave was caused when warm air flowed from the Gulf of Mexico for almost 2 weeks, pushing out across the Mississippi Valley. Few clouds were present to break the sun's radiation. The heat wave finally broke when the flow of air from the south became so weak that cooler air from the north pushed down over the area.

Other high death statistics from heat waves recorded by the Public Health Service include 196 deaths in 1963, 195 deaths in 1964, and 154 in 1962.

916. What studies have been made on heat deaths in New York City? Several studies have been made of specific heat waves in cities and the resulting increase in deaths.

In New York City, during a heat wave in June, 1952, the Health Department reported the following statistics:

Average number of deaths per day, June 1–24, before heat wave—213

Average number of deaths per day, June 25–27, at 90 degrees—376

Another study of a heat wave in August, 1948, showed a 300 percent increase in deaths in New York City.

Even in winter, a rise in temperature or heat wave can cause more deaths than usual. About 10 percent more people died in the New York region following a 2-day rise of 16 degrees Fahrenheit in winter than on days with a corresponding drop in temperature.

917. What is a heat island? As populations increase, cities expand, and the numbers of buildings, streets, and vehicles increase, a phenomenon occurs that meteorologists call a heat island—a defined area where temperatures are higher than those of the surrounding atmosphere or neighborhoods—sometimes as much as 15 degrees or more. This heat accumulates from the city activities—exhaust fumes, burning furnaces and heating units, smokestacks, and people. During the day, concrete buildings and asphalt streets absorb the sun's heat, and radiate it at night. High buildings block off paths of cooling winds. Only when the wind blows faster than 15 to 20 miles an hour can it blow away the heat and dissipate the heat island.

918. What are dog days? Dog days are days of extremely hot, humid, sultry weather, traditionally occurring in the Northern Hemisphere for 4 to 6 weeks in July and August.

The hot-weather period received its name from the bluish-white Sirius, the brightest visible star in the sky, known as the dog star of the constellation Canis Major, the Greater Dog.

At this time of year, Sirius rises in the east at the same time as the sun. Ancient Egyptians believed the brilliant star added its "heat" to the heat of the mid-summer sun. Sirius was blamed for withering droughts, mid-summer sickness, discomfort, and death.

Although Sirius contributes no heat to the parched earth, dog days are well known for their heat and humidity. At this time of year, the sun has already reached its highest point over the Northern Hemisphere and is starting back again toward the equator. The great continents have been well warmed, and oceans have lost their winter chill. The rays of the sun beat directly down on the temperate zone, drawing up moisture from oceans, streams, rivers, and ponds at the rate of some 16 million tons a second. This moisture is retained as water vapor in the warm atmosphere, creating uncomfortable, muggy weather.

Dog days traditionally begin on July 3, about 20 days before the simultaneous rising of the sun and Sirius. They end August 11, about 20 days after the joint rising.

Superstitions that dogs become mad during dog days are unfounded. Medical records show that more cases of rabies occur in spring and autumn than in summer.

Advice to people suffering during these days was offered centuries ago by the Greek poet Hesiod: "When Sirius parches head and knees, and the body is dried up by reason of the heat, then sit in the shade and drink."

919. How does heat affect a human body? The human body constantly works to keep a balance of the heat produced internally from combustion of food and that lost to the outside atmosphere by conduction, convection, and radiation.

When a person generates more heat by being overly active, or when the air temperatures rise, the body starts a series of activities to rid itself of the excess heat. Blood vessels dilate in order to bring more blood circulating to the skin areas where it can lose heat to the air. Heat is also expelled from the body by some 2 million sweat glands. As the sweat evaporates, the skin area cools.

If a person cannot lose enough heat by these normal mechanisms, the body temperature rises, the pulse rate increases, and breathing quickens. The blood, having lost water, begins to thicken, and the person starts having violent cramps and fits of vomiting. The central nervous system becomes affected, and convulsions, paralysis, comas, stupors, somnolences, and deliriums set in. If the body temperature continues to rise, numerous heat ailments result.

Doctors believe that heat illness occurs more often than is generally known and is responsible for many deaths attributed to other causes.

920. What is heat exhaustion? Heat exhaustion is a physiologic state brought on when the body is overexposed to high temperatures and high humidity. The difficulty in trying to get rid of excess heat from the body causes an overly sharp curtailment of heat production within the body. Symptoms include subnormal body temperature, clammy skin, dizziness, headache, vomiting, and rapid pulse rate.

921. What is heat stroke? A more lethal result of too much heat is heat stroke, when perspiration stops and body heat accumulates rapidly, raising the body temperature to near 110 degrees. The skin becomes hot and dry, the pulse becomes fast and irregular. The victim falls into a coma and sometimes is delirious. Then come convulsions and death by asphyxia may follow. Death takes about 80 percent of those suffering from heat stroke. A heat stroke caused by direct exposure to the sun is sometimes called sun stroke.

922. How much heat can a person stand? With normal human body temperature at 98.6 degrees Fahrenheit, a person cannot stand internal body temperatures of more than 108 degrees for an extended length of time. Few patients have survived body temperatures of more than 112 degrees.

Unofficial reports state that people have survived environmental heat as high as 250 degrees for 14 minutes, 32 seconds, without serious aftereffects.

923. What is meant by the phrase "It's not the heat, it's the humidity"? A person feels more uncomfortable in heat with high humidity because perspiration does not evaporate fast enough to cool the body.

The process of sweating is automatically regulated by sweat glands. As soon as skin temperatures rise to about 95 degrees Fahrenheit, sweat glands pour out a salty fluid that evaporates in the surrounding air. Evaporation needs heat, which is removed from the body, hence cooling it. If the air is already saturated, or humid, the perspired water fails to evaporate, and the body retains its heat. A prolonged state of this condition could lead to heat stroke.

● **924. What is the discomfort index?** The index of discomfort to a human being can be measured by a formula based on the relationship between air temperature and relative humidity. This formula has been developed into a temperature-humidity index—the higher the reading, the more uncomfortable people are. At an index of 72, for instance, 10 percent of the people are uncomfortable; at 77, more than half the people are uncomfortable. At 82 almost all people are uncomfortable, and, over 90, people may be in physical danger.

The use of the discomfort index has been discontinued because problems arose as employees in public and private offices became overly aware of discomfort and would request leave of absence.

925. Does heat affect the rate of crime? Increasing studies are being made of effects of heat and humidity and other environmental factors upon man's behavior. The files of police records are becoming more significant when studied in relation to weather.

In some areas, accumulated evidence indicates that the heat of summer, enhanced perhaps by periods of vacation and inactivity, brings on an increased crime rate. Rape, theft, murder, housebreaking, and other crimes increase noticeably as the hot summer continues, particularly in August. There are more riots, more angry speeches, more irritability, quarrels, moodiness, and acts of aggression during this season.

926. What may be the effect of summer heat on race rioting? Many race riots in highly congested U.S. cities have occurred during the summer. The violence of the 1967 summer brought some of the worst riots in the history of the nation. Riots occurred in the muggy night air of Tampa, Florida, and shooting and rioting in Cincinnati. Death and maimings grew out of civil disorders in Atlanta, and many youths battled police in Boston. In Newark, 24 people were killed, 1,100 injured, and 1,300 arrested. Property damage and the riot costs rose to more than $5 million. A 4-day riot in Detroit toward the end of July was the worst slum violence in American history, with 39 dead, 1,500 injured, and about 3,400 arrested.

The summer of 1968 brought more riots and destruction. During riots in Cleveland, Ohio, 10 persons were killed, 23 wounded, and damages to insured property were estimated at $1,500,000.

Police records of crime during the winter months, however, show that crimes abate, and there is less need for arrests, particularly in January.

927. What other noteworthy riots have broken out in summer? Other riots have boiled over during the heat of the summer . . . caused by hot tempers and brewing psychological resentments as well as by exterior hot weather and climate.

On August 24, 1572, the Massacre of St. Bartholomew erupted in Paris, with massive slaughter of the Huguenots by the Catholics, and contributed to the start of 8 wars of religion.

On July 6, 1892, the Homestead, Pennsylvania, strike at Carnegie steel mills resulted in 7 guards and 11 strikers and spectators being shot and many others wounded.

Violence broke out June 22 and 23, 1922, during a coal-mine strike at Herrin, Illinois, with 36 dead.

The race riot of Detroit June 21, 1943, caused 34 people to die and 700 to be injured. At the same time, 6 Negroes of Harlem, New York, were killed.

On June 16, 1953, a demonstration of workers in East Berlin against increased work quotas stirred up anti-communistic riots with some 50,000 people participating and involving a general strike of some 200,000 people in East Germany. Soviet troops, in an attempt to restore order, killed 16 people.

Sufficient data has not been accumulated to indicate definitely that more riots and crimes are committed in the hot seasons than in the cold. Many severe acts of violence have occurred also in the cold seasons.

928. What other weather phenomena affect living organisms?

Other atmospheric hazards besides temperature extremes can affect living beings—sometimes suddenly and dramatically, sometimes indirectly and subtly in ways we do not yet understand.

For centuries people have been aware of the effects of winds, temperatures, storms, and other weather phenomena upon their health and spirits. Changes in weather can bring on attacks of nausea, heart ailments, arthritis, and asthma; or make a person moody, depressed, or even violent. The effects have seemed particularly noticeable among older people, sick patients, or among emotionally highstrung or hypersensitive people.

In past centuries, these effects were duly recognized and accepted. In the past few decades, however, these factors and their effects have come under more precise scientific studies, and a relatively new field called biometeorology (see Question 929) has been formed.

Effects of weather such as winds, heat, cold, and humidity upon people have been mentioned in previous questions and chapters pertaining to those aspects of meteorology. The next few questions of this chapter will briefly describe the more obscure effects of pressures upon man. These phenomena can be termed disasters, for a disaster is a sudden calamity, whether inflicted upon thousands of people or upon only one person.

929. What is biometeorology? Biometeorology is a science dealing with the relation between living things and weather—including storms, winds, heat, cold, humidity, pressures, and electricity. This science is called human biometeorology when it applies specifically to human beings. Studies include atmospheric environments ranging from deep levels beneath the earth's soils to the highest atmospheric levels, as well as artificial atmospheres such as those found in buildings, shelters, submarines, jetliners, and satellites.

Information about this new science is still quite scarce, found in isolated research reports, and as yet not well centralized. Scientific viewpoints often conflict even in relatively small research projects.

930. What is atmospheric pressure? Atmospheric pressure is the force exerted on a surface from the weight of air above that surface. The atmosphere is a fluid layer of gases surrounding the earth, some 20 miles thick, with a total weight of at least 5,600 trillion tons. At sea level, a vertical column of air one inch square extending to the outer reaches of the atmosphere weighs 14.7 pounds.

At the bottom of this fluid ocean of air live the air-breathing organisms of the earth, bearing pressures of about 2,116 pounds on each square foot of their bodies. Men and other living creatures survive this enormous pressure just as fish survive at the bottom of the sea—by exerting an equal pressure from the inner body out.

Whenever man moves out of the thin layer of his accustomed atmospheric pressure—whether up into mountain heights where air is thinner and lighter or down into mines and beneath oceans where pressures of air and water are heavier and crush in on him—he must be equipped with pressure-regulating devices such as oxygen masks, pressurized suits, space suits, and diving suits.

931. How is pressure measured? Atmospheric pressure is measured essentially by two types of barometers.

The mercurial barometer, used in most weather stations, is a curved glass tube filled with mercury. One end is closed, and the other open. The weight of air pushing upon the mercury at the open end forces mercury up the closed end. The greater the air pressure, the higher the column of mercury is pushed up. This height is read and converted into units of millibars for meteorologists; of inches for laymen.

The aneroid (a word derived from *a*, meaning not, and the Greek *nero*, meaning wet or moist) barometer is an instrument with no fluid, consisting of a corrugated metal container from which the air has been removed. The pressure of air above bends in the top of the box. As it increases, the top bends down; as pressure decreases, the top bends out. Gears and levers transmit these changes to a pointer on a dial.

932. What happens when a person is suddenly exposed to low pressures? When someone is suddenly exposed to low pressures, the internal gases and liquids within his body begin to expand because the pressure inside is greater than that outside. The same principle applies when a balloon expands as it ascends to higher altitudes. The body's water vapor and blood may literally boil and capillaries may burst, in a condition called explosive decompression.

A phenomenon called the bends is sometimes encountered when deep sea divers are pulled out of ocean depths too fast for the body to adjust from high pressures to normal pressures. These men experience severe pain in the muscles and joints of arms and legs. More severe symptoms may develop, including itching and skin rash, vertigo, nausea, vomiting, fatigue, choking, and shock—and sometimes death. All this occurs because the sudden decompression permits the formation and enlargement of nitrogen bubbles in the body tissues. Bends are also known as caisson disease, tunnel disease, and diver's paralysis.

933. What dies it feel like to suffer sudden low pressures? A vivid description was given by Marine Corps Lt. Col. Rankin, who suddenly jumped from his jet at an altitude of 47,000 feet—with no pressure suit. This was the highest emergency ejection on record. (See Questions 306 and 307.)

Meanwhile, the pain of explosive decompression was unbearable. I could feel my abdomen distending, stretching, stretching, stretching, until I thought it would burst. My eyes felt as though they were being ripped from their sockets, my head as if it were splitting into several parts, my ears bursting inside, and throughout my entire body there were severe cramps. . . .

I was preoccupied with the pain of decompression. It was nature's cruelest torture, the screw and rack of space, the body crushes, the body stretches, each second another turn of the screw, another wrench of the rack, another interminable shot of pain. Once I caught a horrified glimpse of my stomach, swollen as though I were in well advanced pregnancy. I had never known such savage pain. I was convinced I would not survive; no human could.*

934. What happens when men encounter low pressures living in mountains? As a person ascends into the higher altitudes of mountains, atmospheric pressures decrease at a general rate of one half the amount for every 18,000 feet.

Visitors from lower altitudes may feel the effect of low pressures and reduced oxygen supply. Symptoms of this condition, called mountain sickness or altitude sickness, include breathlessness, palpitations, loss of appetite, and nosebleeds.

In the sixteenth century, the Spanish conquistadores invading the Inca civilization in the Andes Mountains encountered this mountain sickness. They called it the soroche, the "sickness of the Andes."

When the Spaniards settled in Potosi, Peru, some 13,000 feet high, no Spanish woman was able to give birth to a child that survived. When animals were transported to these elevations, their reproductive organs were so affected they became temporarily or even permanently sterile.

935. What does it feel like to have mountain sickness? Father José d'Acosta describes vividly the effect of the high altitude and decreased oxygen on Spaniards in the sixteenth century as they climbed the Peruvian Andes:

There is in Peru a high mountain which they call Pariacaca. . . . When I came to mount the stairs, as they call them, which is the top of this mountain, I was suddenly surprised with so mortal and strange a pang that I was ready to fall from my beast to the ground. . . . I was surprised with such pangs of straining and casting

---

* Lt. Col. William H. Rankin, U.S.M.C., *The Man Who Rode the Thunder* (Englewood Cliffs, N.J.: Prentice-Hall, Inc., 1960), pp. 152, 153.

as I thought to cast up my soul too: for having cast up meat, phlegm, and choler (bile), both yellow and green, in the end I cast up blood, with the straining of my stomach. . . . I persuade myself, that the element of the air is there so subtle and delicate, as it is not proportionable with the breathing of man, which requires a more gross and temperate air, and I believe it is the cause that doth so much alter the stomach and trouble all the disposition.*

936. Can people adapt to low pressures of the high mountains? Certain people in the Andes mountains of South America, in the Himalayas of Asia, and in other high mountains have gradually adapted to the rarefied air of the high altitudes. Residents of the mining community of Cerro de Pasco in Peru live at altitudes of more than 14,000 feet above sea level. Another community lives about 18,000 feet high near the sulfur mines at Mount Aucanquilcha in Chile.

These people have developed long, thick hearts, and enlarged chests. Veins are distended and they have a greater number of corpuscles than people living at lower altitudes. Their heart beats slower than normal. At the very high altitudes, women have difficulty in bearing babies, although when they descend to lower altitudes they are able to bear them.

937. How do changes in atmospheric pressure affect men psychosomatically? Fascinating statistics are being accumulated from studies made on effects of atmospheric pressure upon human beings.

As a low-pressure, storm-bearing system approaches and atmospheric pressure is falling, people often complain of asthma attacks, stiffness in joints and muscles, and heart ailments. Some persons feel generally depressed, grouchy, and ill-tempered without knowing why. Also, people have more difficulty concentrating, and general efficiency is reduced. On the other hand, some persons may become overly excited as a storm approaches. They may not be able to sleep, their heart beats faster, and headaches develop.

As the storm breaks and passes on, however, a person's mood and sense of well being can change completely. As the air clears and the barometer rises, spirits also rise, and people feel bright, mentally alert, and able to do difficult jobs with relative ease.

* David I. Blumenstock, *The Ocean of Air* (New Brunswick: Rutgers University Press, 1959), p. 5, from *Historia natural y moral de las Indias*, 1590.

938. What specific studies have been made on pressure effects?
A 3-year study at the Pennsylvania Health Department noted a "significant" rise in the number of suicides on days when the barometer dropped 0.35 or more inches. The studies covered 527 suicides that occurred in Philadelphia from 1960 through 1962.

Extensive studies on barometric effects on human beings have been made in Germany and Switzerland. Some of the results have been published in *A Survey of Human Biometeorology*, Technical Note No. 65, of the World Meteorological Organization at Geneva. For instance, a person's reaction decreases at times of unfavorable weather conditions when a low-pressure storm is moving in. In experiments conducted on 118,000 visitors at the International Communications Exhibits at Munich, Germany, in 1953, researchers noted that even healthy people took a longer time to respond to a light signal under adverse weather conditions. In experiments at an industrial plant in Munich, researchers found that work accidents almost doubled in unfavorable weather, and many persons complained about their health.

In a study of 43,000 traffic accidents in Munich, researchers found that accidents increased 15 to 20 percent in unfavorable weather. Some experts believe that such weather hazards as fog, rain, and glazed or slippery roads have less effect in causing accidents than changes in pressures. In Hamburg a study showed that accidents increased 6.4 percent on days of glazed frost; 5.2 percent on days of fog; and 40 percent on days of decreasing atmospheric pressure.

939. What other phenomenon accompanies low-pressure cyclones?
Some scientists think the frequency of atmospheric electricity tends to rise in cyclonic, low-pressure, storm-bearing systems and tends to decrease in high-pressure weather.

Whole populations may be affected by high-frequency electricity, according to scientists with the World Meteorological Organization. In a statistical study in Germany of 1 million individual reactions, it was found that when atmospheric electricity was high, births increased 11 percent and deaths increased 20 percent. Traffic accidents soared by 70 percent, and work accidents in factories and companies increased by 20 percent. In hospitals, the complaints of brain patients increased 30 percent, those of patients with amputations increased by 50 percent, and the complaints of chronically ill patients rose a full 100 percent.

Electricity itself may not primarily affect the mood of a human being, some scientists believe. The current may be stimulating the brain blood flow or it may be affecting chemicals in the brain. Or it may be slightly changing the positive-negative electrical polarity in the head.

940. What happens as pressures increase? As men descend into the ocean, or into deep caves and mines, atmospheric pressures increase. These higher pressures can also be felt by aircraft passengers when a plane descends to land. In the oceans, pressure increases at the rate of one atmosphere (14.7 pounds) for each 33 feet of descent below sea level. At 33 feet below sea level, for instance, pressure is 29.4 pounds per square inch—at 66 feet below, it is 44.1 pounds.

If a person's Eustachian tubes and openings of his air sinuses are blocked, the differences in the inner and outer pressures are greater, and may result in severe pain, with congestion, edema, and hemorrhage within the cavities. Sometimes the eardrums can rupture.

Normally, pressure in the middle ear can be equalized by repeated swallowing, yawning, or by trying to breath out when the nose and mouth are kept closed.

941. What other kinds of pressure affect people? Pressure waves that travel outward from any vibrating object are compression waves, or sound waves. These waves are actually pulsations of alternating higher and lower pressures.

More attention is being given to the growing nuisance value of man-made sound, particularly the growing noise-making activities in cities and in the airways. Increasing noise is coming from city sources such as traffic, construction, machinery, boiler factories, and various industries. Other distressful noises are being created by shock waves of guns or projectiles, sonic booms and supersonic sounds of high-speed jets, and screams of rockets.

942. What organizations include heat and cold research studies? The World Meteorological Organization has encouraged research and information exchange in meteorological studies of heat and cold.

Other organizations conducting temperature research include the U.S. Weather Bureau, the National Academy of Sciences, the American Academy of Arts and Sciences, and many atmospheric science departments such as the Massachusetts Institute of Technology, the University of Washington, and Colorado State University.

Medical studies on effects of extreme temperatures on living organisms have been increasing throughout the world. The World Health Organization (WHO) has furthered many such studies, as have branches of medical military establishments—including the Air Force Cambridge Research Center, the U.S. Naval Research Laboratory, the U.S. Army Medical Corps—and medical and biological departments in such Universities as Duke, Michigan, Arizona, Rice, and California. Valuable contributions have been made in other countries, for instance, from the Liverpool School of Tropical Medicine and the Chemical Defense Experimental Establishment in England, the University of Edinburgh in Scotland, the University of Khartoum in the Sudan, and the National Research Center in Paris.

In England, the British Glaciological Society and the Scott Polar Research Institute carry on research work. In Germany, research studies are under way at universities such as those of Mainz, Munich, and Munster. The Polar Institute at Oslo, Norway, fosters studies of snow and ice. In Japan, the Japanese Society of Snow and Ice Research in Tokyo and the Institute of Low Temperature Science of Hokkaido University in Sapporo have conducted thorough studies on snow and ice.

Research on cold and the Antarctic is conducted under the terms of the 1959 Antarctic Treaty, signed by 12 nations which conducted research there during the International Geophysical Year: Argentina, Australia, Belgium, Chile, France, Japan, New Zealand, Norway, Union of South Africa, Union of Soviet Socialist Republics, United Kingdom, and United States. The program is coordinated by the ICSU Scientific Committee on Antarctic Research (SCAR).

Arctic research programs are being carried out by Canada, Denmark, Finland, Iceland, Norway, Sweden, Union of Soviet Socialist Republics, and the United States.

943. Where is research being conducted in biometeorology? Individual scientists and researchers throughout the world are beginning to collect information on this relatively new field and to offer recognized studies. They are so numerous, however, that it is difficult to assemble or compile them in a list. A partial list of organizations includes the World Meteorological Organization of the United Nations, the International Society of Biometeorology, with headquarters in Oegstgiest-Leiden, Holland, the American Institute of Medical Climatology in Philadelphia, and the Massachusetts General Hospital. Other organizations involved more in electrical aspects include the Institute of Electrical and Electronic Engineers and private companies such as the General Electric Company and Westinghouse Electric Corporation.

IX. STORMS OF SPACE

Introduction. Scientists and technicians are beginning to probe deeper into the space storms that rage throughout the solar system, perhaps beyond—the great solar flares, the bursts of plasma clouds, the streams of cosmic rays, and the solar winds, all of which constantly sweep past our planet earth. With extraordinarily precise and imaginative instruments, space scientists are now beginning to probe and learn more about the nature of the radiation waves and particles that stream through space—light waves, x-rays, low energy protons, electrons, cosmic rays, and many more, including meteorites. Masses of these particles shoot past or onto our planet every instant—some are deflected back again into space; some are caught in the magnetic fields that surround the earth; and some shatter atoms and molecules in the earth's atmosphere as they cascade over its surface.

Although pieces of matter and radiation have been zig-zagging, speeding, and crashing around the universe for eons, man is just beginning to detect and identify them and define their destructive or beneficial effects upon him and his world. As astronauts prepare to journey farther out toward the stars, the nature and danger of these space storms will be increasingly analyzed, and more precise methods of prediction and protection will emerge.

The subject of space storms and space weather is so complex, and the research projects, instrumentation, and methods of study so numerous and constantly changing, that latest information is often uneven and incomplete. This chapter merely touches on a few basic aspects of space storms in relation to living creatures of the planet earth.

944. What are solar storms? The violent storms that seethe and roil on the sun are of incomparably greater energy and magnitude than any tornado, hurricane, or maelstrom on earth. This is apparent when one realizes that only a minute fraction of the sun's massive energy output reaches earth, and only a small percentage of that energy is used to generate this planet's windstorms.

Solar storms may take the form of enormous eruptions of solar flares, forming plasma clouds, solar prominences, winds, and other disturbances. Some of these storms emit streams of radiation waves and high-velocity particles out through the solar system and beyond. Like outward-radiating seismic waves, or waves from a rock tossed in a pond, these radiation waves and particles break upon every object they encounter in their paths—upon each planet, satellite, asteroid, or debris particle in turn. Only a small fraction of these high-energy particles strike the relatively tiny earth, but even these can cause huge disruptions and damages to living creatures and man's evolving civilizations.

945. What is a solar flare? A solar flare is a sudden energetic eruption of hot gaseous material—mostly protons which are the cores of hydrogen atoms—thrown spaceward by violently turbulent processes within the sun. Usually associated with sunspots, solar flares may appear within minutes, and then disappear within an hour. They cover a wide range of intensity and size, sometimes extending over as many as 2 billion square miles of the sun's surface—10 times as large as the total surface of the earth.

These bright massive eruptions spew out streams of radiation waves and particles such as x-rays, radio waves, light waves, plasma clouds, and high-energy protons that spread out through the solar system. Some of these bombard the earth in a few seconds, 93,000,000 miles away; others take several days to arrive. Much of this energy is halted by the earth's magnetic field and atmosphere before it can reach and damage the earth's living creatures.

About once a year, gigantic flares erupt, throwing out greater amounts of particles and x-rays faster than normal flares. The particles of these giant flares are considered dangerous for manned space flights, and journeys are planned not to coincide with times the flares are expected. Some thought has been given to the construction of "storm cellars" within the spacecraft, into which astronauts could retreat when flares become particularly intense.

946. What are sunspots? Sunspots are transient, apparently dark spots on the sun's surface, sometimes a few hundred miles in diameter, sometimes as large as 100,000 miles across. They have intense magnetic fields and are accompanied by very bright areas or mottlings known as faculae. They usually occur in pairs with opposite magnetic polarities, or in paired groups with several spots in each group. Scien-

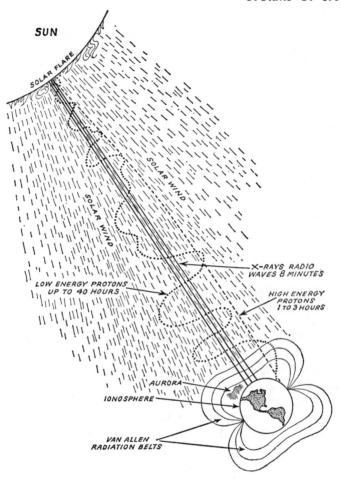

General manifestations of a solar flare

tists believe the interaction of these magnetic fields may help create solar flares.

It is generally believed that sunspots are formed when local magnetic fields on the sun's surface become so strong that they immobilize the upward transference of heat in a bubbling convective heat spot. The area then becomes cooler, and appears darker, than its hotter surroundings.

947. What is the 11-year sunspot cycle? Each year, a certain number of sunspots appear and disappear on the surface of the sun. At times there may be no sunspots observed, then a few may appear. The number may rise from a minimum of 1 to 10 to a maximum of 50 to 140 sunspots about 4 years later, and then slowly decline during the next 7 years.

The periods of minimum activity occur on an average of about every 11.3 years—hence the name 11-year cycle. Periods of maximum activity, however, do not occur so regularly and may vary anywhere between 7 and 17 years.

Sunspots rarely appear on the sun's equator and never at the poles. They do not migrate across the face of the sun, but are carried around with the sun's rotation. The life span of an individual sunspot may be a few days, or months.

At the beginning of the sunspot maximum activity cycle, sunspots appear nearer the poles. As the cycle progresses, more spots appear at lower latitudes. Toward the end of a cycle they appear near the equator.

Although men have been observing sunspots for several hundred years and sunspot cycles for the past three centuries, no one yet knows why the cycles occur so regularly. Many people have made long detailed studies attributing cycles of drought and rainfall, auroras and other earth phenomena to the sunspot cycle—but none has been substantiated.

948. What is a plasma cloud? Sometimes a mass of ionized gas bursts out from the sun. This electrically conductive gas, called a plasma cloud, is composed of ionized particles, free electrons, and neutral particles. Taken as a whole, a plasma cloud is electrically neutral, but it contains a large amount of stored energy due to the interactions between particles.

Plasma clouds emitted from solar flares have been known to strike earth 1 to 3 days later. Sometimes this cloud is preceded by a stream of high-energy protons that may reach earth in 20 minutes.

949. What is a solar prominence? A solar prominence is a conspicuous geyser of bright gas that erupts in huge arcs or fountains from the sun's surface. Observed on the edge of the sun, these eruptions are often associated with flares. They sometimes arch as high as 500,000 miles or more and descend back onto the sun. Tremendous magnetic forces are involved in these eruptions which can take various forms—sometimes as relatively small, thin flames; sometimes as broad, luminous, low clouds; sometimes as huge arches.

Scientists believe the prominences are expelled from the edges of sunspots, and they are drawn in again toward those sunspot centers that show a downward action.

950. What is the solar wind? The solar wind is a stream of electrically charged gas particles constantly boiling out from the sun's surface at the rate of one million tons a second. The wind consists primarily of protons (positive ions of hydrogen). Recent studies show there are also charged helium-4 and helium-3 ions, as well as oxygen ions. Other heavy ions exist, but as yet have not been exactly identified. These energized particles stream out from the sun at speeds of nearly 2 million miles an hour.

Spreading through the solar system like a fluid, bumping and sweeping around objects in its path, the wind extends from 10 to 100 astronomical units into space—more than 9 billion miles. (An astronomical unit is defined as the mean distance of the earth from the sun—about 93,000,000 miles.)

951. What effects does the solar wind have? The constantly blowing solar wind spreading out in all directions throughout the solar system has many effects, scientists are finding.

The pressure of this wind pushes against the earth's magnetosphere (see Question 994) in such a way that it flattens the sphere to windward on the sunny side of the earth and causes it to stream millions of miles to leeward.

Solar wind pressure also blows against the gaseous tails of comets entering the solar system, causing them to point away from the sun.

On earth, the solar wind is a constant source of particles becoming trapped in the magnetic zone around the earth and contributing to the brightly colored auroras. It also causes variations in cosmic rays entering the solar system.

952. What is a magnetic storm? A magnetic storm is a world-wide disturbance of the earth's magnetic field caused when a burst of energy and particles from the sun strike the earth's atmosphere. Several of these storms may occur during a year. Frequently distinct magnetic changes occur in less than an hour, after which the magnetic field returns to normal during a period of several days.

Although scientists know the storms are caused by solar disturbances, they do not yet know exactly how. The storms seem to occur most frequently at times when sunspot activity is high and are experienced on earth 1 to 3 days after a solar flare erupts on the sun.

953. What happens on earth during a magnetic storm? Magnetic storms can cause magnetic, electromagnetic, and other interruptions in many man-made systems. Radio-telephone communications may be interrupted, radios can black out, teletype machines may start typing gibberish, and television sets may get out of focus.

At times of these bursts of excess energy, many particles spiral downward toward the polar regions, causing the auroras to grow larger and brighter.

954. What is radiation? Radiation is energy emitted through space or through matter exhibiting properties of waves and particles.

Radiation can be defined as the process by which energy is transferred by means of oscillations or waves in the electrical and magnetic fields. The length of these waves, which is related to their frequencies, depends primarily on the temperature of the radiating body. For instance, intensely cold bodies with temperatures approaching absolute zero radiate the longest waves. The hotter the body, the shorter the wavelengths. The chemical composition and physical state of the body also determine frequencies and intensities.

Every object in the universe—rocks, ice, people, stars, galaxies—radiates some sort of energy, produced by the motions of variously charged particles in atoms.

Since an inseparable relationship exists between the two types of energy, electricity and magnetism, radiation is also called electromagnetic energy or electromagnetic radiation.

955. What damage on earth is caused by radiation waves and particles? As streams of radiation waves and particles blast over the earth, many crash into and interact with molecules in the earth's atmosphere, forming showers of electrons, gamma rays, mesons, and other radiations that scientists are learning more about.

Many of these particles and solar ultraviolet and x-rays can cause damage in various ways. As they bombard the ionosphere, they can increase the ionization in the ionosphere and decrease its height, severely disrupting radio systems that depend on this ceiling to bounce back transmissions. They can cause fading or complete black-out of radio communications, particularly in the polar regions. For instance, in 1966, communications were severed for about 7 days at Antarctic research stations. Radio emissions from the sun can create so much noise that earth radio systems are interrupted.

Radiation waves and particles can injure delicate solar sensors, solar cells, and other precision instruments. They have fogged up photographic film on satellites. The effect of solar flares on the winds of the earth, on pressures and temperatures, and on polar highs and tropical lows is uncertain, although some scientists believe they may help cause more variety and violence in the weather. It is relatively certain that solar flares create atmospheric magnetic storms on earth and bring about brilliant displays of aurora storms.

956. What biological effects do cosmic particles and radiations have on living organisms? Cosmic particles and radiations can affect living organisms on earth in strange and powerful ways that scientists and doctors are beginning to analyze in detail. The living body cells of human beings, animals, and plants can be injured. Germ cells in particular are damaged by radiation which can break up chromosomes by which genes are carried and hence cause mutations and affect future generations.

After an intense exposure to ionizing radiations and high energy particles, a person can suffer radiation sickness, including nausea and vomiting, sometimes diarrhea, hemorrhage, and a decrease in the blood-cell level.

As man advances higher and deeper into space, he will become more subjected to radiations. Occasional showers of radiations and particles can even affect passengers in supersonic transports flying at altitudes of 60,000 feet to 80,000 feet.

957. What are two diverse theories concerning radiation? In the latter part of the seventeenth century, some scientists considered certain radiation phenomena, such as light, as corpuscles; others thought they were waves. Those men who postulated the wave theory said that waves may be transmitted through an "ether" that served as a medium much as gaseous, liquid, or solid material transfers acoustic or sound waves.

As more data accumulated and theories advanced, the duality of radiation as having the properties both of waves and of particles was recognized.

Today modern scientists consider electromagnetic radiation, called EMR for short, to be both continuous waves traveling at a constant speed through "free" space and distinct quanta or packets of energy. (See Questions 958 and 959.) In other words, they believe that some particles behave as if they had wave-like properties and that some wave forms have the capacity to exist in packets of particles.

958. What is the quantum theory of radiated energy? In 1900 the German physicist Max Karl Planck postulated that all forms of radiated energy could be viewed as existing in packets called quanta (plural for quantum). These packets or particles of radiation are called photons. They are electromagnetic particles.

Albert Einstein substantiated and elaborated upon this quantum theory of radiation.

959. What is the quantum theory of material particles? The mechanical laws developed by the British mathematician Sir Isaac Newton, the French astronomer Pierre La Place, and others, and further developed by Einstein in his theory of relativity cover material particles moving with velocities near the velocity of light (186,000 miles per second)—such as protons and electrons.

The quantum theory developed by the German physicists Erwin Schroedinger, Werner Heisenberg, and others around 1925 showed that some material particles behave as if they had wave-like properties.

960. How is radiation characterized? Electromagnetic radiation can be characterized by frequency or wavelength, intensity, and polarization. Some forms of radiation, such as x-rays, are often characterized by their energy and momentum.

Frequency and wavelength are closely related to each other. Frequency is actually the number of wavelengths produced per second.

Characterizations commonly used by scientists include:

gamma-rays	energy and momentum	photons
x-rays	energy and momentum	photons
ultraviolet	wavelength	waves
visible	wavelength	waves
infrared	wavelength	waves
microwaves	wavelength	waves
radio and TV bands	frequency	waves
longer waves	frequency	waves

961. What is solar radiation? The sun generates nearly every form of radiation and emits energetic particles in addition to particles constituting the solar wind. These energies thrown off from the sun include radio waves with wavelengths many miles long, infrared rays, light waves, ultraviolet waves, and x-rays with wavelengths only a few billionths of an inch long. The greatest part of the sun's radiation is in the form of light.

The radiant solar energy received on the earth's surface is called insolation, a word contracted from *in*coming *sol*ar radi*ation*. This is not to be confused with insulation, which refers to materials that do not conduct heat or electricity.

Less than half the sun's radiation passing through the earth's atmosphere reaches the earth's surface. Many waves are absorbed and reflected back into space by dust and clouds. Also waves are absorbed by atomic and molecular constituents of the earth's upper atmosphere, for example, oxygen and nitrogen. All solar radiation that strikes the earth's surface, no matter what wavelength, is converted into heat and long-wave infrared waves.

962. What is terrestrial radiation? The radiant energy received by the earth is in turn radiated from the earth or transferred into the atmosphere. But for this terrestrial radiation, the earth would become increasingly hotter and perhaps burn up. Because outgoing radiation keeps in balance with incoming radiation, the world neither freezes nor burns up. Of all solar radiation falling on the planet earth, scientists estimate that about 42 percent is reflected back into space by clouds and atmospheric dust; 15 percent is absorbed directly in the atmosphere; and 43 percent actually reaches the earth. Of the 43 percent reaching earth, about 8 percent is radiated back into space. This fraction of earth's reflection, called the earth's albedo, is actually measured in terms of a ratio—the ratio of light reflected to light received by the earth. Albedo varies widely from place to place. Snow and ice have a very high albedo, as well as sand deserts. Yet cultivated fields and forests radiate practically nothing back to space.

963. How are radiation waves classified? The electromagnetic waves transferring energy outward from an object can exist in any length. They are classified in an orderly arrangement of radiation according to their wavelengths—the distances from crest to crest of succeeding waves. This array, called the electromagnetic spectrum, is continuous.

Starting from the very long radio waves and microwaves that have waves several miles long, radiations extend through infrareds measured in microns; into the visible, ultraviolet, x-rays, and gamma rays, measured in angstroms, which is a unit about one 250-millionth of an inch, named after the nineteenth century Swedish astronomer Anders Jonas Angstrom, who first studied solar radiation with a spectroscope. X-rays and gamma rays can be described in energy units. Their energies range upward from about 10 kilovolts (1 kilovolt equals 1,000 electron volts).

964. What is the electromagnetic spectrum? The electromagnetic spectrum is the array of electromagnetic radiations found in space arranged in order of their wavelengths. These classifications have been given arbitrary divisions by scientists. The boundaries are not definite, but often merge into one another.

965. What are some of the longer waves of the electromagnetic spectrum? Some of the longest waves of the electromagnetic spectrum, known as radio waves, have wavelengths measured in terms of megameters, kilometers, and meters. Radio waves, first called Hertzian waves after the German physicist Heinrich Hertz, are used to transmit power, telephone and radiotelegraphic communications—i.e., communication not by electric currents along regular telegraph wires but by waves radiating through space. The great long whistlers generated by lightning discharges and apparently projected along a geomagnetic line of force that carries them between the Northern and Southern Hemisphere are in this category. These can be heard on radio receivers.

Microwaves, with wavelengths shorter than radio waves, measure about 1 meter to 1 millimeter. These waves include the ultra high frequency (UHF), super high frequency (SHF), and extremely high frequency (EHF) radio frequency bands. Radar is operated within these bands.

966. What are infrared rays? Infrared (below the red) rays are an important part of solar energy. Generated mostly from heat sources by large-scale intramolecular processes that are mainly molecular rotations and internal vibrations, they have wavelengths from 1 millimeter to 1 micron.

Most of the solar infrared rays are absorbed in the atmosphere's troposphere, but some reach the earth's surface.

These rays cannot be seen, but they can be felt. Excessive amounts can produce severe burns on living flesh.

967. What is light? Light is the phenomenon of luminous radiation from the sun and other stars and bright objects, radiating with wavelengths from 7,600 angstroms to 3,800 angstroms. The waves are produced by vibrations of electrons in a large number of randomly oriented atoms and molecules. They are visible radiations that stimulate the sense of sight. Altogether they make up white light, which is ordinary sunlight. Separated, as with a prism, they break into colored spectral bands ranging from the longer wavelengths of the reds, through the oranges, yellows, greens, and blues, into the violets which have the shortest wavelengths of the visible waves.

Blue light is scattered much more than red light. That is, blue light is knocked out of a direct beam more than red light.

When we look at the sun on the horizon during sunrises and sunsets, we are looking through much more atmosphere than when we see the sun overhead. Thus the sun appears red when near the horizon, because more blue light has been knocked out of the beam, while the longer rays of red light penetrate through. The sky looks blue by day because a larger portion of blue light than any other light is scattered throughout the sky.

968. How is light essential to life on earth? Without light, life could not exist on earth. Light waves are essential in the process of photosynthesis, by which plants manufacture carbohydrates, proteins, and other basic foods of animals and human beings. Light also gives men and animals the gift of sight.

Yet too much sunlight can bring harmful effects in the form of sunburn blisters, inflammation of the eye's cornea, and total blindness when a person looks directly at the sun.

969. What is ultraviolet radiation? Ultraviolet (literally meaning, beyond the violet) radiation has wavelengths shorter than those of visible light—from 1,000 to 10 angstroms. Emanating from disturbances of electrons, it can neither be seen nor felt without special equipment.

Many of these rays are effectively absorbed in the higher atmosphere, where they are responsible for many complex photochemical reactions—such as the formation of the ozone layer by dissociating oxygen molecules followed by the recombination to form the ozone. Some ultraviolet rays reach earth where they can cause harm to living material by having a toxic effect on cellular structure. In excess, ultraviolet rays can cause burns on exposed skin, and eventually cancer.

Ultraviolet rays can also be used for beneficial purposes, such as killing harmful germs and helping convert ergosterol and other materials in the body into vitamin D.

970. What are x-rays? X-rays are non-nuclear electromagnetic radiations of very short wavelengths, from 10 to 0.1 angstroms. They are usually characterized by energy in kilovolts.

$$10 \text{ angstroms} = \text{approximately} \quad 1 \text{ kilovolt}$$
$$1 \text{ angstrom} = \text{approximately} \quad 10 \text{ kilovolts}$$
$$0.1 \text{ angstrom} = \text{approximately} \quad 100 \text{ kilovolts}$$

Named x-rays because of their unknown quality at the time of discovery in 1895, they are also called Roentgen rays, after the German physicist, Wilhelm Konrad Roentgen.

These powerful rays can penetrate various thicknesses of all solids on earth. They act upon photographic plates in much the same manner as light. Overexposure to these rays can seriously damage living tissues, burning and destroying cells of flesh, blood, and bone; and causing permanent changes in genetic cells that produce mutations or new species of living organisms. When used under control, x-rays have proved useful tools to mankind in medical and industrial fields.

971. What are gamma rays? Gamma rays are products of secondary cosmic ray showers. (See Question 974.) Emitted from nuclei of atoms undergoing nuclear rearrangement, these rays are like x-rays but have a greater amount of energy and a shorter wavelength— from about 0.1 angstrom through .001 milliangstrom and shorter. These rays are also characterized by energy in kilovolts. Their energy starts at about 100 kilovolts and goes up indefinitely.

Like x-rays, gamma rays can penetrate various materials, including tissues of plants and animals. They can cause genetic changes in reproductive cells.

Gamma rays are 1 of 3 varieties of radiation from radioactive substances, named after the first 3 letters of the Greek alphabet— *alpha, beta, gamma*. Many are emitted from atomic explosions and by atomic nuclei in the course of certain radioisotopic disintegrations. Today scientists are creating their own gamma rays from high voltage generators, linear accelerators, betatrons, and synchrotrons.

972. What is black body radiation? Scientists use the concept of black body radiation as an ideally radiating body or surface that reradiates all the radiation that it receives. Theoretically it is the highest amount of energy radiated at each wavelength at any given temperature.

973. What are cosmic rays? Cosmic rays are highly charged nuclei of many different kinds of atoms possessing enormous energies of motion (kinetic energies).

Possibly originating from outer space, they move through the solar system at velocities near the speed of light.

Some scientists have calculated that about 90 percent of these particles are hydrogen nuclei or protons; about 9 percent are helium nuclei or alpha particles; and about 1 percent are charged nuclei of heavier atoms such as lithium or iron. These particles have enormous amounts of energies—some may have more than 10 billion billion electron volts, whereas particles in the solar wind have energies of 500 electron volts.

Scientists believe these cosmic rays originate in stars, including our own sun. The majority may come from outside the solar system. They have been found to be densest in the galaxies, in the Milky Way for instance, and are the dominant gas pressure in these galaxies. Cosmic rays might be debris from immense thermonuclear explosions of stars called supernovae, picking up extra speed from being bounced around between magnetic fields accompanying interstellar plasma clouds. Cosmic radiation of lesser energy level is believed to come from our own sun.

974. What kinds of cosmic rays exist? Scientists group cosmic rays into two basic classes: primary and secondary.

Primary cosmic rays occur mostly in outer space, before they enter the earth's atmosphere. They are submicroscopic electrically charged particles, mainly protons, that travel in space at speeds near the speed of light. Other primary-ray particles are helium-4 nuclei and the remainder are nuclei of still heavier atoms. Stripped of electrons, they are highly charged nuclei of atoms. As they travel, they cause ionization all along their paths.

As these primary cosmic rays rip through the earth's atmosphere, they sometimes collide with and shatter atoms or nuclei in the air, creating showers of secondary rays that have less energy than primary rays, but still have very high energies. These secondary cosmic rays are neutrons, protons, mesons, and residual nuclei such as alpha particles.

975. What harm do cosmic rays cause? Space scientists have been concerned with primary cosmic rays that may strike a space ship and hurl secondary showers inward.

Radiation doses may destroy body cells, particularly sensitive eye cells and reproductive cells. Some primary particles may go through the human body so fast they have no time to do damage. Other slower and heavier particles may slow down as they pass through a body, possibly pulling out streams of electrons from the body cells.

976. What are auroras? Sometimes beautiful displays of flickering colored lights appear in the upper atmosphere over middle and high latitudes. These often assume various shapes such as rays, arcs, curtains, crowns, bands, streamers, and diffuse luminous glows. The more common lower glows may be mostly apple-green arcs and bands that flicker at heights of 60 to 100 miles. The higher glows, mostly blue, violet, and gray-violet, occur some 200 to 400 miles high.

The reason auroras occur is as follows: ordinarily, low energy particles in the solar winds are deflected around the earth by the earth's magnetic field. When the sun is active, as when flares occur, the energy of particles emitted by the sun is increased. These penetrate farther into the earth's atmosphere. Since the earth's field is weakest at the magnetic poles, the particles penetrate farthest into the atmosphere at these places, speeding toward the poles at several thousand miles per second. As they reach the atmosphere, they collide with molecules and atoms, "exciting" them and causing them to glow. As the amount and energies of the space particles change, the aurora changes and flickers.

In northern regions, these spectacular lights are called aurora borealis, or northern lights; in the Southern Hemisphere, they are called aurora australis, or southern lights.

977. What is solar debris? Solar, cosmic, interplanetary debris—all are names for countless pieces of dust, metal, and rock flying through space, presumably broken off from planets, satellites, comets, and other objects. Some scientists believe this debris may be some of the primeval matter from which the solar system may have been formed.

As the earth sweeps around the sun in orbit, it sometimes collides with some of this material which, if caught in the gravity field of earth, may fall to the earth's surface as a meteorite. (See Question 979.)

978. What are meteoroids? A meteoroid is a fragment of stony or metallic material traveling in space as an independent body in the solar system. Meteoroids may be pieces of comets or asteroids. (See Questions 988 and 989.) Most meteoroids are no larger than a particle of dust or a head of a pin, but some are larger, weighing thousands of tons. As they are caught in the earth's gravitational field, they fall toward earth at speeds of 10 to 40 miles per second, braked by the friction of the atmosphere which causes them to generate tremendous heat, to glow, and then to burn up. They are first visible about 70 miles above the earth's surface and flash downward as streaks of light before being burned up in half a second some 30 miles above the earth. In its fall, the meteoroid excites and disassociates air molecules to such a point that energized glowing particles are created along its path, leaving a trail of light behind it called a meteor. (See Question 985.)

Some scientists estimate that about 80 to 100 million meteoroids strike the earth's atmosphere every 24 hours, nearly all of which burn themselves out in the sky. They have erroneously been called shooting stars or falling stars. They are not stars at all.

Scientists do not exactly know where meteoroids come from, but many believe they are fragments of asteroids, remnants of comets, or part of the general debris sweeping through the solar system.

979. What are meteorites? Some of the larger meteoroids do not completely disintegrate in their fall through the atmosphere, but land with a crash on the earth's surface—either in one piece or in fragments. They are then called meteorites and range in size from particles so tiny they fall unnoticed, to tear-drop-sized pieces called tektites, to mammoth pieces weighing several tons.

980. What happens as meteorites strike the earth? The energy of the motion of meteorites falling through the atmosphere changes to heat on collision with the earth and is released as an explosion. If the meteorite is large, the blast can be enormous. Several large craters on the earth's surface, many miles in diameter, have been created this way.

Most meteorites fall into the sea or on uninhabited or sparsely populated regions. As communities and cities continue to expand across the face of the earth, however, chances become greater that a meteorite can hit a populated area.

981. Has anyone been killed by a meteorite? As far as is known, no human being has been killed by a meteorite.

There are many ancient stories reporting the deaths of people by what might have been meteorites, but none of these have been substantiated. Authentic accounts have been given of people and animals being struck or injured by a falling meteorite, or of coming close to being hit—but not of being killed. For instance, in 1847, a 40-pound meteorite crashed through a house in Braunau, Bohemia, and injured

3 children in bed. A Japanese girl was reported hit by a small meteorite in the late 1930's. On June 24, 1938, a meteorite, later called Chicora, crashed near Pittsburgh, Pennsylvania, rocking the city and terrifying thousands of people, but the only damage reported was an injured cow. Scientists estimate that at a height of 12 miles above the earth, this meteorite weighed about 500 tons. Had it fallen at a different angle, it might have hit Pittsburgh and killed half a million people.

982. What large meteorites have been found? The largest meteorite known is about 9 by 8 feet and weighs more than 70 tons. This is known as Hoba West and lies near Grootfontein, southwest Africa, slowly disintegrating in the weather. It may have fallen a million years ago.

The largest meteorite on display is in the Hayden Planetarium in New York City. This weighs 34 tons, 85 pounds.

A huge meteorite may lie buried beneath Meteor Crater, also known as Crater Mound or Barringer Crater, 19 miles west of Winslow, Arizona. This crater is about one mile in diameter at the top and almost 500 feet deep, edged by a parapet some 130 to 160 feet high. Pieces of meteorites have been found around the crater, the largest weighing more than 1,000 pounds. Scientists estimate the original meteorite may have weighed about 63,000 tons and was traveling at speeds of 10 miles a second. As it hit the earth, it gouged out some 300 million tons of rock. Efforts to find the meteorite body have failed, after drilling to depths of more than 1,000 feet.

A large meteorite fell February 17, 1930, near Paragould, Arkansas, and broke into several fragments weighing 80 to 820 pounds.

The Black Stone of the Kaba, in a shrine in Mecca, Arabia, and considered the holiest of holies by people of Mohammedan faith, may be a meteorite with a black crust.

983. What devastation occurred in Siberia in 1908? On June 30, 1908, about 7 A.M. out of a clear southwestern sky, a large incandescent mass crashed upon a lonely forest near Vanavara in central Siberia.

Men and horses were jarred off their feet, 300 miles away; a great thud was heard by people near Turokhansk, 600 miles away; freak atmospheric air waves were felt in Britain 5 hours later, 3,500 miles away, and in Batavia, Djakarta, 4,410 miles away; seismic waves were recorded in Germany and parts of Russia.

Not until many years later did scientific expeditions investigate the site of the explosion. Here men found wreckage of some 80 million trees over an area of nearly 2,000 square miles. At the center of the area the ground was shattered and rocks folded. For nearly 80 square miles around the impact center almost all trees had been killed by tremendous heat. Beyond the burning center, mile after mile of trees had been felled, all with their tops pointing away from the center.

Several puzzling discoveries have led scientists to believe this was no meteorite. No crater was found comparable to the vast destruction the object caused. No meteorites or fragments have been recovered from the area. Some scientists believe it may have been the nucleus of a comet—a mass of frozen substances at extremely low temperatures that thawed to water on hitting earth.

Had this object landed on or near a populous city, millions of people could have been wiped out.

984. What was a recent spectacular meteorite shower? Early Saturday morning, a few minutes after 1 A.M. on February 8, 1969, a brilliant meteorite shower began to fall over a ten-square-mile area in central Mexico.

According to reports, a blinding blue-white fireball streaked across the sky above Pueblito de Allende and other villages of the area. The fireball exploded with a flash of light and showered meteorite fragments throughout the countryside. Some of the stones were pebble size, and some weighed 40 pounds. This was an unusual meteor shower because more than a thousand pounds of fragments were collected in one of the largest, quickest, and most intense meteorite research and recovery operations on record. Scientists from many U.S. and Mexican organizations made immediate trips to the area to gather samples of the extraterrestrial material for future study. They believe the fragments could have come from an asteroid.

985. What is a meteor? A meteor is actually the brief streak of light that occurs as a meteoroid falls through the earth's atmosphere and is heated to incandescence by friction. The word meteor comes from the Greek term meaning "something in the air." The Greeks believed an occasional "shooting star" was actually a star falling from heaven and considered the phenomenon atmospheric such as falling rain or snow.

Meteorology is the study of the atmosphere and the weather, not of meteors. The study of meteors is now termed meteoritics.

986. What is a meteor shower? When a large number of meteors are seen, the event is called a meteor shower.

The earth in its orbit passes through two particularly large comet tails each year, causing large meteor showers. The meteors are named after the constellation from which they seem to radiate—for instance, the Perseid meteors, once called the "Tears of St. Lawrence," seem to come from the constellation Perseus. The earth always passes through these comet tails in mid-August. Another large shower occurs in mid-November, called the Leonid meteors which seem to shoot down from the constellation of Leo.

987. What was a spectacularly large meteor shower? One of the brightest and most spectacular meteor showers on record was the Leonid meteor shower that fell the night of November 12, 1833. The meteors flared so brightly that people were wakened from sleep, and the sky was reported literally ablaze with fire. Meteors were described larger and brighter than Jupiter or Venus, and one was nearly as large as the moon. One location estimated that 250,000 meteors were counted between midnight and dawn.

Scientists calculate that a major meteor shower can be expected every 33 years, when the main body of meteoroids is encountered at the intersection of the earth's orbit and the comet's orbit. Although the Leonid meteor showers are some of the brightest, none has been so spectacular as that of 1833. Another bright, heavy shower fell in 1966.

988. What is an asteroid? An asteroid is a small fragment of a planet revolving in a belt around the sun between the orbits of Mars and Jupiter. Some scientists believe that millions of years ago, a planet orbiting around the sun exploded or collided and burst into millions of fragments that still circle the sun. No one knows how many asteroids there are, although estimates run as high as 50,000. Some asteroids are tiny irregular pieces of rock less than a mile in diameter;

others are several hundred miles in diameter. Ceres, the first asteroid discovered, in 1801, is the largest, with a diameter about 480 miles.

Scientists believe that some smaller asteroids may come near the earth's gravitational field and are pulled to earth as meteoroids.

989. What is a comet? A comet is a huge lump of frozen gases in which many small solid particles lie buried. A comet is a regular member of the solar system and moves in a definite path or orbit, usually elliptical, around the sun and then far out beyond the most distant planet. Some comets have small orbits and return relatively frequently around the sun—at periods about every 3 to 9 years. Some comets appear every 10 to 1,000 years. Others seem to come from incalculable distances, make a sharp turn around the sun, then travel straight out of the solar system. One comet is estimated to take more than 2 million years to complete one orbit.

As a comet nears the sun, its frozen gases become vaporized, releasing gas and solid particles that continue traveling as a group in the comet's orbit. This is the tail of the comet, which is driven out from the comet's head, away from the sun, by the solar wind. Sometimes these tails stretch millions of miles into space. This long streaming tail reminded ancient observers of hair, and they called the phenomenon comet, from the Latin word *coma*, meaning hair.

Comets shine partly with an intense light that is reflected sunlight. As they approach the sun and become more vaporized, comets grow tremendously in size without gaining mass.

990. What were some memorable comets? The appearance of a huge and threatening comet in the sky has often brought fear and hysteria to people, who for centuries considered them as omens of death and destruction.

Throughout history spectacular comets have been recorded: one appeared in 344 B.C. "like a flaming torch"; in 146 B.C. a comet appeared "as bright as the sun"; in 1811 a comet rode across the sky with a tail some 130 million miles long; and in 1843, the Great Comet loomed with a tail 200 million miles long. One of the most famous comets is Halley's Comet which has been circling the sun at intervals of about 76 years ever since 240 B.C. Named after British astronomer Edmund Halley who calculated this and other comets' journeys around the sun, Halley's Comet last appeared in 1910 and is due to appear again about 1986.

991. What layers of atmosphere help shield the earth from space dangers? Many particles and radiations flying toward the orbiting earth are trapped, shattered, or bounced off by various layers of the earth's magnetic fields and atmosphere. The atmosphere is composed of several distinct but merging layers that are constantly changing in height and thickness.

The troposphere (from the Greek word *tropos*, meaning turn or mix) is the bottom layer, nearest the surface of the earth. In this turning, moving, mixing layer of air is the weather as we know it—the winds, storms, clouds, lightning, and other phenomena. Here molecules of air are closely packed together, and the air as a whole is heavy. About 80 percent of the air by weight, and nearly all water vapor, is packed into this troposphere, which varies in thickness from 5 miles high at the north and south poles, to about 10 miles high at the equator.

The stratosphere (from the Latin meaning layer or covering) lies above the troposphere, extending to about 50 or 60 miles above the earth. Here there are no clouds, no convection air currents. There is little water vapor or dust, and skies are clear. Strong, steady winds blow parallel to the earth's surface. Here temperatures increase with height. Within the stratosphere, about 20 miles above sea level, is the ozone layer that absorbs most of the sun's powerful short radiations hurtling toward earth. Life on earth would probably be destroyed if this layer were suddenly dispersed.

The ionosphere extends out to 650 miles above earth. Here electrons are bombarded off molecules by incoming radiation and particles, forming a large supply of ions and electrified particles. Like mirrors, certain highly electrified zones reflect long waves back to earth, a basic factor in radio communications. Short waves pass through the ionosphere into space. This is why shortwave length (microwave) communications require relay towers, strung across the country. During solar disturbances and magnetic storms, these zones rapidly change position and intensity, causing communications to be disrupted.

The exosphere is the outermost layer of the earth's atmosphere, extending above the ionosphere into interplanetary space.

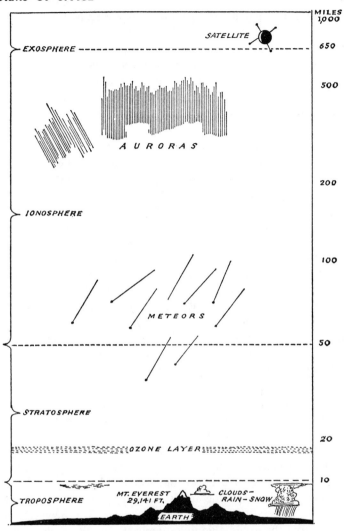

Layers of earth's atmosphere

992. What are temperatures in these various layers of atmosphere?
Temperatures vary in the different layers of atmosphere. The lowest
layer, the troposphere, is characterized by decreasing temperatures
with increasing height. Temperatures at the bottom of this layer on a

hot summer day may be about 95 degrees Fahrenheit, but at the top they may be minus 70 degrees. Temperatures of this air layer drop at a relatively steady rate of about 3.5 degrees for every 1,000 feet of altitude. The reason the air becomes cooler is simply because the sun-warmed earth radiates its heat into the atmosphere.

In the tropopause, a transition zone between the troposphere and the next layer, the stratosphere, temperatures remain relatively constant at minus 70 degrees.

In the lower regions of the stratosphere, strong winds blow and temperatures are cold. But about 20 miles high, the winds weaken and die, and temperatures take a turn and gradually rise to about 25 or 30 degrees. This happens because of the ozone layer which stops the ultraviolet rays from reaching earth in full impact. Beyond the ozone layer, about 32 miles high, the temperature again drops.

Temperature variations in the ionosphere are being determined with more accuracy as space instruments continually accumulate more data. In the mesosphere, the ionospheric layer of air about 20 miles deep above the stratosphere, temperatures fall. At its base, temperatures are about 30 degrees, but at the top they are about minus 130 degrees. In the mesopause, another boundary zone, the temperature remains constant about minus 130 degrees. Above this, above 70 miles, the temperatures rise again, and at altitudes of more than 20,000 miles, they are more than 2,000 degrees.

993. What is the magnetosphere? The magnetosphere is a region of the earth's atmosphere thousands of miles above the earth's surface where the geomagnetic field ends.

Essentially the earth is a giant magnet surrounded by a magnetic field that permeates the earth, the oceans, and the atmosphere around it. Between the magnetic north and south poles, magnetic lines of force curve outward into the atmosphere, spreading widely above the tropics and descending to earth over the poles. The magnetic poles, different from the geographic poles, are not opposite each other, and often change locations. They have reversed their magnetic charge several times during the eons the planet has been spinning around the sun. The north magnetic pole lies near Bathurst Island, north of central Canada. The south magnetic pole is in the Antarctic.

The solar wind consists of charged particles. Charged particles in a directed flow constitute an electric current. Associated with all

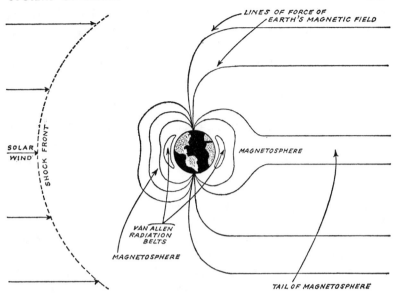

Magnetic spheres of the earth

electric currents are magnetic fields. Thus, the solar wind creates a magnetic field.

This magnetic field and the earth's magnetic field add together in such a way as to create the magnetosphere. This net magnetic field, in turn, reacts back on the current that helped create it—the solar wind—causing the current to flow around the earth. In essence, the magnetosphere acts as a shield against which the solar-wind particles blow and are dispersed around the planet earth.

994. What is the shape of the magnetosphere? Were there no solar wind, the earth's magnetic field would extend far out into space in all directions. The solar wind shapes the limits of the magnetosphere into a streamlined tear drop. On the side facing the sun, the shield is bluntly rounded, some 40,000 miles high. On the other side, to the lee of the solar wind, the magnetosphere stretches into a long tail on the dark side of the earth to distances of 3½ million miles. This tail is pushed back and forth by the wind, sometimes breaking off in great pieces that whirl into space.

995. What is the Van Allen belt? Charged particles from space, mostly protons and electrons, are trapped in the earth's magnetic field and spiral back and forth from near the north magnetic pole to near the south magnetic pole. These trapped particles are concentrated in a zone called the Van Allen belt, shaped somewhat like a huge doughnut surrounding the earth high above the magnetic equator.

This zone of trapped particles, calculated and suggested previously by a number of scientists, is named after the American physicist Dr. James A. Van Allen who discovered its existence in 1958 with data from an early space probe.

996. What is happening to these magnetic shields? Some scientists believe that heat from the earth's inner core may be flowing into fluid layers near the polar regions, causing motions that may weaken and destroy the earth's magnetic field. Since 1670 A.D., scientists believe the strength of these poles has decreased by about 15 percent. In another 100 years, the magnetic fields may not be strong enough in the outer atmospheres to shield the earth's surface against incoming radiation.

997. Can solar storms be predicted? The science and technology of predicting solar and cosmic disturbances is in an early stage of development, but with increasing numbers and precision of space instruments, more accuracy is being obtained.

Since 1965 daily space weather forecasts have been made at the Space Disturbance Laboratory in Boulder Colorado, part of the Environmental Science Services Administration. Here close watch is kept on the sun's activities through cooperation of more than 20 world-wide observatories.

By constant information obtained through earth-based observatories, satellites, and space probes, space scientists are beginning to analyze the cycles of solar storms in an attempt to forecast when they might occur.

998. What was the International Geophysical Year? The International Geophysical Year (IGY) was an 18-month cooperative program from July, 1957, to December, 1958, involving scientists from some 70 nations. Conducted during the height of the sunspot activity, it was designed for systematic study of the earth and its cosmic environment. Studies in meteorology included analysis of weather and climate changes and of cosmic rays and ionization in the atmosphere. During this time the sun was intensively observed in both visual and radio wave regions of the electromagnetic spectrum and data was obtained on the number and size of sunspots, solar flares, and other solar disturbances.

This was the third formal international cooperative program in a series, the first two of which were the First International Polar Year (1882–1883) and the Second International Polar Year (1932–1933).

999. What was the program of the International Years of the Quiet Sun? The program of the International Years of the Quiet Sun (IQSY), an outgrowth of the International Geophysical Year, was undertaken in 1964–1965, when solar activity was at its expected minimum. At this time when the solar wind was less dense and fewer solar particles were being emitted from the sun, scientists fired thousands of rockets into the upper atmosphere to gather astronomical, geophysical, and meteorological data.

Each of more than 60 participating countries planned its own programs to study solar-terrestrial relationships, in consultation with other countries. The U.S. research program included studies on auroras, airglow, ionospheric physics, radio cosmic rays, and trapped radiation.

1000. What is the World Magnetic Survey? The World Magnetic Survey (WMS), a continuing project of the International Geophysical Year, is a project to map the world's magnetic field during the period of minimum solar activity. The survey is coordinated internationally by the World Magnetic Survey Board of the International Association of Geomagnetism and Aeronomy, International Union of Geodesy and Geophysics.

1001. What organizations are conducting research in space storms? With many available devices of aircraft, satellites, space probes, and high-speed computers, many organizations are conducting vast studies on the mysterious worlds and regions beyond the earth. It is not possible to list every laboratory and office engaged in such fascinating work, but a few of the more outstanding organizations include the

Goddard Space Flight Center, of the National Aeronautics and Space Administration, in Greenbelt, Maryland; the Institute for Environmental Research, of the Environmental Science Services Administration, in Boulder, Colorado; the Office of Aerospace Research, of the U.S. Air Force, in Arlington, Virginia; the Naval Research Laboratory, the Air Force Cambridge Research Laboratories; the Jungfraujoch Scientific Station in Berne, Switzerland; the World Meteorological Organization in Geneva, Switzerland; the European Space Research Organization in Paris, France; University College in London, England; and the University of Paris.

APPENDIX

The following 27 questions are identical with those bearing the same numbers in the preceding text. The answers, however, have been entirely updated, incorporating new research data and information available from 1969 through 1985. The hurricane map on page 44 has been similarly supplemented by a new map, on page 361, depicting major Atlantic hurricanes from 1969 through 1985.

6. What was the worst tragedy caused by a hurricane? On November 13, 1970, a storm with winds of 100 miles per hour and waves 10 to 15 feet above normal surged across the low-lying islands offshore from Bangladesh. It struck between the mouth of the Haringhata River and the lower Meghna estuary and along vast stretches of coastline near Chittagong, sweeping away houses, crops, animals, and human beings. More than 1.1 million acres of rice paddies disappeared, a million livestock animals drowned, and human losses, first reported as between 200,000 and 500,000, may have reached one million.

13. What was another destructive hurricane? Tropical hurricane Fifi blew across the Caribbean from September 15 through 20, 1974, flooding Honduras with winds of more than 110 miles an hour and heavy rains, leaving 5,000 dead and 60,000 homeless.

26. What was one of the worst hurricane years? After the powerful hurricane Allen churned harmlessly through the Caribbean in 1980, there was a five-year respite from large hurricanes along the Atlantic and Gulf coasts, except for hurricane Alicia, which swept across Galveston in 1983 with winds of 120 miles per hour and a death toll of 22. In 1985, however, a slew of hurricanes was spawned, 3 of which were memorable. From August 28 to September 4, Elena moved over Florida, Mississippi, and Alabama, causing some 2 million people to be evacuated but bringing no fatalities. Before Hurricane Gloria (September 16 to October 2) crossed over Long Island, it created the lowest atmospheric pressure in its center measured in the southwest North Atlantic since aircraft reconnaissance began in the mid-1940's. Hurricane Juan (October 26 through November 1) looped 2 times around the Louisiana coastline, bringing death to 12 people and causing high floods for 3 days.

29. What U.S. hurricane caused the most property damage on record?
Hurricane Agnes was tracked from June 14 to 23, 1972, and has been to date
the costliest hurricane in American history—an estimated $3 billion in
immediate property damage. Moving across the Gulf of Mexico and over
Florida, Agnes turned inland and dumped billions of tons of water over five
mid-Atlantic states: Maryland, Virginia, New Jersey, northern Pennsylvania, and southern New York. Bringing the worst flooding in history,
Agnes caused severe flood damage in 25 cities and 142 counties of these
states, leaving 5,000 square miles under flood waters, 122 people dead, and
some 330,000 homeless.

30. What was another expensive hurricane? On September 12, 1979,
hurricane Frederic smashed across the Gulf coast near Mobile Bay and
caused damages of some $2.3 billion.

36. How are hurricanes named today? In the late 1970's the system of
calling hurricanes after girls became too unpopular to continue. New alphabetical lists alternating men's and women's names were set up in a six-year
cycle. After the lists have been used, they are used again. The names now
also have an international flavor in recognition of the fact that hurricanes
affect other nations and are tracked by the public and weather services of
countries other than the United States.

37. What are names of future Atlantic hurricanes? The following listing
identifies the names of hurricanes for the next six years, after which the sets
will be used again.

The Six-Year List of Names for Atlantic Storms

1987	*1988*	*1989*
Arlene	Alberto	Alicia
Bret	Beryl	Barry
Cindy	Chris	Chantal
Dennis	Debby	Dean
Emily	Ernesto	Erin
Floyd	Florence	Felix
Gert	Gilbert	Gabrielle
Harvey	Helene	Hugo
Irene	Isaac	Iris
Jose	Joan	Jerry
Katrina	Keith	Karen

1987 (cont.)	*1988 (cont.)*	*1989 (cont.)*
Lenny	Leslie	Luis
Maria	Michael	Marilyn
Nate	Nadine	Noel
Ophelia	Oscar	Opal
Philippe	Patty	Pablo
Rita	Rafael	Roxanne
Stan	Sandy	Sebastien
Tammy	Tony	Tanya
Vince	Valerie	Van
Wilma	William	Wendy

1990	*1991*	*1992*
Arthur	Ana	Allen
Bertha	Bob	Bonnie
Cesar	Claudette	Charley
Diana	Danny	Danielle
Edouard	Elena	Earl
Fran	Fabian	Frances
Gustav	Gloria	Georges
Hortense	Henri	Hermine
Isidore	Isabel	Ivan
Josephine	Juan	Jeanne
Klaus	Kate	Karl
Lili	Larry	Lisa
Marco	Mindy	Mitch
Nana	Nicholas	Nicole
Omar	Odette	Otto
Paloma	Peter	Paula
Rene	Rose	Richard
Sally	Sam	Shary
Teddy	Teresa	Tomas
Vicky	Victor	Virginie
Wilfred	Wanda	Walter

38. What are the names of future hurricanes (also called typhoons) in the eastern North Pacific Ocean? Names of men and women are also used to identify hurricanes in the Pacific Ocean. A set of 6 alphabetical lists is used and, as in the Atlantic, the lists are used again when the 6-year set is completed. For instance, the 1987 list will be used again in 1993.

The Six-Year List of Names for Eastern Pacific Storms

1987	*1988*	*1989*
Adrian	Aletta	Adolph
Beatriz	Bud	Barbara
Calvin	Carlotta	Cosme
Dora	Daniel	Dalilia
Eugene	Emilia	Erick
Fernanda	Fabio	Flossie
Greg	Gilma	Gil
Hilary	Hector	Henriette
Irwin	Iva	Ismael
Jova	John	Juliette
Knut	Kristy	Kiko
Lidia	Lane	Lorena
Max	Miriam	Manuel
Norma	Norman	Narda
Otis	Olivia	Octave
Pilar	Paul	Priscilla
Ramon	Rosa	Raymond
Selma	Sergio	Sonia
Todd	Tara	Tico
Veronica	Vicente	Velma
Wiley	Willa	Winnie

1990	*1991*	*1992*
Alma	Andres	Agatha
Boris	Bianca	Blas
Cristina	Carlos	Celia
Douglas	Dolores	Darby
Elida	Enrique	Estelle
Fausto	Fefa	Frank
Genevieve	Guillermo	Georgette
Hernan	Hilda	Howard
Iselle	Ignacio	Isis
Julio	Jimena	Javier
Kenna	Kevin	Kay
Lowell	Linda	Lester
Marie	Marty	Madeline
Norbert	Nora	Newton
Odile	Olaf	Orlene

1990 (cont.)	*1991 (cont.)*	*1992 (cont.)*
Polo	Pauline	Paine
Rachel	Rick	Roslyn
Simon	Sandra	Seymour
Trudy	Terry	Tina
Vance	Vivian	Virgil
Wallis	Waldo	Winifred

40. Why are hurricanes named for girls? Hurricanes are no longer named for girls. That quaint custom lost its humor and popularity in the feminist movement of the late 1970's. They now are named for men and women. (See Question 37.)

182. What were some other outstanding tornado disasters? During an 18-hour period on April 3 and 4, 1974, a swarm of 148 tornadoes speeded across 13 states from Georgia to Canada, killing 315 people and injuring nearly 5,500.

Another series of tornadoes killed 117 people on February 21, 1971, as it swept through Mississippi and Louisiana.

On May 31, 1985, a series of tornadoes blew through towns in Ohio and Pennsylvania, killing 76 people.

In tribute to people's swift response to the U.S. Weather Service's warning system, only 3 persons were killed on May 6, 1975, when a tornado cut an 11-mile swath through a residential section in Omaha, destroying 500 homes, damaging another 4,500, and causing $100 million in damage.

186. On the average, how many people have been killed each year by tornadoes? During the years 1953 through 1985, tornadoes have killed 102 persons per year on an average.

188. When was the death rate the lowest? Lowest death rate figures were given for the years 1972, when 27 died, and 1981, when only 24 people died.

227. What was the highest number of tornadoes reported in one year? In 1973, 1,102 tornadoes were spotted throughout the United States; in 1982, 1,046 were recorded.

228. What states have been hardest hit each year, on an average? The U.S. Weather Service has reported that during the years 1953 through 1985, Texas had the highest average number of tornadoes per year—124.

Oklahoma was next with 56 a year; Florida and Kansas both averaged 43. Other hard-hit states during this period were Nebraska with 35, and Missouri and Iowa each with 29.

229. What states have been hit least, on an average? During the years 1953 through 1985, all 50 states were hit, the District of Columbia being the only area in the U.S. not hit by a tornado. Alaska was hit by only one—in 1959. Connecticut, Delaware, Hawaii, Idaho, Nevada, Oregon, Rhode Island, Utah, Vermont, Washington, and Puerto Rico and the Virgin Islands were hit by at least one tornado per year on an average during those years.

230. What are the averages for the other states? During the years 1953 through 1985, other states averaged the following yearly numbers:

Alabama	22	Montana	4
Arizona	4	New Hampshire	2
Arkansas	21	New Jersey	2
California	4	New Mexico	8
Colorado	19	New York	4
Georgia	21	North Carolina	12
Illinois	27	North Dakota	18
Indiana	20	Ohio	14
Kentucky	8	Pennsylvania	9
Louisiana	22	South Carolina	9
Maine	2	South Dakota	26
Maryland	3	Tennessee	12
Massachusetts	3	Virginia	6
Michigan	16	West Virginia	2
Minnesota	18	Wisconsin	19
Mississippi	21	Wyoming	9

270. How are tornadoes being spotted and studied? A tornado classification scale that gives a numerical representation of a tornado's intensity was developed by Professor T. Theodore Fujita of the University of Chicago and by Allen Pearson, former director of the Severe Storms Forecast Center of the National Weather Service in Kansas City. Known in short as the Fujita scale, the following shows the ranking of tornadoes according to wind speed.

F-0. Very weak, 40–72 mph. Tree branches broken off, shallow-rooted or old trees toppled; signboards damaged.

F-1. Weak, 73–112 mph. A wind velocity of 73 miles per hour is the beginning of hurricane wind speed; roof surfaces peeled off, windows broken, trailer houses overturned.

F-2. Strong, 113–157 mph. Roofs torn off frame houses but strong upright walls still stand; weak structures and trailer houses demolished; railroad boxcars pushed over; large trees snapped or uprooted; cars blown off highways.

F-3. Severe, 158–206 mph. Some rural buildings flattened or demolished, trains overturned, cars lifted off the ground, most trees in a forest snapped or leveled.

F-4. Devastating, 207–260 mph. Well-constructed frame houses leveled, structures with weak foundations lifted and blown some distance, cars thrown some distance.

F-5. Incredible, 261–318 mph. Strong frame houses lifted clear off foundations for considerable distance, steel-reinforced concrete structures badly damaged.

F-6. Inconceivable, 319 mph to sonic speed. Should a tornado with a maximum wind speed in the range of F-6 occur, the extent and types of damage may not be conceived. Assessment of tornadoes in these categories is feasible only through detailed surveys involving engineering and aerodynamical calculations as well as meteorological models of tornadoes.

350. How many people are killed each year by lightning? The *General Summary of Lightning* of 1985, a report from the National Climatic Data Center of NOAA (National Oceanic and Atmospheric Administration) showed that an average of 98 people died per year from 1959 through 1985.

351. When are people hit most often by lightning? NOAA's National Climatic Data Center reports that 27 percent of deaths from lightning in the years 1959 through 1984 occurred when people were in open fields, ball fields, or other open spaces. Only 16 percent of deaths occurred when people stood under trees.

441. Has anyone been killed in the United States by hail? Updated research and cataloguing have turned up other hail fatalities reported from time to time in local newspapers throughout the country: on June 10, 1742, at Amwell, New Jersey; on September 11, 1863, in St. Charles County, Missouri; on May 17, 1909, near Uvalde, Texas.

463. What was the largest U.S. hailstone found? On September 3, 1970, a hailstone dropped on Coffeyville, Kansas, measuring 5.57 inches in diameter and weighing 1.671 pounds.

763. What was the blizzard of 1966? A snowfall of 20 inches or more in a single storm places cities in the exclusive Snowstorm Club. Several cities have earned that distinction: *New York City* on December 26, 1947, received 25.8 inches officially—26.4 inches at Central Park (see also Question 780, original text). *Chicago* in a 2-day storm January 26 and 27, 1967, was blanketed with 23 inches, with a record 19.8 inches falling in 24 hours. *Boston* was filled with snow January 20, 1978, when 21 inches were recorded at Logan Airport and 24 inches in the suburban districts. On February 7, an additional 22 inches fell, bringing the total snowfall in some places to 27 inches. *Buffalo* received 28 inches of snow on January 11, 1982, with the same record-breaking cold wave that swept the rest of the country. A record of 25.3 inches in 24 hours was set.

818. What was a great New England ice storm? Another severe ice storm hit most of southern New England on December 16 and 17, 1973, when more than an inch of freezing precipitation coated branches, wires, and other objects with ½ inch of ice. The greatest power outages to date in New England history resulted. Extensive tree damage and broken electric and telephone wires left many homes without power or phone for a week.

844. Where are the coldest temperatures on record? On July 21, 1983, the Vostok station in Antarctica recorded a new low of minus 128.6 degrees Fahrenheit. In the United States, extremes of cold have been recorded in the following places: Moran, Wyoming, February 9, 1933—minus 63 degrees Fahrenheit; Parshall, North Dakota, February 15, 1936—minus 60 degrees Fahrenheit; Rogers Pass, Montana, January 20, 1954—minus 70 degrees Fahrenheit; and Prospect Creek Camp, Alaska, January 23, 1971—minus 80 degrees Fahrenheit.

849. What were some record cold waves? The first week of January, 1982, produced bitter cold masses of air that moved from the arctic regions of Alaska and northwest Canada and spilled across the United States, with 41 weather stations reporting new record-breaking cold temperatures. During this week, International Falls, Minnesota, reported temperatures as low as minus 36 degrees Fahrenheit; north-central Montana reported minus 21 degrees Fahrenheit; and eastern Oregon minus 18 degrees Fahrenheit. In

Illinois, Chicago broke records with minus 26 degrees Fahrenheit at O'Hare and Midway airports. Minus 23 degrees was reported at Moline, and minus 18 degrees at Peoria. The cold air swept into Florida, where on January 12 vegetables were killed and the citrus crops damaged.

914. What was the severest heat wave in the United States? During a severe heat wave in 1980, more than 1,250 people died. Records for the duration of high temperatures were broken over much of Texas during a span of 42 days from June 23 to August 3 of that year, when temperatures rose to 100 degrees Fahrenheit or more, crescendoing to 113 degrees Fahrenheit on June 26 and 27, and reaching 119 degrees Fahrenheit at Weathersford, Texas. From June 23 to July 3 in the Wichita Falls area, along the Red River, the mercury reached 110 degrees Fahrenheit every day from June 23 through July 3, with a high of 117 degrees on June 28.

In the 20th century, extremes of heat have been recorded in the following places in the United States: Death Valley, California, July 10, 1913—134 degrees Fahrenheit (see also Question 908, original text); Parker, Arizona, July 7, 1905—127 degrees Fahrenheit; Overton, Nevada, June 23, 1954—122 degrees Fahrenheit; Steele, North Dakota, and a spot near Alton, Kansas, both on July 24, 1936—121 degrees Fahrenheit.

924. What is the discomfort index? The National Weather Service of NOAA has devised the "heat index," sometimes referred to as the "apparent temperature." This index is an accurate measure of how hot a person really feels when relative humidity is added to the actual air temperature.

(Continued)

Heat Index Chart

*Air Temperature and Relative Humidity
versus Apparent Temperature*

**Heat Index
(or Apparent
Temperature)**

AIR TEMPERATURE (°F)	\ RELATIVE HUMIDITY (%) 0	5	10	15	20	25	30	35	40	45	50	55	60	65	70	75	80	85	90	95	100
140	125																				
135	120	128																			
130	117	122	131																		
125	111	116	123	131	141																
120	107	111	116	123	130	139	148														
115	103	107	111	115	120	127	135	143	151												
110	99	102	105	108	112	117	123	130	137	143	150										
105	95	97	100	102	105	109	113	118	123	129	135	142	149								
100	91	93	95	97	99	101	104	107	110	115	120	126	132	138	144						
95	87	88	90	91	93	94	96	98	101	104	107	110	114	119	124	130	136				
90	83	84	85	86	87	88	90	91	93	95	96	98	100	102	106	109	113	117	122		
85	78	79	80	81	82	83	84	85	86	87	88	89	90	91	93	95	97	99	102	105	108
80	73	74	75	76	77	77	78	79	79	80	81	82	83	85	86	86	87	88	89	91	
75	69	69	70	71	72	72	73	73	74	74	75	75	76	76	77	78	78	79	79	80	
70	64	64	65	65	66	66	67	67	68	68	69	69	70	70	70	71	71	71	71	72	

NOAA

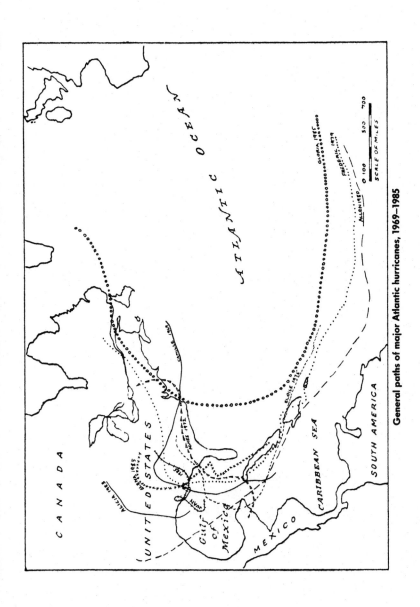

General paths of major Atlantic hurricanes, 1969–1985

BIBLIOGRAPHY

General Works

The following works are a few references relevant to all chapters:

Aubert de la Rue, Edgar. *Man and the Winds*. New York: Philosophical Library, 1955.

Battan, Louis J. *The Nature of Violent Storms*. Garden City, N.Y.: Doubleday & Co., Inc., 1961.

Blumenstock, David I. *The Ocean of Air*. New Brunswick, N.J.: Rutgers University Press, 1959.

Brown, Slater. *World of the Wind*. Indianapolis: Bobbs-Merrill Co., 1961.

Edinger, James G. *Watching for the Wind: The Seen and Unseen Influences on Local Weather*. Garden City, N.Y.: Doubleday & Co., Inc., 1967.

"Federal Aviation Agency-Flight Standards Service; and Department of Commerce-Weather Bureau." *Aviation Weather*. Washington, D.C.: U.S. Government Printing Office, 1965.

Forrester, Frank H. *1001 Questions Answered About the Weather*. New York: Dodd, Mead & Co., 1961.

Humphreys, W. J. *Physics of the Air*. New York: Dover Publications, 1964.

Kimble, George H. *Our American Weather*. New York: McGraw-Hill Book Co., Inc., 1955.

Koeppe, Clarence E.; and de Long, George C. *Weather and Climate*. New York: McGraw-Hill Book Co., Inc., 1958.

Lane, Frank W. *The Elements Rage*. Philadelphia: Chilton Books, 1965.

Lehr, Paul E.; Burnett, R. Will; and Zim, Herbert. *Weather*. New York: Golden Press, 1963.

Loebsack, Theo. *Our Atmosphere*. New York: The New American Library of World Literature, Inc., 1961.

Malone, T., ed. *Compendium of Meteorology*. Boston: American Meteorological Society, 1951.

Nesbitt, Paul H.; Pond, Alonzo W.; and Allen, William H. *The Survival Book*. Princeton, N.J.: D. Van Nostrand Co., Inc., 1959.

Petterssen, Sverre. *Introduction to Meteorology*. New York: McGraw-Hill Book Co., Inc., 1958.

Pilkington, Roger. *The Ways of the Air*. New York: Criterion Books, 1962.

Riehl, Herbert. *Introduction to the Atmosphere.* New York: McGraw-Hill Book Co., Inc., 1965.

Ross, Frank. *Weather: The Science of Meteorology from Ancient Times to the Space Age.* New York: Lothrop, Lee & Shepard Co., Inc., 1965.

Sutton, Ann and Myron. *Nature on the Rampage.* Philadelphia: J. B. Lippincott Co., 1962.

Sutton, Oliver G. *Understanding Weather.* Baltimore, Md.: Penguin Books, 1960.

Tannehill, Ivan B. *Weather Around the World.* Princeton, N.J.: Princeton University Press, 1943.

Thompson, Philip D.; O'Brien, Robert; and editors of *Life. Weather.* New York: Life Science Library, Time, Inc., 1965.

Trewartha, Glenn T. *An Introduction to Climate.* New York: McGraw-Hill Book Co., Inc., 1965. 3rd edition.

Wenstrom, William H. *Weather and the Ocean of Air.* Boston: Houghton Mifflin Co., 1942.

Hurricanes: Chapter I

Dunn, Gordon E.; and Miller, Banner. *Atlantic Hurricanes.* Baton Rouge: Louisiana State University Press, 1960.

Helm, Thomas. *Hurricanes: Weather at Its Worst.* New York: Dodd, Mead & Co., 1967.

Tannehill, Ivan Ray. *Hurricanes: Their Nature and History—Particularly Those of the West Indies and the Southern Coasts of the United States.* Princeton, N.J.: Princeton University Press, 1956. 9th edition.

Tannehill, Ivan Ray. *The Hurricane.* Washington, D.C.: Weather Bureau, U.S. Government Printing Office, 1956. Revised.

Weather Bureau (ESSA). *Hurricane Information and Atlantic Tracking Chart.* P.I. 680006. Washington, D.C.: U.S. Government Printing Office, 1968.

Tornadoes: Chapter II

Changnon, Stanley, Jr.; and Semonin, Richard G. "Tri-State Tornado: A Great Tornado Disaster in Retrospect." *Weatherwise,* Vol. 19, No. 2, pp. 56–65. Boston: American Meteorological Society, 1966.

Colgate, Stirling A. "Tornadoes: Mechanism and Control." *Science,* Vol. 157, pp. 1431–1434. Washington, D.C.: American Association for the Advancement of Science, Sept. 22, 1967.

Flora, Snowden D. *Tornadoes of the United States.* Norman, Oklahoma: University of Oklahoma Press, 1953.

Weather Bureau (ESSA). *Tornado*. P.I. 660028. Washington, D.C.: U.S. Government Printing Office, 1969.

Thunderstorms: Chapter III

Battan, Louis J. *The Thunderstorm*. New York: Signet Science Library Book, New American Library, 1964.
Bell, Thelma H. *Thunderstorm*. New York: The Viking Press, 1960.
Kessler, Edwin. "Purposes and Program of the U.S. Weather Bureau. National Severe Storms Laboratory, Norman, Oklahoma." *Transactions*, Vol. 46, No. 2. Washington, D.C.: American Geophysical Union, June, 1965.
Viemeister, Peter E. *The Lightning Book*. Garden City, N.Y.: Doubleday & Co., 1961.
Weather Bureau (ESSA). *Lightning*. P.I. 660024. Washington, D.C.: U.S. Government Printing Office, 1968.
Weather Bureau (ESSA). *Thunderstorm*. P.I. 670004. Washington, D.C.: U.S. Government Printing Office, 1969.

Hailstorms: Chapter IV

Chagnon, Stanley A., Jr. *25 Most Severe Summer Hailstorms in Illinois During 1915–59*. Research Report No. 4. Chicago: 1960.
Flora, Snowden D. *Hailstorms of the United States*. Norman, Oklahoma: University of Oklahoma Press, 1956.
Hull, Blanche B. *Hail Size and Distribution*. Technical Report, EP–83. Natick, Mass.: U.S. Quartermaster Research & Engineering Center. Environmental Protection Research Division, Feb., 1957.
Koch, Charles R. "Hail Is Hell." *The Farm Quarterly*. Cincinnati, Ohio: F & W Publishing Corp., Spring, 1959.
Roth, Richard. "Hailstones and Hailstorms." *Weatherwise*. Boston, Mass.: American Meteorological Society, June, 1952.
Schaefer, Vincent J. *Hailstorms and Hailstones of the Western Great Plains*. Annual Report of the Board of Regents of the Smithsonian Institution, Publication 4435, p. 341. Washington, D.C.: U.S. Government Printing Office, 1961.

Winds: Chapter V

Bodley, Ronald V. C. *Wind in the Sahara*. New York: Creative Age Press, Inc., 1944.
Namias, Jerome. "The Jet Stream," *Scientific American*, Vol. 187, No. 4. New York: Oct., 1952.
Snow, Edward Rowe. *Great Gales and Dire Disasters*. New York: Dodd, Mead & Co., 1960.

Snow, Edward Rowe. *New England Sea Tragedies*. New York: Dodd, Mead & Co., 1960.

Stick, David. *Graveyard of the Atlantic: Shipwrecks of the North Carolina Coast*. Chapel Hill: The University of North Carolina Press, 1952.

Vaeth, J. Gordon. *Weather Eyes in the Sky: America's Meteorological Satellites*. New York: The Ronald Press Co., 1965.

Fogs: Chapter VI

Myers, Joel N. "Fog," *Scientific American*, Vol. 219, No. 6, pp. 75–82. New York: Dec., 1968.

Rainstorms, Snowstorms, and Icestorms: Chapter VII

Byers, Horace Robert. *Elements of Cloud Physics*. Chicago: University of Chicago Press, 1965.

Mason, B. J. *Clouds, Rain and Rainmaking*. Cambridge: Cambridge University Press, 1962.

Mason, B. J. *The Physics of Clouds*. Oxford: The Clarendon Press, 1957.

Extremes of Cold, Heat, and Pressure: Chapter VIII

"Biometeorology Today and Tomorrow." *Bulletin of the American Meteorological Society*, Vol. 48, No. 6, pp. 378–393. Boston: June, 1967.

Brooks, Charles E. P. *Climate in Everyday Life*. London: Ernest Benn Ltd., 1950.

Dyson, James L. *The World of Ice*. New York: Alfred A. Knopf, 1962.

Huntington, Ellsworth. *Civilization and Climate*. New Haven: Yale University Press, 1915.

Huxley, Leonard, arranger. *Scott's Last Expedition*. 2 vols. New York: Dodd, Mead & Co., 1913.

Ice Observations. Hydrographic Office Observers Manual (H.O. Pub. No. 606–d). Washington, D.C.: U.S. Navy Department, 1954.

Leithead, C. S.; and Lind, A. R. *Heat Stress and Heat Disorders*. London: Cassell & Co., Ltd., 1964.

Mills, Clarence A. *Climate Makes the Man*. New York: Harper & Bros., 1942.

Sargent, F., II; and Tromp, S. W., eds. *A Survey of Human Biometeorology*. Technical Note 65. Geneva: World Meteorological Organization, 1964.

Storms of Space: Chapter IX

Asimov, Isaac. *The Universe: From Flat Earth to Quasar*. New York: Walker & Co., 1966.

Pickering, James S. *1001 Questions Answered About Astronomy*. New York: Dodd, Mead & Co., 1966.

Von Braun, Wernher. *Space Frontier*. New York: Holt, Rinehart & Winston, 1967. Revised.

Whipple, Fred L. *Earth, Moon, and Planets*. Cambridge: Harvard University Press, 1968.

Supplement to the Dover Edition

Climatological Data, National Summary (of NOAA). U.S. Dept. of Commerce, 1980–85.

Cornell, James. *The Great International Disaster Book*. New York: Charles Scribner's Sons, 1976.

Ludlum, David M. *The American Weather Book*. Boston: Houghton Mifflin Co., 1982.

Storm Data. Asheville, N.C.: National Climatic Data Center, NOAA, 1983–85.

Vigansky, Henry N. *General Summary of Lightning*. Asheville, N.C.: National Climatic Data Center, NOAA, 1980–85.

Vigansky, Henry N. *General Summary of Tornadoes*. Asheville, N.C.: National Climatic Data Center, NOAA, 1981–83.

INDEX

References are to question numbers

Supplement to the Dover Edition

References are to question numbers in the Appendix

A CATALOG OF SELECTED
DOVER BOOKS
IN ALL FIELDS OF INTEREST

A CATALOG OF SELECTED DOVER
BOOKS IN ALL FIELDS OF INTEREST

CONCERNING THE SPIRITUAL IN ART, Wassily Kandinsky. Pioneering work by father of abstract art. Thoughts on color theory, nature of art. Analysis of earlier masters. 12 illustrations. 80pp. of text. 5⅜ × 8½.　　　　23411-8 Pa. $3.95

ANIMALS: 1,419 Copyright-Free Illustrations of Mammals, Birds, Fish, Insects, etc., Jim Harter (ed.). Clear wood engravings present, in extremely lifelike poses, over 1,000 species of animals. One of the most extensive pictorial sourcebooks of its kind. Captions. Index. 284pp. 9 × 12.　　　　23766-4 Pa. $10.95

CELTIC ART: The Methods of Construction, George Bain. Simple geometric techniques for making Celtic interlacements, spirals, Kells-type initials, animals, humans, etc. Over 500 illustrations. 160pp. 9 × 12. (USO)　　　　22923-8 Pa. $8.95

AN ATLAS OF ANATOMY FOR ARTISTS, Fritz Schider. Most thorough reference work on art anatomy in the world. Hundreds of illustrations, including selections from works by Vesalius, Leonardo, Goya, Ingres, Michelangelo, others. 593 illustrations. 192pp. 7⅛ × 10¼.　　　　20241-0 Pa. $8.95

CELTIC HAND STROKE-BY-STROKE (Irish Half-Uncial from "The Book of Kells"): An Arthur Baker Calligraphy Manual, Arthur Baker. Complete guide to creating each letter of the alphabet in distinctive Celtic manner. Covers hand position, strokes, pens, inks, paper, more. Illustrated. 48pp. 8¼ × 11.

24336-2 Pa. $3.95

EASY ORIGAMI, John Montroll. Charming collection of 32 projects (hat, cup, pelican, piano, swan, many more) specially designed for the novice origami hobbyist. Clearly illustrated easy-to-follow instructions insure that even beginning papercrafters will achieve successful results. 48pp. 8¼ × 11.　　　　27298-2 Pa. $2.95

THE COMPLETE BOOK OF BIRDHOUSE CONSTRUCTION FOR WOOD-WORKERS, Scott D. Campbell. Detailed instructions, illustrations, tables. Also data on bird habitat and instinct patterns. Bibliography. 3 tables. 63 illustrations in 15 figures. 48pp. 5¼ × 8½.　　　　24407-5 Pa. $1.95

BLOOMINGDALE'S ILLUSTRATED 1886 CATALOG: Fashions, Dry Goods and Housewares, Bloomingdale Brothers. Famed merchants' extremely rare catalog depicting about 1,700 products: clothing, housewares, firearms, dry goods, jewelry, more. Invaluable for dating, identifying vintage items. Also, copyright-free graphics for artists, designers. Co-published with Henry Ford Museum & Green-field Village. 160pp. 8¼ × 11.　　　　25780-0 Pa. $8.95

HISTORIC COSTUME IN PICTURES, Braun & Schneider. Over 1,450 costumed figures in clearly detailed engravings—from dawn of civilization to end of 19th century. Captions. Many folk costumes. 256pp. 8⅜ × 11¾.　　　　23150-X Pa. $10.95

STICKLEY CRAFTSMAN FURNITURE CATALOGS, Gustav Stickley and L. & J. G. Stickley. Beautiful, functional furniture in two authentic catalogs from 1910. 594 illustrations, including 277 photos, show settles, rockers, armchairs, reclining chairs, bookcases, desks, tables. 183pp. 6½ × 9¼.   23838-5 Pa. $8.95

AMERICAN LOCOMOTIVES IN HISTORIC PHOTOGRAPHS: 1858 to 1949, Ron Ziel (ed.). A rare collection of 126 meticulously detailed official photographs, called "builder portraits," of American locomotives that majestically chronicle the rise of steam locomotive power in America. Introduction. Detailed captions. xi + 129pp. 9 × 12.   27393-8 Pa. $12.95

AMERICA'S LIGHTHOUSES: An Illustrated History, Francis Ross Holland, Jr. Delightfully written, profusely illustrated fact-filled survey of over 200 American lighthouses since 1716. History, anecdotes, technological advances, more. 240pp. 8 × 10¾.   25576-X Pa. $10.95

TOWARDS A NEW ARCHITECTURE, Le Corbusier. Pioneering manifesto by founder of "International School." Technical and aesthetic theories, views of industry, economics, relation of form to function, "mass-production split" and much more. Profusely illustrated. 320pp. 6⅛ × 9¼. (USO)   25023-7 Pa. $8.95

HOW THE OTHER HALF LIVES, Jacob Riis. Famous journalistic record, exposing poverty and degradation of New York slums around 1900, by major social reformer. 100 striking and influential photographs. 233pp. 10 × 7⅞.
                                                                        22012-5 Pa $10.95

FRUIT KEY AND TWIG KEY TO TREES AND SHRUBS, William M. Harlow. One of the handiest and most widely used identification aids. Fruit key covers 120 deciduous and evergreen species; twig key 160 deciduous species. Easily used. Over 300 photographs. 126pp. 5⅜ × 8½.   20511-8 Pa. $2.95

COMMON BIRD SONGS, Dr. Donald J. Borror. Songs of 60 most common U.S. birds: robins, sparrows, cardinals, bluejays, finches, more—arranged in order of increasing complexity. Up to 9 variations of songs of each species.
                                                            Cassette and manual 99911-4 $8.95

ORCHIDS AS HOUSE PLANTS, Rebecca Tyson Northen. Grow cattleyas and many other kinds of orchids—in a window, in a case, or under artificial light. 63 illustrations. 148pp. 5⅜ × 8½.   23261-1 Pa. $3.95

MONSTER MAZES, Dave Phillips. Masterful mazes at four levels of difficulty. Avoid deadly perils and evil creatures to find magical treasures. Solutions for all 32 exciting illustrated puzzles. 48pp. 8¼ × 11.   26005-4 Pa. $2.95

MOZART'S DON GIOVANNI (DOVER OPERA LIBRETTO SERIES), Wolfgang Amadeus Mozart. Introduced and translated by Ellen H. Bleiler. Standard Italian libretto, with complete English translation. Convenient and thoroughly portable—an ideal companion for reading along with a recording or the performance itself. Introduction. List of characters. Plot summary. 121pp. 5¼ × 8½.
                                                                        24944-1 Pa. $2.95

TECHNICAL MANUAL AND DICTIONARY OF CLASSICAL BALLET, Gail Grant. Defines, explains, comments on steps, movements, poses and concepts. 15-page pictorial section. Basic book for student, viewer. 127pp. 5⅜ × 8½.
                                                                        21843-0 Pa. $3.95

BRASS INSTRUMENTS: Their History and Development, Anthony Baines. Authoritative, updated survey of the evolution of trumpets, trombones, bugles, cornets, French horns, tubas and other brass wind instruments. Over 140 illustrations and 48 music examples. Corrected and updated by author. New preface. Bibliography. 320pp. 5⅜ × 8½. 27574-4 Pa. $9.95

HOLLYWOOD GLAMOR PORTRAITS, John Kobal (ed.). 145 photos from 1926–49. Harlow, Gable, Bogart, Bacall; 94 stars in all. Full background on photographers, technical aspects. 160pp. 8⅜ × 11¼. 23352-9 Pa. $9.95

MAX AND MORITZ, Wilhelm Busch. Great humor classic in both German and English. Also 10 other works: "Cat and Mouse," "Plisch and Plumm," etc. 216pp. 5⅜ × 8½. 20181-3 Pa. $5.95

THE RAVEN AND OTHER FAVORITE POEMS, Edgar Allan Poe. Over 40 of the author's most memorable poems: "The Bells," "Ulalume," "Israfel," "To Helen," "The Conqueror Worm," "Eldorado," "Annabel Lee," many more. Alphabetic lists of titles and first lines. 64pp. 5³⁄₁₆ × 8¼. 26685-0 Pa. $1.00

SEVEN SCIENCE FICTION NOVELS, H. G. Wells. The standard collection of the great novels. Complete, unabridged. First Men in the Moon, Island of Dr. Moreau, War of the Worlds, Food of the Gods, Invisible Man, Time Machine, In the Days of the Comet. Total of 1,015pp. 5⅜ × 8½. (USO) 20264-X Clothbd. $29.95

AMULETS AND SUPERSTITIONS, E. A. Wallis Budge. Comprehensive discourse on origin, powers of amulets in many ancient cultures: Arab, Persian, Babylonian, Assyrian, Egyptian, Gnostic, Hebrew, Phoenician, Syriac, etc. Covers cross, swastika, crucifix, seals, rings, stones, etc. 584pp. 5⅜ × 8½. 23573-4 Pa. $10.95

RUSSIAN STORIES/PYCCKNE PACCKA3bl: A Dual-Language Book, edited by Gleb Struve. Twelve tales by such masters as Chekhov, Tolstoy, Dostoevsky, Pushkin, others. Excellent word-for-word English translations on facing pages, plus teaching and study aids, Russian/English vocabulary, biographical/critical introductions, more. 416pp. 5⅜ × 8½. 26244-8 Pa. $7.95

PHILADELPHIA THEN AND NOW: 60 Sites Photographed in the Past and Present, Kenneth Finkel and Susan Oyama. Rare photographs of City Hall, Logan Square, Independence Hall, Betsy Ross House, other landmarks juxtaposed with contemporary views. Captures changing face of historic city. Introduction. Captions. 128pp. 8¼ × 11. 25790-8 Pa. $9.95

AIA ARCHITECTURAL GUIDE TO NASSAU AND SUFFOLK COUNTIES, LONG ISLAND, The American Institute of Architects, Long Island Chapter, and the Society for the Preservation of Long Island Antiquities. Comprehensive, well-researched and generously illustrated volume brings to life over three centuries of Long Island's great architectural heritage. More than 240 photographs with authoritative, extensively detailed captions. 176pp. 8¼ × 11. 26946-9 Pa. $14.95

NORTH AMERICAN INDIAN LIFE: Customs and Traditions of 23 Tribes, Elsie Clews Parsons (ed.). 27 fictionalized essays by noted anthropologists examine religion, customs, government, additional facets of life among the Winnebago, Crow, Zuni, Eskimo, other tribes. 480pp. 6⅛ × 9¼. 27377-6 Pa. $10.95

FRANK LLOYD WRIGHT'S HOLLYHOCK HOUSE, Donald Hoffmann. Lavishly illustrated, carefully documented study of one of Wright's most controversial residential designs. Over 120 photographs, floor plans, elevations, etc. Detailed perceptive text by noted Wright scholar. Index. 128pp. 9¼ × 10¾.
27133-1 Pa. $10.95

THE MALE AND FEMALE FIGURE IN MOTION: 60 Classic Photographic Sequences, Eadweard Muybridge. 60 true-action photographs of men and women walking, running, climbing, bending, turning, etc., reproduced from rare 19th-century masterpiece. vi + 121pp. 9 × 12.
24745-7 Pa. $10.95

1001 QUESTIONS ANSWERED ABOUT THE SEASHORE, N. J. Berrill and Jacquelyn Berrill. Queries answered about dolphins, sea snails, sponges, starfish, fishes, shore birds, many others. Covers appearance, breeding, growth, feeding, much more. 305pp. 5¼ × 8¼.
23366-9 Pa. $7.95

GUIDE TO OWL WATCHING IN NORTH AMERICA, Donald S. Heintzelman. Superb guide offers complete data and descriptions of 19 species: barn owl, screech owl, snowy owl, many more. Expert coverage of owl-watching equipment, conservation, migrations and invasions, etc. Guide to observing sites. 84 illustrations. xiii + 193pp. 5⅜ × 8½.
27344-X Pa. $7.95

MEDICINAL AND OTHER USES OF NORTH AMERICAN PLANTS: A Historical Survey with Special Reference to the Eastern Indian Tribes, Charlotte Erichsen-Brown. Chronological historical citations document 500 years of usage of plants, trees, shrubs native to eastern Canada, northeastern U.S. Also complete identifying information. 343 illustrations. 544pp. 6½ × 9¼.
25951-X Pa. $12.95

STORYBOOK MAZES, Dave Phillips. 23 stories and mazes on two-page spreads: Wizard of Oz, Treasure Island, Robin Hood, etc. Solutions. 64pp. 8¼ × 11.
23628-5 Pa. $2.95

NEGRO FOLK MUSIC, U.S.A., Harold Courlander. Noted folklorist's scholarly yet readable analysis of rich and varied musical tradition. Includes authentic versions of over 40 folk songs. Valuable bibliography and discography. xi + 324pp. 5⅜ × 8½.
27350-4 Pa. $7.95

MOVIE-STAR PORTRAITS OF THE FORTIES, John Kobal (ed.). 163 glamor, studio photos of 106 stars of the 1940s: Rita Hayworth, Ava Gardner, Marlon Brando, Clark Gable, many more. 176pp. 8⅜ × 11¼.
23546-7 Pa. $10.95

BENCHLEY LOST AND FOUND, Robert Benchley. Finest humor from early 30s, about pet peeves, child psychologists, post office and others. Mostly unavailable elsewhere. 73 illustrations by Peter Arno and others. 183pp. 5⅜ × 8½.
22410-4 Pa. $4.95

YEKL and THE IMPORTED BRIDEGROOM AND OTHER STORIES OF YIDDISH NEW YORK, Abraham Cahan. Film Hester Street based on Yekl (1896). Novel, other stories among first about Jewish immigrants on N.Y.'s East Side. 240pp. 5⅜ × 8½.
22427-9 Pa. $5.95

SELECTED POEMS, Walt Whitman. Generous sampling from *Leaves of Grass*. Twenty-four poems include "I Hear America Singing," "Song of the Open Road," "I Sing the Body Electric," "When Lilacs Last in the Dooryard Bloom'd," "O Captain! My Captain!"—all reprinted from an authoritative edition. Lists of titles and first lines. 128pp. 5³⁄₁₆ × 8¼.
26878-0 Pa. $1.00

THE BEST TALES OF HOFFMANN, E. T. A. Hoffmann. 10 of Hoffmann's most important stories: "Nutcracker and the King of Mice," "The Golden Flowerpot," etc. 458pp. 5⅜ × 8½.   21793-0 Pa. $8.95

FROM FETISH TO GOD IN ANCIENT EGYPT, E. A. Wallis Budge. Rich detailed survey of Egyptian conception of "God" and gods, magic, cult of animals, Osiris, more. Also, superb English translations of hymns and legends. 240 illustrations. 545pp. 5⅜ × 8½.   25803-3 Pa. $10.95

FRENCH STORIES/CONTES FRANÇAIS: A Dual-Language Book, Wallace Fowlie. Ten stories by French masters, Voltaire to Camus: "Micromegas" by Voltaire; "The Atheist's Mass" by Balzac; "Minuet" by de Maupassant; "The Guest" by Camus, six more. Excellent English translations on facing pages. Also French-English vocabulary list, exercises, more. 352pp. 5⅜ × 8½. 26443-2 Pa. $8.95

CHICAGO AT THE TURN OF THE CENTURY IN PHOTOGRAPHS: 122 Historic Views from the Collections of the Chicago Historical Society, Larry A. Viskochil. Rare large-format prints offer detailed views of City Hall, State Street, the Loop, Hull House, Union Station, many other landmarks, circa 1904–1913. Introduction. Captions. Maps. 144pp. 9⅜ × 12¼.   24656-6 Pa. $12.95

OLD BROOKLYN IN EARLY PHOTOGRAPHS, 1865–1929, William Lee Younger. Luna Park, Gravesend race track, construction of Grand Army Plaza, moving of Hotel Brighton, etc. 157 previously unpublished photographs. 165pp. 8⅞ × 11¾.   23587-4 Pa. $12.95

THE MYTHS OF THE NORTH AMERICAN INDIANS, Lewis Spence. Rich anthology of the myths and legends of the Algonquins, Iroquois, Pawnees and Sioux, prefaced by an extensive historical and ethnological commentary. 36 illustrations. 480pp. 5⅜ × 8½.   25967-6 Pa. $8.95

AN ENCYCLOPEDIA OF BATTLES: Accounts of Over 1,560 Battles from 1479 B.C. to the Present, David Eggenberger. Essential details of every major battle in recorded history from the first battle of Megiddo in 1479 B.C. to Grenada in 1984. List of Battle Maps. New Appendix covering the years 1967–1984. Index. 99 illustrations. 544pp. 6½ × 9¼.   24913-1 Pa. $14.95

SAILING ALONE AROUND THE WORLD, Captain Joshua Slocum. First man to sail around the world, alone, in small boat. One of great feats of seamanship told in delightful manner. 67 illustrations. 294pp. 5⅜ × 8½.   20326-3 Pa. $4.95

ANARCHISM AND OTHER ESSAYS, Emma Goldman. Powerful, penetrating, prophetic essays on direct action, role of minorities, prison reform, puritan hypocrisy, violence, etc. 271pp. 5⅜ × 8½.   22484-8 Pa. $5.95

MYTHS OF THE HINDUS AND BUDDHISTS, Ananda K. Coomaraswamy and Sister Nivedita. Great stories of the epics; deeds of Krishna, Shiva, taken from puranas, Vedas, folk tales; etc. 32 illustrations. 400pp. 5⅜ × 8½. 21759-0 Pa. $8.95

BEYOND PSYCHOLOGY, Otto Rank. Fear of death, desire of immortality, nature of sexuality, social organization, creativity, according to Rankian system. 291pp. 5⅜ × 8½.   20485-5 Pa. $7.95

A THEOLOGICO-POLITICAL TREATISE, Benedict Spinoza. Also contains unfinished Political Treatise. Great classic on religious liberty, theory of government on common consent. R. Elwes translation. Total of 421pp. 5⅜ × 8½.
                                                                        20249-6 Pa. $7.95

MY BONDAGE AND MY FREEDOM, Frederick Douglass. Born a slave, Douglass became outspoken force in antislavery movement. The best of Douglass' autobiographies. Graphic description of slave life. 464pp. 5⅜ × 8½. 22457-0 Pa. $7.95

FOLLOWING THE EQUATOR: A Journey Around the World, Mark Twain. Fascinating humorous account of 1897 voyage to Hawaii, Australia, India, New Zealand, etc. Ironic, bemused reports on peoples, customs, climate, flora and fauna, politics, much more. 197 illustrations. 720pp. 5⅜ × 8½. 26113-1 Pa. $15.95

THE PEOPLE CALLED SHAKERS, Edward D. Andrews. Definitive study of Shakers: origins, beliefs, practices, dances, social organization, furniture and crafts, etc. 33 illustrations. 351pp. 5⅜ × 8½. 21081-2 Pa. $7.95

THE MYTHS OF GREECE AND ROME, H. A. Guerber. A classic of mythology, generously illustrated, long prized for its simple, graphic, accurate retelling of the principal myths of Greece and Rome, and for its commentary on their origins and significance. With 64 illustrations by Michelangelo, Raphael, Titian, Rubens, Canova, Bernini and others. 480pp. 5⅜ × 8½. 27584-1 Pa. $9.95

PSYCHOLOGY OF MUSIC, Carl E. Seashore. Classic work discusses music as a medium from psychological viewpoint. Clear treatment of physical acoustics, auditory apparatus, sound perception, development of musical skills, nature of musical feeling, host of other topics. 88 figures. 408pp. 5⅜ × 8½. 21851-1 Pa. $8.95

THE PHILOSOPHY OF HISTORY, Georg W. Hegel. Great classic of Western thought develops concept that history is not chance but rational process, the evolution of freedom. 457pp. 5⅜ × 8½. 20112-0 Pa. $8.95

THE BOOK OF TEA, Kakuzo Okakura. Minor classic of the Orient: entertaining, charming explanation, interpretation of traditional Japanese culture in terms of tea ceremony. 94pp. 5⅜ × 8½. 20070-1 Pa. $2.95

LIFE IN ANCIENT EGYPT, Adolf Erman. Fullest, most thorough, detailed older account with much not in more recent books, domestic life, religion, magic, medicine, commerce, much more. Many illustrations reproduce tomb paintings, carvings, hieroglyphs, etc. 597pp. 5⅜ × 8½. 22632-8 Pa. $9.95

SUNDIALS, Their Theory and Construction, Albert Waugh. Far and away the best, most thorough coverage of ideas, mathematics concerned, types, construction, adjusting anywhere. Simple, nontechnical treatment allows even children to build several of these dials. Over 100 illustrations. 230pp. 5⅜ × 8½. 22947-5 Pa. $5.95

DYNAMICS OF FLUIDS IN POROUS MEDIA, Jacob Bear. For advanced students of ground water hydrology, soil mechanics and physics, drainage and irrigation engineering, and more. 335 illustrations. Exercises, with answers. 784pp. 6⅛ × 9¼. 65675-6 Pa. $19.95

SONGS OF EXPERIENCE: Facsimile Reproduction with 26 Plates in Full Color, William Blake. 26 full-color plates from a rare 1826 edition. Includes "The Tyger," "London," "Holy Thursday," and other poems. Printed text of poems. 48pp. 5¼ × 7. 24636-1 Pa. $3.95

OLD-TIME VIGNETTES IN FULL COLOR, Carol Belanger Grafton (ed.). Over 390 charming, often sentimental illustrations, selected from archives of Victorian graphics—pretty women posing, children playing, food, flowers, kittens and puppies, smiling cherubs, birds and butterflies, much more. All copyright-free. 48pp. 9¼ × 12¼. 27269-9 Pa. $5.95

PERSPECTIVE FOR ARTISTS, Rex Vicat Cole. Depth, perspective of sky and sea, shadows, much more, not usually covered. 391 diagrams, 81 reproductions of drawings and paintings. 279pp. 5⅜ × 8½. 22487-2 Pa. $6.95

DRAWING THE LIVING FIGURE, Joseph Sheppard. Innovative approach to artistic anatomy focuses on specifics of surface anatomy, rather than muscles and bones. Over 170 drawings of live models in front, back and side views, and in widely varying poses. Accompanying diagrams. 177 illustrations. Introduction. Index. 144pp. 8⅜ × 11¼. 26723-7 Pa. $7.95

GOTHIC AND OLD ENGLISH ALPHABETS: 100 Complete Fonts, Dan X. Solo. Add power, elegance to posters, signs, other graphics with 100 stunning copyright-free alphabets: Blackstone, Dolbey, Germania, 97 more—including many lower-case, numerals, punctuation marks. 104pp. 8⅛ × 11. 24695-7 Pa. $6.95

HOW TO DO BEADWORK, Mary White. Fundamental book on craft from simple projects to five-bead chains and woven works. 106 illustrations. 142pp. 5⅜ × 8. 20697-1 Pa. $4.95

THE BOOK OF WOOD CARVING, Charles Marshall Sayers. Finest book for beginners discusses fundamentals and offers 34 designs. "Absolutely first rate . . . well thought out and well executed."—E. J. Tangerman. 118pp. 7¾ × 10⅝. 23654-4 Pa. $5.95

ILLUSTRATED CATALOG OF CIVIL WAR MILITARY GOODS: Union Army Weapons, Insignia, Uniform Accessories, and Other Equipment, Schuyler, Hartley, and Graham. Rare, profusely illustrated 1846 catalog includes Union Army uniform and dress regulations, arms and ammunition, coats, insignia, flags, swords, rifles, etc. 226 illustrations. 160pp. 9 × 12. 24939-5 Pa. $10.95

WOMEN'S FASHIONS OF THE EARLY 1900s: An Unabridged Republication of "New York Fashions, 1909," National Cloak & Suit Co. Rare catalog of mail-order fashions documents women's and children's clothing styles shortly after the turn of the century. Captions offer full descriptions, prices. Invaluable resource for fashion, costume historians. Approximately 725 illustrations. 128pp. 8⅜ × 11¼. 27276-1 Pa. $10.95

THE 1912 AND 1915 GUSTAV STICKLEY FURNITURE CATALOGS, Gustav Stickley. With over 200 detailed illustrations and descriptions, these two catalogs are essential reading and reference materials and identification guides for Stickley furniture. Captions cite materials, dimensions and prices. 112pp. 6½ × 9¼. 26676-1 Pa. $9.95

EARLY AMERICAN LOCOMOTIVES, John H. White, Jr. Finest locomotive engravings from early 19th century: historical (1804–74), main-line (after 1870), special, foreign, etc. 147 plates. 142pp. 11⅜ × 8¼. 22772-3 Pa. $8.95

THE TALL SHIPS OF TODAY IN PHOTOGRAPHS, Frank O. Braynard. Lavishly illustrated tribute to nearly 100 majestic contemporary sailing vessels: Amerigo Vespucci, Clearwater, Constitution, Eagle, Mayflower, Sea Cloud, Victory, many more. Authoritative captions provide statistics, background on each ship. 190 black-and-white photographs and illustrations. Introduction. 128pp. 8⅞ × 11¾. 27163-3 Pa. $12.95

EARLY NINETEENTH-CENTURY CRAFTS AND TRADES, Peter Stockham (ed.). Extremely rare 1807 volume describes to youngsters the crafts and trades of the day: brickmaker, weaver, dressmaker, bookbinder, ropemaker, saddler, many more. Quaint prose, charming illustrations for each craft. 20 black-and-white line illustrations. 192pp. 4⅝ × 6. 27293-1 Pa. $4.95

VICTORIAN FASHIONS AND COSTUMES FROM HARPER'S BAZAR, 1867–1898, Stella Blum (ed.). Day costumes, evening wear, sports clothes, shoes, hats, other accessories in over 1,000 detailed engravings. 320pp. 9⅜ × 12¼. 22990-4 Pa. $12.95

GUSTAV STICKLEY, THE CRAFTSMAN, Mary Ann Smith. Superb study surveys broad scope of Stickley's achievement, especially in architecture. Design philosophy, rise and fall of the Craftsman empire, descriptions and floor plans for many Craftsman houses, more. 86 black-and-white halftones. 31 line illustrations. Introduction. 208pp. 6½ × 9¼. 27210-9 Pa. $9.95

THE LONG ISLAND RAIL ROAD IN EARLY PHOTOGRAPHS, Ron Ziel. Over 220 rare photos, informative text document origin (1844) and development of rail service on Long Island. Vintage views of early trains, locomotives, stations, passengers, crews, much more. Captions. 8⅞ × 11¾. 26301-0 Pa. $13.95

THE BOOK OF OLD SHIPS: From Egyptian Galleys to Clipper Ships, Henry B. Culver. Superb, authoritative history of sailing vessels, with 80 magnificent line illustrations. Galley, bark, caravel, longship, whaler, many more. Detailed, informative text on each vessel by noted naval historian. Introduction. 256pp. 5⅜ × 8½. 27332-6 Pa. $6.95

TEN BOOKS ON ARCHITECTURE, Vitruvius. The most important book ever written on architecture. Early Roman aesthetics, technology, classical orders, site selection, all other aspects. Morgan translation. 331pp. 5⅜ × 8½. 20645-9 Pa. $8.95

THE HUMAN FIGURE IN MOTION, Eadweard Muybridge. More than 4,500 stopped-action photos, in action series, showing undraped men, women, children jumping, lying down, throwing, sitting, wrestling, carrying, etc. 390pp. 7⅞ × 10⅝. 20204-6 Clothbd. $24.95

TREES OF THE EASTERN AND CENTRAL UNITED STATES AND CANADA, William M. Harlow. Best one-volume guide to 140 trees. Full descriptions, woodlore, range, etc. Over 600 illustrations. Handy size. 288pp. 4½ × 6⅜. 20395-6 Pa. $4.95

SONGS OF WESTERN BIRDS, Dr. Donald J. Borror. Complete song and call repertoire of 60 western species, including flycatchers, juncoes, cactus wrens, many more—includes fully illustrated booklet. Cassette and manual 99913-0 $8.95

GROWING AND USING HERBS AND SPICES, Milo Miloradovich. Versatile handbook provides all the information needed for cultivation and use of all the herbs and spices available in North America. 4 illustrations. Index. Glossary. 236pp. 5⅜ × 8½. 25058-X Pa. $5.95

BIG BOOK OF MAZES AND LABYRINTHS, Walter Shepherd. 50 mazes and labyrinths in all—classical, solid, ripple, and more—in one great volume. Perfect inexpensive puzzler for clever youngsters. Full solutions. 112pp. 8⅛ × 11. 22951-3 Pa. $3.95

PIANO TUNING, J. Cree Fischer. Clearest, best book for beginner, amateur. Simple repairs, raising dropped notes, tuning by easy method of flattened fifths. No previous skills needed. 4 illustrations. 201pp. 5⅜ × 8½. 23267-0 Pa. $4.95

A SOURCE BOOK IN THEATRICAL HISTORY, A. M. Nagler. Contemporary observers on acting, directing, make-up, costuming, stage props, machinery, scene design, from Ancient Greece to Chekhov. 611pp. 5⅜ × 8½. 20515-0 Pa. $10.95

THE COMPLETE NONSENSE OF EDWARD LEAR, Edward Lear. All nonsense limericks, zany alphabets, Owl and Pussycat, songs, nonsense botany, etc., illustrated by Lear. Total of 320pp. 5⅜ × 8½. (USO) 20167-8 Pa. $5.95

VICTORIAN PARLOUR POETRY: An Annotated Anthology, Michael R. Turner. 117 gems by Longfellow, Tennyson, Browning, many lesser-known poets. "The Village Blacksmith," "Curfew Must Not Ring Tonight," "Only a Baby Small," dozens more, often difficult to find elsewhere. Index of poets, titles, first lines. xxiii + 325pp. 5⅜ × 8¼. 27044-0 Pa. $7.95

DUBLINERS, James Joyce. Fifteen stories offer vivid, tightly focused observations of the lives of Dublin's poorer classes. At least one, "The Dead," is considered a masterpiece. Reprinted complete and unabridged from standard edition. 160pp. 5³⁄₁₆ × 8¼. 26870-5 Pa. $1.00

THE HAUNTED MONASTERY and THE CHINESE MAZE MURDERS, Robert van Gulik. Two full novels by van Gulik, set in 7th-century China, continue adventures of Judge Dee and his companions. An evil Taoist monastery, seemingly supernatural events; overgrown topiary maze hides strange crimes. 27 illustrations. 328pp. 5⅜ × 8½. 23502-5 Pa. $7.95

THE BOOK OF THE SACRED MAGIC OF ABRAMELIN THE MAGE, translated by S. MacGregor Mathers. Medieval manuscript of ceremonial magic. Basic document in Aleister Crowley, Golden Dawn groups. 268pp. 5⅜ × 8½. 23211-5 Pa. $7.95

NEW RUSSIAN-ENGLISH AND ENGLISH-RUSSIAN DICTIONARY, M. A. O'Brien. This is a remarkably handy Russian dictionary, containing a surprising amount of information, including over 70,000 entries. 366pp. 4½ × 6⅛. 20208-9 Pa. $8.95

HISTORIC HOMES OF THE AMERICAN PRESIDENTS, Second, Revised Edition, Irvin Haas. A traveler's guide to American Presidential homes, most open to the public, depicting and describing homes occupied by every American President from George Washington to George Bush. With visiting hours, admission charges, travel routes. 175 photographs. Index. 160pp. 8¼ × 11. 26751-2 Pa. $10.95

NEW YORK IN THE FORTIES, Andreas Feininger. 162 brilliant photographs by the well-known photographer, formerly with *Life* magazine. Commuters, shoppers, Times Square at night, much else from city at its peak. Captions by John von Hartz. 181pp. 9¼ × 10¾. 23585-8 Pa. $12.95

INDIAN SIGN LANGUAGE, William Tomkins. Over 525 signs developed by Sioux and other tribes. Written instructions and diagrams. Also 290 pictographs. 111pp. 6⅛ × 9¼. 22029-X Pa. $3.50

CATALOG OF DOVER BOOKS

ANATOMY: A Complete Guide for Artists, Joseph Sheppard. A master of figure drawing shows artists how to render human anatomy convincingly. Over 460 illustrations. 224pp. 8⅜ × 11¼. 27279-6 Pa. $9.95

MEDIEVAL CALLIGRAPHY: Its History and Technique, Marc Drogin. Spirited history, comprehensive instruction manual covers 13 styles (ca. 4th century thru 15th). Excellent photographs; directions for duplicating medieval techniques with modern tools. 224pp. 8⅜ × 11¼. 26142-5 Pa. $11.95

DRIED FLOWERS: How to Prepare Them, Sarah Whitlock and Martha Rankin. Complete instructions on how to use silica gel, meal and borax, perlite aggregate, sand and borax, glycerine and water to create attractive permanent flower arrangements. 12 illustrations. 32pp. 5⅜ × 8½. 21802-3 Pa. $1.00

EASY-TO-MAKE BIRD FEEDERS FOR WOODWORKERS, Scott D. Campbell. Detailed, simple-to-use guide for designing, constructing, caring for and using feeders. Text, illustrations for 12 classic and contemporary designs. 96pp. 5⅜ × 8½. 25847-5 Pa. $2.95

OLD-TIME CRAFTS AND TRADES, Peter Stockham. An 1807 book created to teach children about crafts and trades open to them as future careers. It describes in detailed, nontechnical terms 24 different occupations, among them coachmaker, gardener, hairdresser, lacemaker, shoemaker, wheelwright, copper-plate printer, milliner, trunkmaker, merchant and brewer. Finely detailed engravings illustrate each occupation. 192pp. 4⅝ × 6. 27398-9 Pa. $4.95

THE HISTORY OF UNDERCLOTHES, C. Willett Cunnington and Phyllis Cunnington. Fascinating, well-documented survey covering six centuries of English undergarments, enhanced with over 100 illustrations: 12th-century laced-up bodice, footed long drawers (1795), 19th-century bustles, 19th-century corsets for men, Victorian "bust improvers," much more. 272pp. 5⅜ × 8¼. 27124-2 Pa. $9.95

ARTS AND CRAFTS FURNITURE: The Complete Brooks Catalog of 1912, Brooks Manufacturing Co. Photos and detailed descriptions of more than 150 now very collectible furniture designs from the Arts and Crafts movement depict davenports, settees, buffets, desks, tables, chairs, bedsteads, dressers and more, all built of solid, quarter-sawed oak. Invaluable for students and enthusiasts of antiques, Americana and the decorative arts. 80pp. 6½ × 9¼. 27471-3 Pa. $7.95

HOW WE INVENTED THE AIRPLANE: An Illustrated History, Orville Wright. Fascinating firsthand account covers early experiments, construction of planes and motors, first flights, much more. Introduction and commentary by Fred C. Kelly. 76 photographs. 96pp. 8¼ × 11. 25662-6 Pa. $7.95

THE ARTS OF THE SAILOR: Knotting, Splicing and Ropework, Hervey Garrett Smith. Indispensable shipboard reference covers tools, basic knots and useful hitches; handsewing and canvas work, more. Over 100 illustrations. Delightful reading for sea lovers. 256pp. 5⅜ × 8½. 26440-8 Pa. $6.95

FRANK LLOYD WRIGHT'S FALLINGWATER: The House and Its History, Second, Revised Edition, Donald Hoffmann. A total revision—both in text and illustrations—of the standard document on Fallingwater, the boldest, most personal architectural statement of Wright's mature years, updated with valuable new material from the recently opened Frank Lloyd Wright Archives. "Fascinating"—*The New York Times*. 116 illustrations. 128pp. 9¼ × 10¾. 27430-6 Pa. $10.95

PHOTOGRAPHIC SKETCHBOOK OF THE CIVIL WAR, Alexander Gardner. 100 photos taken on field during the Civil War. Famous shots of Manassas, Harper's Ferry, Lincoln, Richmond, slave pens, etc. 244pp. 10⅝ × 8¼.
22731-6 Pa. $9.95

FIVE ACRES AND INDEPENDENCE, Maurice G. Kains. Great back-to-the-land classic explains basics of self-sufficient farming. The one book to get. 95 illustrations. 397pp. 5⅜ × 8½.
20974-1 Pa. $6.95

SONGS OF EASTERN BIRDS, Dr. Donald J. Borror. Songs and calls of 60 species most common to eastern U.S.: warblers, woodpeckers, flycatchers, thrushes, larks, many more in high-quality recording.
Cassette and manual 99912-2 $8.95

A MODERN HERBAL, Margaret Grieve. Much the fullest, most exact, most useful compilation of herbal material. Gigantic alphabetical encyclopedia, from aconite to zedoary, gives botanical information, medical properties, folklore, economic uses, much else. Indispensable to serious reader. 161 illustrations. 888pp. 6½ × 9¼. 2-vol. set. (USO)
Vol. I: 22798-7 Pa. $9.95
Vol. II: 22799-5 Pa. $9.95

HIDDEN TREASURE MAZE BOOK, Dave Phillips. Solve 34 challenging mazes accompanied by heroic tales of adventure. Evil dragons, people-eating plants, bloodthirsty giants, many more dangerous adversaries lurk at every twist and turn. 34 mazes, stories, solutions. 48pp. 8¼ × 11.
24566-7 Pa. $2.95

LETTERS OF W. A. MOZART, Wolfgang A. Mozart. Remarkable letters show bawdy wit, humor, imagination, musical insights, contemporary musical world; includes some letters from Leopold Mozart. 276pp. 5⅜ × 8½.
22859-2 Pa. $6.95

BASIC PRINCIPLES OF CLASSICAL BALLET, Agrippina Vaganova. Great Russian theoretician, teacher explains methods for teaching classical ballet. 118 illustrations. 175pp. 5⅜ × 8½.
22036-2 Pa. $3.95

THE JUMPING FROG, Mark Twain. Revenge edition. The original story of The Celebrated Jumping Frog of Calaveras County, a hapless French translation, and Twain's hilarious "retranslation" from the French. 12 illustrations. 66pp. 5⅜ × 8½.
22686-7 Pa. $3.50

BEST REMEMBERED POEMS, Martin Gardner (ed.). The 126 poems in this superb collection of 19th- and 20th-century British and American verse range from Shelley's "To a Skylark" to the impassioned "Renascence" of Edna St. Vincent Millay and to Edward Lear's whimsical "The Owl and the Pussycat." 224pp. 5⅜ × 8½.
27165-X Pa. $3.95

COMPLETE SONNETS, William Shakespeare. Over 150 exquisite poems deal with love, friendship, the tyranny of time, beauty's evanescence, death and other themes in language of remarkable power, precision and beauty. Glossary of archaic terms. 80pp. 5³⁄₁₆ × 8¼.
26686-9 Pa. $1.00

BODIES IN A BOOKSHOP, R. T. Campbell. Challenging mystery of blackmail and murder with ingenious plot and superbly drawn characters. In the best tradition of British suspense fiction. 192pp. 5⅜ × 8½.
24720-1 Pa. $5.95

THE WIT AND HUMOR OF OSCAR WILDE, Alvin Redman (ed.). More than 1,000 ripostes, paradoxes, wisecracks: Work is the curse of the drinking classes; I can resist everything except temptation; etc. 258pp. 5⅜ × 8½. 20602-5 Pa. $4.95

SHAKESPEARE LEXICON AND QUOTATION DICTIONARY, Alexander Schmidt. Full definitions, locations, shades of meaning in every word in plays and poems. More than 50,000 exact quotations. 1,485pp. 6½ × 9¼. 2-vol. set.
Vol. 1: 22726-X Pa. $15.95
Vol. 2: 22727-8 Pa. $15.95

SELECTED POEMS, Emily Dickinson. Over 100 best-known, best-loved poems by one of America's foremost poets, reprinted from authoritative early editions. No comparable edition at this price. Index of first lines. 64pp. 5³⁄₁₆ × 8¼.
26466-1 Pa. $1.00

CELEBRATED CASES OF JUDGE DEE (DEE GOONG AN), translated by Robert van Gulik. Authentic 18th-century Chinese detective novel; Dee and associates solve three interlocked cases. Led to van Gulik's own stories with same characters. Extensive introduction. 9 illustrations. 237pp. 5⅜ × 8½.
23337-5 Pa. $5.95

THE MALLEUS MALEFICARUM OF KRAMER AND SPRENGER, translated by Montague Summers. Full text of most important witchhunter's "bible," used by both Catholics and Protestants. 278pp. 6⅝ × 10. 22802-9 Pa. $10.95

SPANISH STORIES/CUENTOS ESPAÑOLES: A Dual-Language Book, Angel Flores (ed.). Unique format offers 13 great stories in Spanish by Cervantes, Borges, others. Faithful English translations on facing pages. 352pp. 5⅜ × 8½.
25399-6 Pa. $7.95

THE CHICAGO WORLD'S FAIR OF 1893: A Photographic Record, Stanley Appelbaum (ed.). 128 rare photos show 200 buildings, Beaux-Arts architecture, Midway, original Ferris Wheel, Edison's kinetoscope, more. Architectural emphasis; full text. 116pp. 8¼ × 11. 23990-X Pa. $9.95

OLD QUEENS, N.Y., IN EARLY PHOTOGRAPHS, Vincent F. Seyfried and William Asadorian. Over 160 rare photographs of Maspeth, Jamaica, Jackson Heights, and other areas. Vintage views of DeWitt Clinton mansion, 1939 World's Fair and more. Captions. 192pp. 8⅞ × 11. 26358-4 Pa. $12.95

CAPTURED BY THE INDIANS: 15 Firsthand Accounts, 1750–1870, Frederick Drimmer. Astounding true historical accounts of grisly torture, bloody conflicts, relentless pursuits, miraculous escapes and more, by people who lived to tell the tale. 384pp. 5⅜ × 8½. 24901-8 Pa. $7.95

THE WORLD'S GREAT SPEECHES, Lewis Copeland and Lawrence W. Lamm (eds.). Vast collection of 278 speeches of Greeks to 1970. Powerful and effective models; unique look at history. 842pp. 5⅜ × 8½. 20468-5 Pa. $12.95

THE BOOK OF THE SWORD, Sir Richard F. Burton. Great Victorian scholar/adventurer's eloquent, erudite history of the "queen of weapons"—from prehistory to early Roman Empire. Evolution and development of early swords, variations (sabre, broadsword, cutlass, scimitar, etc.), much more. 336pp. 6⅛ × 9¼. 25434-8 Pa. $8.95

CATALOG OF DOVER BOOKS

AUTOBIOGRAPHY: The Story of My Experiments with Truth, Mohandas K. Gandhi. Boyhood, legal studies, purification, the growth of the Satyagraha (nonviolent protest) movement. Critical, inspiring work of the man responsible for the freedom of India. 480pp. 5⅜ × 8½. (USO) 24593-4 Pa. $6.95

CELTIC MYTHS AND LEGENDS, T. W. Rolleston. Masterful retelling of Irish and Welsh stories and tales. Cuchulain, King Arthur, Deirdre, the Grail, many more. First paperback edition. 58 full-page illustrations. 512pp. 5⅜ × 8½.
26507-2 Pa. $9.95

THE PRINCIPLES OF PSYCHOLOGY, William James. Famous long course complete, unabridged. Stream of thought, time perception, memory, experimental methods; great work decades ahead of its time. 94 figures. 1,391pp. 5⅜ × 8½. 2-vol. set.
Vol. I: 20381-6 Pa. $12.95
Vol. II: 20382-4 Pa. $12.95

THE WORLD AS WILL AND REPRESENTATION, Arthur Schopenhauer. Definitive English translation of Schopenhauer's life work, correcting more than 1,000 errors, omissions in earlier translations. Translated by E. F. J. Payne. Total of 1,269pp. 5⅜ × 8½. 2-vol. set. Vol. 1: 21761-2 Pa. $10.95
Vol. 2: 21762-0 Pa. $11.95

MAGIC AND MYSTERY IN TIBET, Madame Alexandra David-Neel. Experiences among lamas, magicians, sages, sorcerers, Bonpa wizards. A true psychic discovery. 32 illustrations. 321pp. 5⅜ × 8½. (USO) 22682-4 Pa. $7.95

THE EGYPTIAN BOOK OF THE DEAD, E. A. Wallis Budge. Complete reproduction of Ani's papyrus, finest ever found. Full hieroglyphic text, interlinear transliteration, word-for-word translation, smooth translation. 533pp. 6½ × 9¼.
21866-X Pa. $9.95

MATHEMATICS FOR THE NONMATHEMATICIAN, Morris Kline. Detailed, college-level treatment of mathematics in cultural and historical context, with numerous exercises. Recommended Reading Lists. Tables. Numerous figures. 641pp. 5⅜ × 8½. 24823-2 Pa. $11.95

THEORY OF WING SECTIONS: Including a Summary of Airfoil Data, Ira H. Abbott and A. E. von Doenhoff. Concise compilation of subsonic aerodynamic characteristics of NACA wing sections, plus description of theory. 350pp. of tables. 693pp. 5⅜ × 8½. 60586-8 Pa. $13.95

THE RIME OF THE ANCIENT MARINER, Gustave Doré, S. T. Coleridge. Doré's finest work; 34 plates capture moods, subtleties of poem. Flawless full-size reproductions printed on facing pages with authoritative text of poem. "Beautiful. Simply beautiful."—*Publisher's Weekly*. 77pp. 9¼ × 12. 22305-1 Pa. $5.95

NORTH AMERICAN INDIAN DESIGNS FOR ARTISTS AND CRAFTS-PEOPLE, Eva Wilson. Over 360 authentic copyright-free designs adapted from Navajo blankets, Hopi pottery, Sioux buffalo hides, more. Geometrics, symbolic figures, plant and animal motifs, etc. 128pp. 8⅜ × 11. (EUK) 25341-4 Pa. $6.95

SCULPTURE: Principles and Practice, Louis Slobodkin. Step-by-step approach to clay, plaster, metals, stone; classical and modern. 253 drawings, photos. 255pp. 8⅛ × 11. 22960-2 Pa. $9.95

THE INFLUENCE OF SEA POWER UPON HISTORY, 1660-1783, A. T. Mahan. Influential classic of naval history and tactics still used as text in war colleges. First paperback edition. 4 maps. 24 battle plans. 640pp. 5⅜ × 8½.
25509-3 Pa. $12.95

THE STORY OF THE TITANIC AS TOLD BY ITS SURVIVORS, Jack Winocour (ed.). What it was really like. Panic, despair, shocking inefficiency, and a little heroism. More thrilling than any fictional account. 26 illustrations. 320pp. 5⅜ × 8½. 20610-6 Pa. $7.95

FAIRY AND FOLK TALES OF THE IRISH PEASANTRY, William Butler Yeats (ed.). Treasury of 64 tales from the twilight world of Celtic myth and legend: "The Soul Cages," "The Kildare Pooka," "King O'Toole and his Goose," many more. Introduction and Notes by W. B. Yeats. 352pp. 5⅜ × 8½. 26941-8 Pa. $7.95

BUDDHIST MAHAYANA TEXTS, E. B. Cowell and Others (eds.). Superb, accurate translations of basic documents in Mahayana Buddhism, highly important in history of religions. The Buddha-karita of Asvaghosha, Larger Sukhavativyuha, more. 448pp. 5⅜ × 8½. 25552-2 Pa. $9.95

ONE TWO THREE . . . INFINITY: Facts and Speculations of Science, George Gamow. Great physicist's fascinating, readable overview of contemporary science: number theory, relativity, fourth dimension, entropy, genes, atomic structure, much more. 128 illustrations. Index. 352pp. 5⅜ × 8½. 25664-2 Pa. $7.95

ENGINEERING IN HISTORY, Richard Shelton Kirby, et al. Broad, nontechnical survey of history's major technological advances: birth of Greek science, industrial revolution, electricity and applied science, 20th-century automation, much more. 181 illustrations. ". . . excellent . . ."—Isis. Bibliography. vii + 530pp. 5⅜ × 8¼.
26412-2 Pa. $13.95

Prices subject to change without notice.

Available at your book dealer or write for free catalog to Dept. GI, Dover Publications, Inc., 31 East 2nd St., Mineola, N.Y. 11501. Dover publishes more than 500 books each year on science, elementary and advanced mathematics, biology, music, art, literary history, social sciences and other areas.